AF883000

Ecology of a Glacial Flood Plain

Ecology of a Glacial Flood Plain

Edited by

J.V. Ward

Swiss Federal Institute of Technology (ETH), Zurich
and former Head of Limnology,
Swiss Federal Institute for Environmental Science and Technology (EAWAG),
Dübendorf, Switzerland

and

U. Uehlinger

Department of Limnology,
Swiss Federal Institute for Environmental Science and Technology (EAWAG),
Dübendorf, Switzerland

KLUWER ACADEMIC PUBLISHERS
DORDRECHT / BOSTON / LONDON

A C.I.P. Catalogue record for this book is available from the Library of Congress.

ISBN 1-4020-1792-8

Published by Kluwer Academic Publishers,
P.O. Box 17, 3300 AA Dordrecht, The Netherlands.

Sold and distributed in North, Central and South America
by Kluwer Academic Publishers,
101 Philip Drive, Norwell, MA 02061, U.S.A.

In all other countries, sold and distributed
by Kluwer Academic Publishers,
P.O. Box 322, 3300 AH Dordrecht, The Netherlands.

Printed on acid-free paper

All Rights Reserved
© 2003 Kluwer Academic Publishers
No part of this work may be reproduced, stored in a retrieval system, or transmitted
in any form or by any means, electronic, mechanical, photocopying, microfilming, recording
or otherwise, without written permission from the Publisher, with the exception
of any material supplied specifically for the purpose of being entered
and executed on a computer system, for exclusive use by the purchaser of the work.

Printed in the Netherlands.

AQUATIC ECOLOGY SERIES

Aquatic ecology is an extraordinarily broad and diverse discipline. Limnology is the study of the functional relationships and productivity of organisms and communities of inland waters as they are regulated by their physical, chemical, and biotic environment. Aquatic ecology encompasses all inland aquatic environments, including streams, rivers, lakes, reservoirs, and wetlands. Research has accelerated in certain areas and has been less active in others. Reassessments and syntheses are stimulating to the discipline as a whole, as well as enormously useful to students and researchers in ecology sciences.

As humans struggle with intensifying environmental crises, understanding of how aquatic ecosystems function is essential to their effective management and indeed to our survival. Information emerging from comprehensive analyses of specific lake, reservoir, and river ecosystems is burgeoning at such a rapid rate, however, that it is difficult to sort out overarching widespread principles.

This series of succinct monographs in aquatic ecology include both in-depth analyses of individual topics, such as hydrodynamics or metabolic processes, as well as comprehensive treatments of intensively studied individual ecosystems from which generic properties can emerge that can be applicable to other ecosystems. Syntheses of information provide interpretations important to our attempts to find order and commonality among ostensible chaos and information overload. Those interpretations are critical not only to students and scientists accepting the challenges of understanding complex and highly dynamic ecosystems. Such syntheses are also important to persons charged with the responsibility of managing aquatic resources under the extreme pressures of multiple uses, valuations, and political stresses of competing sociological demands. We anticipate that the Aquatic Ecology Series will contribute toward enhanced understanding of these resources so critical to the well-being of humankind.

Robert G. Wetzel, Series Editor
Kenan Professor of Environmental Sciences
University of North Carolina

This book is dedicated to Vernon Grant Ward and Hilde Uehlinger

CONTRIBUTORS

PETER BURGHERR is a research scientist with the Paul Scherrer Institute. He completed his PhD research at the Swiss Federal Institute of Technology (ETH) in Zürich on spatio-temporal community patterns of lotic zoobenthos in the Val Roseg. Present address: Paul Scherrer Institute, CH-5232 Villigen-PSI, Switzerland.

HANSRUDOLF BÜRGI is a senior lecturer with the Swiss Federal Institute of Technology (ETH) in Zürich. His research focusses on the biodiversity and ecology of algae and micro-invertebrates, especially planktonic forms. Present address: Department of Limnology, EAWAG/ETH, Ueberlandstrasse 133, CH-8600 Duebendorf, Switzerland.

MARK O. GESSNER is a research scientist at the Swiss Federal Institute of Environmental Science and Technology (EAWAG) and adjunct lecturer at ETH in Zürich. His research addresses ecosystem processes in streams and wetlands with a focus on the roles of fungi, bacteria and invertebrates on litter decomposition.. Present address: Department of Limnology, EAWAG/ETH, Limnological Research Center, CH-6047 Kastanienbaum, Switzerland.

RICHARD ILLI is a technical assistant with the Department of Limnology, EAWAG, Ueberlandstrasse 133, CH-8600 Duebendorf, Switzerland.

BIRGITT KLEIN is a PhD student at the University of Jena. She completed her Masters of Science at the University of Giessen, Germany and at the Department of Limnology, EAWAG, Zürich, Switzerland on the ecology of alluvial springs in the Val Roseg. Present address: Institut für Allgemeine Botanik, Friedrich-Schiller-Universität, D-07743 Jena, Germany

MAX MAISCH is a geomorphologist and lecturer at the University of Zurich. His research focusses on late glacial and Holocene glacier history and alpine geomorphology. Present address: Physical Geography Division, Glaciology and Geomorphodynamics Group, Department of Geography, University of Zürich–Irchel, Winterthurerstrasse 190, CH-8057 Zurich, Switzerland.

FLORIAN MALARD is a groundwater ecologist in the Freshwater and River Ecology Laboratory in Lyon. He occupied a Post-doctoral position within the Department of Limnology at EAWAG from 1996 to 1998, working on surface water - groundwater interactions and hyporheic fauna in the Val Roseg. His research

J.V. Ward & U. Uehlinger (eds.), Ecology of a Glacial Flood Plain, ix-x.
© 2003 *Kluwer Academic Publishers. Printed in the Netherlands.*

interests include invertebrate community dynamics and large scale biodiversity patterns in subsurface water. Present address: Université Claude Bernard - Lyon 1, UMR CNRS 5023, Ecologie des Hydrosystèmes Fluviaux, 43 Boulevard du 11 Novembre 1918, F-69622 Villeurbanne, Cedex, France.

CHRISTOPHER T. ROBINSON is an Alpine stream ecologist at the EAWAG and a senior lecturer at the ETH in Zürich, Present address: Department of Limnology, EAWAG/ETH, Ueberlandstrasse 133, CH-8600 Duebendorf, Switzerland.

CHRISTINE ROTHENBÜHLER completed her Masters of Science in geomorphology and GIS-application at the University of Zürich and presently is a PhD student within the GISALP-project (NFP 48). Present address: Institute of Tourism and Landscape, Academia Engiadina, Quadratscha 18, CH-7503 Samedan, Switzerland.

KLEMENT TOCKNER is an ecologist with the Department of Limnology at the EAWAG and a lecturer at the ETH in Zurich. His research focusses on the interaction between aquatic and terrestrial ecosystems and on the biocomplexity of riverine flood plains. Present address: Department of Limnology, EAWAG/ETH, Ueberlandstrasse 133, CH-8600 Duebendorf, Switzerland.

URS UEHLINGER is a stream ecologist with the Department of Limnology at the EAWAG. His research interests include organic matter dynamics and ecosystem metabolism in alpine and prealpine rivers. Present address: Department of Limnology, EAWAG/ETH, Ueberlandstrasse 133, CH-8600 Duebendorf, Switzerland.

CAMILLA WELLSTEIN is a PhD student at Justus-Liebig University, Germany. She completed her Masters of Science at the University of Freiburg, Germany on vegetation dynamics in the glacier foreland and the flood plain of the Val Roseg. Present address: Institute for Landscape Ecology and Resource Management, Interdisciplinary Research Center (IFZ), Justus-Liebig University, Heinrich-Buff-Ring 26, D- 35392 Giessen, Germany.

RAINER ZAH is research scientist with the Department of Sustainable Information Technology at the Swiss Federal Institute for Material Testing (EMPA), St. Gallen. He carried out his PhD research at the Swiss Federal Institute of Technology in Zürich on organic matter pathways in the Val Roseg. Present address: Department Sustainable Information Technology, Swiss Federal Institute for Material Testing, Lerchenfeldstrasse 5, CH-9014 St.Gallen.

PREFACE

This book is a monograph reporting the major findings from a comprehensive study of a glacial flood plain in the Swiss Alps, based on an intensive research program conducted year-round over several years. Until recently, very little was known regarding the ecology of glacial streams. Previous studies typically focused on one or a few aspects and were limited to the summer period. Moreover, this is the first ecological study of a glacial flood plain with a dynamic, multi-thread channel network. Year-round sampling of a system with a complex channel network spawned unanticipated results and new insights into the ecology of glacial streams.

The book begins with the landscape features, glacial history, and floodplain evolution of the Val Roseg. This is followed by chapters on channel typology, groundwater-surfacewater interactions, thermal heterogeneity, and nutrient dynamics. Chapters on the biota deal with terrestrial and aquatic flora, hyphomycete fungi, surface zoobenthos, and the interstitial fauna. Functional processes are addressed in chapters on organic matter dynamics, litter decomposition, nutrient limitation, and drift and colonization patterns. The final chapter provides a synthesis of our current understanding of the ecology of Val Roseg.

We are indebted to many individuals and organizations for assistance and support of the research program on the Val Roseg and the production of this book. First and foremost we thank the chapter authors for their contributions. Enthusiastic field assistance was provided by Donna Anderson, David Arscott, Christine Dambone, Barbara Fasnacht, David Ferreira, Fadri Guidon, Mäggi Hieber, Barbara Känel, Birgitt Klein, Mike Monaghan, Carsten Rinke, Phil Schenkel, Robert Skvarc, and Todd Wellnitz. Simone Blaser, Christine Dambone, and Christa Jolidon sorted and counted numerous invertebrate samples. Benthic algal samples were processed by Regula Illi, Esther Keller, and Sieglinde Gäbel. Gabi Meier, Silke Meyns, Richard Illi, Bruno Ribi, and David Kistler analyzed numerous water, chlorophyll a and ergosterol samples. Kirsten Callies assisted with the analyses of fungal community structure. We thank. Plinio Testa and his crew at the Hotel Roseggletscher for their hospitality, and the communes of Pontresina and Samedan for access to the study sites.

The following individuals provided taxonomic expertise: Michel Lafont (Oligochaeta), Reinhard Gerecke (Hydrachnellae) Pierre Marmonier (Ostracoda), Diana Galassi, (Copepoda), Johann Waringer (Trichoptera), Rolf Glatthaar (Simuliidae), Berthold Janecek (Chironomidae), Brigitte Lods-Crozet

J.V. Ward & U. Uehlinger (eds.), Ecology of a Glacial Flood Plain, xi-xii.
© 2003 *Kluwer Academic Publishers. Printed in the Netherlands.*

(Chironomidae), Philippe Richoux (Coleoptera), Wolfgang Riss (Limoniidae), and Enrique Descals (aquatic hyphomycetes).

We thank Werner Stumm and Mike Sturm for their advice regarding the identification of chemical tracers of water sources, Edi Hoehn for his expertise on the hydrogeological functioning of the Val Roseg, and Urs Beyerle for dating ground water samples. Jürg Zobrist provided valuable advice for the calculation of nutrient and organic matter export rates. Peter Huggenberger, employing ground penetrating radar, supplied critical information on the structure of the floodplain alluvium.

The Coordination Center for Aerial Survey and Aerial Photos of the Federal Office of Topography provided high resolution aerial photographs of the Roseg flood plain in 1998 and 1999, as well as historic b/w aerial pictures. Markus Niederöst of the Institute of Geodesy and Photogrammetry at ETH Zurich generated the Digital Elevation Model (DEM) and Philippe Meuret of the Remote Sensing Laboratory of the University of Zurich provided support for the ortho-rectification of historic aerial images. The Federal Office for Water and Geology provided discharge records of the Roseg River and MeteoSwiss made data available on air temperature, global radiation, and snow cover. The Swiss Federal Institute for Snow and Avalanche Research in Davos helped with advice and equipment to sample snow cover. Robert Berger and the members of the EAWAG workshop manufactured a variety of field equipment that ranged from logger housings and instrument shelters to respiration chambers, piezometers, and litter traps. Michael Döring helped with editing of the manuscript.

Research support was provided by the Swiss National Science Foundation (Grant 21-49243.96) and the EAWAG.

We sincerely thank Professor Robert Wetzel (Series Editor) and the professional staff at Kluwer, especially Judith Terpos, and Dr. Anne Besse for making this book a reality.

J. V. Ward and U. Uehlinger, Editors

CONTENTS

CHAPTER 1

U. UEHLINGER, M. MAISCH, C. ROTHENBÜHLER & R. ZAH

VAL ROSEG: A HIGH ALPINE CATCHMENT

1. INTRODUCTION

The Alps, covering an area of 240,000 km^2, extend from southwest to northeast with a length of about 1,000 km and a width between 130 and 250 km. The highest peak is Mont Blanc at 4807 m a.s.l. Mountains exceeding 3000 m in altitude usually have snow and ice covered summits; the glacierized area totals about 2900 km^2 (Haeberli et al., 1989). The Alps are a young mountain range with rugged topography and steep slopes. The Roseg valley (46° 29' 28" N, 9° 53' 57" E) is located in the Bernina massif, within the eastern part of the Swiss Alps (Fig. 1). The valley is drained by the Roseg River, a second order glacial stream that flows in a north-northeast direction before it joins the Bernina River near the village of Pontresina. The Roseg valley is part of the catchment of the Inn River, which is the major Alpine tributary of the upper Danube.

This first chapter is a general overview of the Roseg catchment. It describes the local and regional settings, which affect the structure and function of the complex stream ecosystem within this catchment. Some of the catchment characteristics are related to the fluctuations of glaciers since the Pleistocene. Glacial history is portrayed in detail in Chapter 2.

2. TECTONICS AND GEOLOGY

The rocks in the Bernina massif are more than 400 million years old, formed during several orogenic periods and subjected to folding, magmatism, metamorphism, intrusions, erosion and sedimentation. The Alpine orogeny began at the end of the Lower Cretaceous and continued until the Pliocene; the shortening of the crust produced folds and nappes. In Pliocene time, the Alps were presumably eroded to a chain of low hills (Trümpy, 1980) but the uplift resumed during the Pleistocene.

The high mountains of the Bernina massif are part of the Lower Austroalpine nappes. Basement rocks, in particular Variscan granitoids, form several sheets such as Sella, Corvatsch, Bernina and Stretta nappes. The granitoids consist of rock of varied chemistry and include diorites, granodiorites, syenits and akaline granites. The south ridge of the Val Roseg belongs to the Sella nappe (Fig. 2). A shallow formation (Coaz Mulde) consisting of breccia and dolomite boulders embedded in a

J.V. Ward & U. Uehlinger (eds.), Ecology of a Glacial Flood Plain, 1-16.
© 2003 *Kluwer Academic Publishers. Printed in the Netherlands.*

limestone-dolomite matrix separates the Sella and Corvatsch nappes. The Bernina nappe forms the northeastern part of the Val Roseg, including Piz Bernina.

The Quaternary geology is characterized by the Pleistocene glaciations, which shaped the morphology of the Roseg catchment. However, the geological record is almost lacking or poor for the Low and Middle Pleistocene, because of intense erosion during and between the younger glaciations.

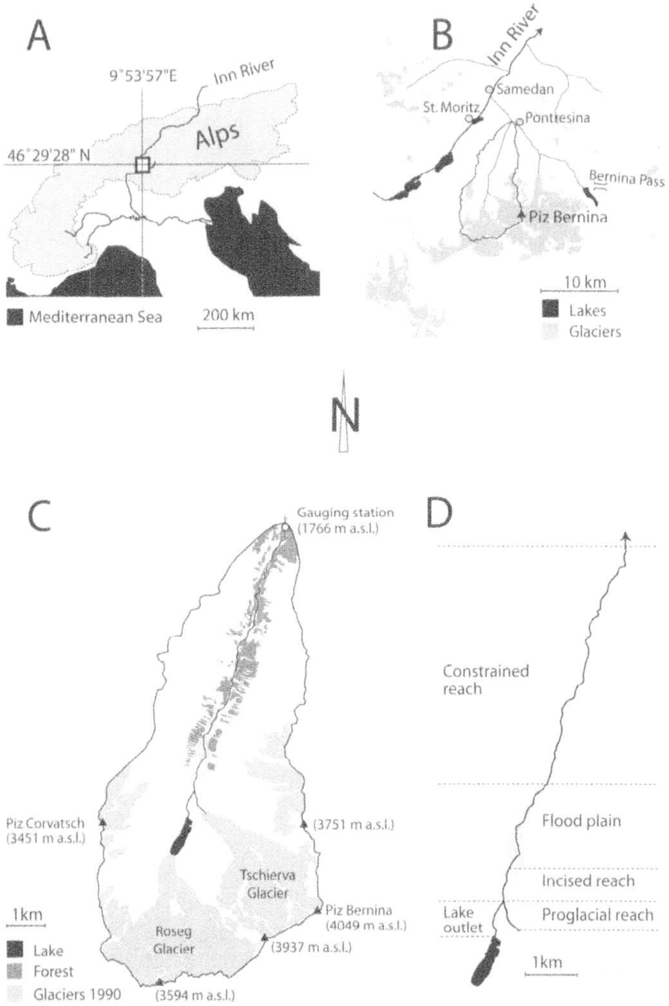

Figure 1. A) Location of the Val Roseg within the Alps, and B) within the upper Inn Valley. C) Roseg catchment, and D) the five main reaches of the Roseg River corridor.

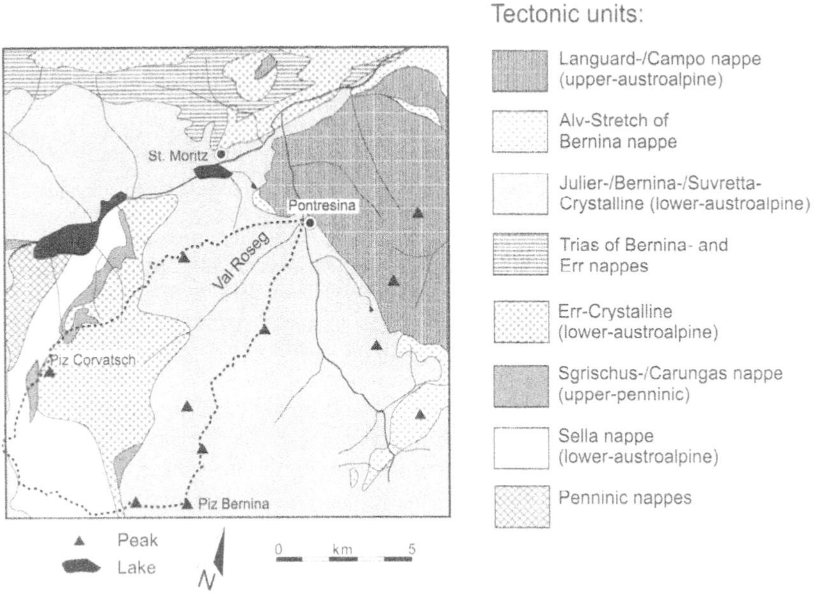

Figure 2. Tectonic map of the Roseg valley. Modified from Maisch et al. (1999), Fig. 62, p. 61.

3. CLIMATE

The Alps are in the transition zones between the temperate latitudes and the Mediterranean subtropics (north-south) and between the Atlantic and continental influence (west-east). Front ranges are humid because moist air from the Atlantic is forced to rise, which enhances precipitation. The climate of the inner Alps is more continental with lower precipitation and high seasonal temperature variation (Barry, 1992).

The Bernina massif separates the dry Inn valley from the wet slopes to the south. Precipitation declines from 1700 mm at the Bernina pass summit (2256 m a.s.l.) to 821 mm at Pontresina (1774 m a.s.l.). Average precipitation in the Roseg catchment equals 1573 mm (GIS-based interpolation, M. Zappa, unpublished data). June is the wettest and February the driest month (Fig. 3A). The probability of snow fall (days with snowfall/days with precipitation) is 0.5 for the upper Inn valley at Bever (1712 m a.s.l.) and 1.0 at elevations >3800 m a.s.l. (Gensler, 1978). Snow cover at the Bernina pass averaged 216 days per year (Gensler, 1978). The Roseg flood plain is usually snow covered from November to May. In the Bernina massif recent snowline[1] ranges from 2800 m a.s.l. (north facing slopes) to 3100 m a.s.l (southwest facing slopes) (Fig. 3B, Maisch, 1992).

[1] Equilibrium Line Altitude (ELA) on glaciers

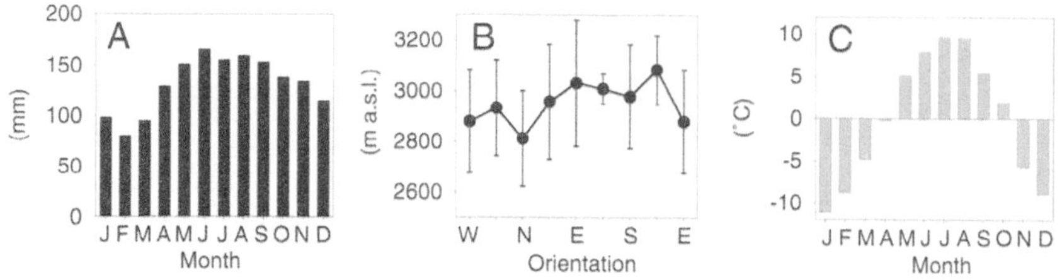

Figure 3. A) Average precipitation in the Roseg catchment (1981-2000, M. Zappa, unpublished data), B) Equilibrium line altitude (ELA) in the Bernina massif (average ± 1 SD). Modified from Maisch (1992), Fig. 8.11, p. 147. C) Estimated monthly average air temperature in the Roseg flood plain (see text).

Mean daily air temperatures in the Roseg flood plain (2025 m a.s.l.) are usually below 0°C until mid April and drop again below 0°C at the beginning of November (Fig. 3C). Mean annual temperature ranged from –0.5°C in 1996 to 0.5°C in 2000. In July, daily temperatures varied from 2.5 to 14.3°C and averaged 9.7°C. In January, monthly temperature averaged -11.1°C. These temperature ranges were obtained by extrapolation of temperatures recorded at the climate station of Samedan (1705 m a.s.l.) using the average surface air temperature lapse rate of the region (–0.56°C/100 m (Urfer et al., 1979).

4. VERTICAL ZONATION IN THE VAL ROSEG

The snow line and the treeline delimit the alpine life zone (Körner, 1999, 1995). In temperate latitudes the alpine treeline coincides with a mean air temperature of the warmest month of about 10°C (Körner, 1999). Elevation and vertical extent of the alpine life zone change from subarctic latitudes to the tropics (Fig. 4). However, superimposed on this global trend are altitudinal variations reflecting regional climate gradients[2].

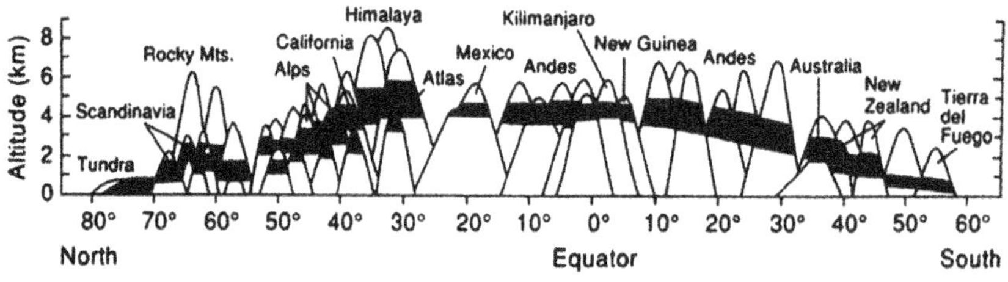

Figure 4. Altitudinal position of the alpine life zone between arctic and antarctic latitudes. From Körner (1999), Fig. 2.1, p. 10.

[2] In the Alps, treeline ranges from 1500-2600 m a.s.l. (Hermes, 1955). In the Swiss Alps, the treeline increases from 1750 m a.s.l. in the northern front range up to 2300 m a.s.l. in the central parts and declines to 1900 m a.s.l. in the southern front ranges.

With elevations ranging from 1760 m a.s.l. (near Pontresina) to 4049 m a.s.l. (Piz Bernina) the Roseg catchment extents from the subalpine to the nival zone. The subalpine zone comprises the transition zone between the upper boundary of the montane zone and the treeline. In the Alps, the lower boundary of the subalpine zone ranges from 1400 m a.s.l. in the front ranges to 1700 m a.s.l. in the inner mountain ranges. Coniferous forests dominate the vegetation of the subalpine zone unless mountain farming or frequent disturbances such as avalanches, debris flow and rock fall restrict tree growth. The elevation of the treeline in the Bernina massif varies between 2100 and 2250 m a.s.l. Environmental conditions above the treeline are harsh. Low temperatures (<10°C in July) and extended snow cover (> 6 months) restrict the growing season to a few months. Within the alpine life zone vegetation cover changes with altitude from shrubs and meadows (low alpine) to prostrate plants (high alpine). About 50% of the Val Roseg catchment belong to the alpine life zone. The snowline in the Val Roseg is 700 to 800 m above the treeline. In the nival zone, i.e. above the snowline, bare rocks, permanent snowfields and glaciers prevail, depending on topographic conditions. About 30% of the catchment belong to the nival zone (>3000 m a.s.l.).

5. LANDFORMS AND LANDSCAPE ELEMENTS

About 65% of the Roseg catchment is covered by glaciers, bedrock, and rock debris and 32% by grasslands and forests (Table 1). The contribution of the river corridor and a proglacial lake are relatively small.

Table 1. Relative areal contribution of different landscape elements in the Roseg catchment.

Catchment cover	*Upper catchment[1)] (%)*	*Entire catchment[2)] (%)*
Glacier	41.7	30.1
Bedrock, rock debris	32.3	35.2
Proglacial lake	0.6	0.5
River corridor incl. flood plain	3.4	2.6
Grassland	18.4	25.3
Subalpine forest	3.6	6.3

[1)] upstream of the lower floodplain end, area = 49.5 km^2 (Tockner et al., 1997)
[2)] area = 66.5 km^2

5.1. Landforms

The diversity of landforms in the Roseg catchment is impressive (Fig. 5). Many of the landforms are relatively young and distinct. The Val Roseg is a U-shaped glacial trough with steep, glacially smoothed side slopes and a valley floor dominated by glacio-fluvial processes. The south ridge of the upper valley forms a cirque of 7.5 km diameter (Fig. 1C) that is >60% glacierized. Superimposed on this macro-scale geomorphology are numerous meso-scale landforms. Late Würmian moraines can be traced along the valley between Pontresina and the lower end of the Roseg flood plain. However, the most impressive features of glacial activity are the long, sharp

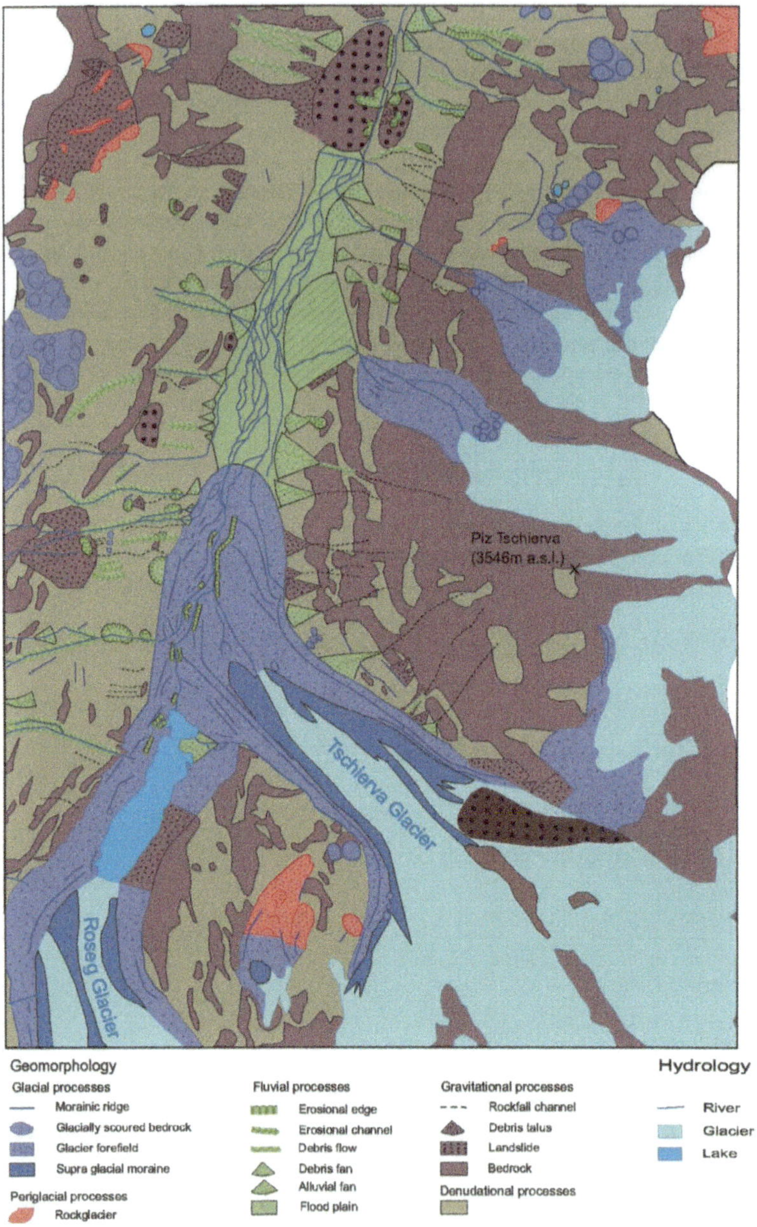

Figure 5. Geomorphology of the upper Roseg valley. Modified from Rothenbühler (2000).

crested lateral moraines of the Tschierva Glacier (Fig. 6). These moraines and those of the Roseg Glacier reflect the cumulative effect of Holocene cold periods including the last glacier advance at the conclusion of the Little Ice Age. Large numbers of talus, rock fall channels, erosional scars, debris flow tracks, alluvial- or debris fans, and the multi-channel flood plain reflect gravitational and fluvial processes. The formation of the flood plain is ultimately linked with a huge landslide, which controls the sediment output of the upstream-located fluvio-glacial system (see Chapter 2 for details). Slow mass movements driven by gravitation and water (denudation processes) result in rock creep features, rotational slips and successive slumping. Rock glaciers are the most apparent periglacial landforms, indicating discontinuous permafrost. Computer simulations that consider climate parameters such as temperature and solar radiation, as well as empirically found ranges of permafrost occurrence predict that the lower limit of permafrost varies between 2400 (north facing slopes) and 3000 m a.s.l. (south facing slopes) (Hoelzle, 1994). In the upper Inn valley the zone of potential permafrost decreased by about 15% between 1850 and 1990 (Hoelzle & Haeberli, 1995).

Figure 6. Lateral moraines and recent foreland of the Tschierva Glacier in 1997.

5.2. Glaciers

Glaciers belong undoubtedly to the most spectacular features of high alpine environments. Glaciers respond to climate variations but the interactions between climate, mass balance and glacier responses (advance and retreat) are complex. The ongoing glacier recession in the Alps is considered to be a clear signal of long-term changes in the energy and mass balance of the cryosphere (e.g., Haeberli & Beniston, 1998; Haeberli et al., 2002).

The Bernina region is the most densely glaciated mountain range in the Eastern Swiss Alps, despite relatively low precipitation. Glacier size and regional glacier distribution depend on topography, which determines aspect, altitude, and extent of the accumulation zones of existing glaciers. Glacierization is dense in the north facing valleys of the Bernina massif (e.g. Val Roseg, Val Morteratsch) and declines towards the outer edges, where the elevation of cirque headwalls is relatively low (about 3000 m a.s.l., Maisch, 1992).

The glacier inventories of 1850 (end of the Little Ice Age) and 1973 (reference year of the glacier inventory by Müller et al., 1976) both listed 87 glaciers located in the Bernina massif. In 1850, the glacierized area equaled 120 km^2 (Table 2, Maisch, 1992). Since then this area decreased by 34 km^2 (-28.5%, Maisch et al., 2000)), which is in accordance with the loss of glacierized areas in the Swiss Alps (-27.2%). However, local effects such as differences in aspect, glacier bed and cirque topography, wind direction and avalanches result in high variation at the level of individual glaciers. The area of small glaciers declined by 40 - 80% and a few tiny glaciers almost disappeared. In contrast, the relatively large and long glaciers lost only between 15 - 23% (Table 2). Detailed glacial history since the late glacial period is given in Chapter 2.

Table 2. Temporal changes of the four largest glaciers in the Bernina massif (Maisch et al., 2000, updated).

Glacier	*Areal extent (km^2)*			*Glacier length (km)*		
	1850/60	*2000*	*loss*	*1850*	*2000*	*loss*
Morteratsch	19.3	16.4	2.9 (-15.0%)	8.9	6.7	2.2 (-24.7%)
Roseg	11.1	8.5	2.6 (-23.4%)	6.8	4.3	2.5 (-36.8%)
Palü	8.0	6.5	1.5 (-18.8%)	5.5	3.8	1.7 (-30.9%)
Tschierva	7.6	6.2	1.4 (-18.4%)	5.8	4.7	1.1 (-19.0%)
Bernina region total [1)]	118.6	84.8	33.8 (-28.5%)			

[1)] 87 glaciers

The Roseg and Tschierva Glaciers are typical "valley glaciers" with pronounced tongue-shaped ice fronts (Maisch, 1992). During the climax of the last glacial advance in 1850 both glaciers formed a single spectacular terminus in the upper flood plain. The photograph of 1910 still shows an impressive ice front (Fig. 7). In 1934, after 80 years of continuous recession, the Roseg Glacier began to separate from the Tschierva Glacier (see also section 5.6 of this chapter).

5.3. Bedrock and sediments

Bedrock and rock debris prevail in ice-free areas of the nival zone and on steep slopes below the snow line. The glacier foreland of the Tschierva Glacier is rapidly growing (see Chapter 2). The receding glacier each year exposes an area of 3000 - 5000 m^2 (Zah et al., 2001), large parts of which consist of loose glacial till (ground moraine) and outwash gravel (Fig. 6). These mobile sediments are subject to fluviatile processes and might be transported into the flood plain.

Figure 7. The Roseg Glacier in 1910 (postcard).

5.4. Alpine grasslands

The vegetation above the treeline is characterized by nutrient-poor alpine grassland dominated by tussock forming sedges. On the siliceous rocks ferric podzols develop (Richter, 1994, see also Chapter 7). Tall-growing herbs (*Adenostylo-Cicerbitetum* association) occur in the riparian zone of rivulets. The prevailing short grass alpine meadow is formed by the sedge *Carex curvula* All. Below 2600 m a.s.l. alpine meadows are dominated by evergreen sedge *Carex sempervirens* Vill. (the *Seslerio-Caricetum sempervirentis* association, Richter, 1994). In the transition zone from the alpine grasslands to the timberline ecotone the vegetation is characterized by dwarf-shrub vegetation such as juniper heath (*Arctostaphylo-Juniperetum nanae* association), rhododendron and blueberry heath (*Rhododendro-Vaccinietum* association). Mat-grass (*Nardus stricta* L.) mainly occurs in areas influenced by cattle grazing (*Aveno Nardetum* association).

5.5. Subalpine forests

The subalpine forest extends along the side slopes of the Roseg Valley to an elevation of about 2250 m a.s.l. The forest is fragmented by erosional channels and avalanche tracks (Fig. 8) usually colonized by green alder (*Alnetum viridis* association). The treeline declines from the lower end to the middle of the flood plain by about 100 m and upstream of the flood plain forests are lacking (Fig. 1C). The dominant tree species of the subalpine forests are larch (*Larix decidua* Miller) and Arolla pine (*Pinus cembra* L.). The larch-Arolla pine forest (*Larici-Pinetum Cembrae* association) on the moraine deposits of the side slopes is the climax community of the supra-subalpine zone of the Central Alps (Ellenberg, 1988). On the siliceous bedrock podzol develop under the humus-rich forest floor (Richter, 1994). A variety of successional vegetation stages can be observed on geomorphologically young landforms <150 year old, see Chapter 7).

Figure 8. Subalpine forest (1998) on the left valley slope along the flood plain.

5.6. The proglacial lake

After 1934, the Roseg Glacier receded behind the left lateral moraine of the Tschierva Glacier. This moraine reduced the export of glacial sediments and gave rise to an unusual low-gradient glacier foreland, where a lake was formed. In 1977, mean depth and volume of the lake equaled 7 m and 1,519,000 m^3, respectively (Kilger, 1979). Since then the lake area increased from 0.22 to about 0.36 km^2 in 1998 (Fig. 9).

5.7. The river corridor

Five morphologically different sections characterize the corridor of the Roseg River (Figs. 1, 9):

1) The 650 m long proglacial reach below the Tschierva Glacier is steep in the uppermost part (slope: >7%) and characterized by braided channels in the middle reach. Elevations range from 2085 – 2145 m a.s.l. The upper end of the proglacial section, where large amounts of unconsolidated sediments prevail, is

rapidly growing (see Chapter 2). Channel stability within the proglacial reach is low, particularly in the uppermost reach.

Figure 9. Views of the Roseg River Corridor. A) Proglacial reach below the Tschierva Glacier (1997). From Burgherr et al (2001), Fig. 1, p. 1707. B) Lake outlet stream and Roseg Lake (1997), C) Incised reach between confluence and flood plain (2002), D) The Roseg flood plain (1997). From Zah & Uehlinger (2001), Fig. 2, p. 1600. E) constrained reach downstream of the Roseg flood plain (2002).

2) The lake outlet system is about 900 m long with elevations ranging from 2085 – 2159 m a.s.l. The upper part of this section is relatively flat (slope 2%), whereas

the middle reach is steep (slope: >8%) and deeply incised in the lateral moraine of the Tschierva Glacier. High channel stability and lack of bedload transport characterize the entire reach.

3) The incised reach consists of a 700 ml long stable single-thread channel incised in glacial till downstream of the confluence of the proglacial stream and the lake outlet (elevation 2055 – 2085 m a.s.l., slope 5.2%).
4) The Roseg flood plain is 2750 m long and 130 –510 m wide with elevations varying between 1981 – 2055 m a.s.l. The area of the flood plain is 0.67 km^2, 85% of which is sparsely vegetated (see Chapter 7). The upper section of the flood plain is steeper (slope: 4.8%) and wider (maximum width = 510 m) than the lower section (slope: 1.9%, width: 130-260m). Several channel types can be distinguished based on water source and connectivity to the main channel (Malard et al., 2000; Tockner et al., 1997, Chapter 4).
5) The 7200 m long constrained reach (slope: 3.3%), where the river flows through a narrow valley, extends from the lower end of the flood plain to the confluence with the Bernina River (1768 m a.s.l.) . Numerous sills have been installed in this reach to control channel incision and a few banks have been artificially stabilized.

6. HYDROLOGY AND CHANNEL MORPHOLOGY

The glacial flow regime is a major factor determining the physical habitat template of the river-floodplain system in the Val Roseg. Glacial meltwater from the Tschierva and Roseg Glaciers is the primary water source of the Roseg River. The Tschierva and Roseg Glaciers contribute about 40% and 60%, respectively, of the water entering the flood plain through the main channel. Valley slope aquifers and floodplain tributaries provide an additional 5 - 7% to the discharge at the end of the flood plain.

6.1. Discharge

Mean annual discharge averaged 2.8 m^3s^{-1} between 1955 and 2000 (FOWG, 2000). The time series of annual discharge reflect decadal climatic variation (Fig. 10A). A series of cool summers between 1965 and 1980 coincided with low annual runoff The water balance equation for the Val Roseg (Table 3) indicates net loss of storage (glacier mass loss) during the period from 1991 to 1998.

Table 3. Average water balance (1991 – 1998) of the Roseg catchment (mm).

Discharge at the end of the catchment	*Precipitation* [1)]	*Evapo-transpiration* [2)]	*Input from Storage*
1465	1573	337	229

[1)] M. Zappa, unpublished data [2)] Zappa et al. (2000).

6.1.1. The glacial flow pulse

Annual precipitation falling as snow increases by 2.5 – 3.5% for every 100 m increase in elevation; in the Alps virtually all precipitation is in solid form at altitudes of 3500 – 4000 m a.s.l. (Röthlisberger & Lang, 1987). Snow and ice form a transient water storage that is partly released during summer. Major drivers of the snow and ice melt process are air temperature and radiation (Hock, 1999), both of which change in a rather predictable way during an annual cycle. As consequence, discharge patterns of glacial streams are characterized by a distinct flow pulse that lasts for several months (Fig. 10B). Uehlinger et al. (2003) defined a flow threshold of 0.75 m^3s^{-1} (5x average March discharge) for the beginning and end of the annual flow pulse of the Roseg River. Based on this criterion the flow pulse begins on 30 April (± 8 days) and ends on 30 October (± 14 days). Average discharge is maximum in July with 9.4 m^3s^{-1} and minimum in February with 0.08 m^3s^{-1} (1990 – 2000). Despite distinct inter-annual variation in discharge (Fig. 10C) the general flow pattern is highly predictable compared to streams of similar size at low elevation.

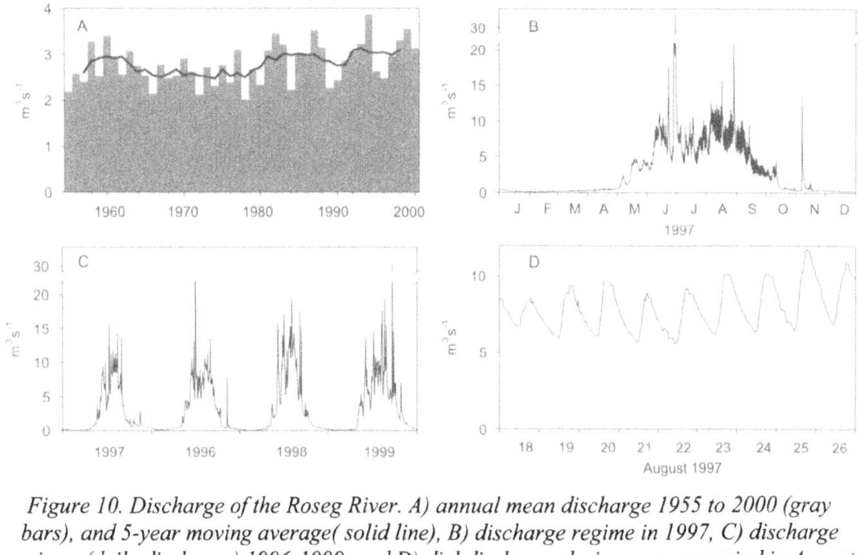

Figure 10. Discharge of the Roseg River. A) annual mean discharge 1955 to 2000 (gray bars), and 5-year moving average(solid line), B) discharge regime in 1997, C) discharge regimes (daily discharge) 1996-1999, and D) diel discharge during a warm period in August 1997.

6.1.2. Short-term variations

Periodic and aperiodic discharge fluctuations are superimposed on the flow pulse. During episodes of high air temperatures and net radiation, enhanced freeze-thaw cycles produce regular diel discharge patterns (Fig. 10D). Precipitation usually coincides with reduced air temperature and incoming radiation. Freshly fallen snow raises the albedo of the glacier surface, which can reduce ice melt and runoff for several days (Collins, 1998; Röthlisberger & Lang, 1987). However, intense rainfall over an entire glaciated catchment result in major floods as e.g. in Val Roseg between 27 and 28 June 1997, Fig. 10B). Heavy rain may also trigger sudden changes in the unstable glacial drainage system and give rise to meltwater outbursts (Collins, 1998). Rainfall induced floods occur more frequently during the second half of the ablation period (Table 4) when the transient snowline is high and the englacial and subglacial drainage system well developed (Collins, 1998; Röthlisberger & Lang, 1987). Flow of this magnitude may immediately modify channel configuration in the flood plain or initiate subsequent lateral channel shift. The partial outbreak of the Roseg Lake during a heavy rainstorm on 21 August 1954 was a single event that resulted in a flow peak of about 120 m^3s^{-1}; within few hours the lake lost about 100,000 m^3 of water causing severe damage in the constrained reach (Töndury, 1954).

Table 4. Monthly frequency of spates with maximum discharge > 20 m^3s^{-1} (1974 –2000). U. Uehlinger, unpublished data.

Month	*Spates per month*
June	0.22
July	0.61
August	1.00
September	0.74
October	0.04

6.2. Channel characteristics

Depending on discharge, the width of the main channel ranged from 2 to 8 m in the proglacial reach (sites 1, 6), 6 to 14 m in the lake outlet (sites10, 11), 0 (dry period) to 20 m in the flood plain (sites 56 and 80), and 6 to 25 m the constrained reach (sites 97, 100 and 101) (Burgherr & Ward, 2000; Burgherr et al., 2002; Uehlinger et al. 2003)[3]. In floodplain channels excluding the main channel, channel width varied between 0 (dry) and 13 m (see also Fig. 9 in Chapter 1 and Fig. 8 in Chapter 4).

Bed sediments, in particular those of the surface layer, were relatively coarse (Table 5). Large boulders occurred in the uppermost steep proglacial reach, in the incised reach and the lake outlet, where the river was deeply cut into the lateral moraine of the Tschierva Glacier. In the main channel, sediment particles were often embedded. Freeze core samples showed a significant increase of fine fractions (<4 mm) with depth (R. Zah, unpublished data).

[3] Zah et al. (2000) reported that the width of the main channel ranged from 2.2 to 5.1 m in the flood plain. However, these values also included the widths of first order side channels in a corridor along the main channel.

Table 5. Size of surface layer bed sediments (B-diameter in cm) in the upper Val Roseg[1].

	Main channel			*Flood plain channels*[2]
	Lake outlet	*Proglacial reach*	*Flood plain*	
No. of sites	1	1	2	4
Median	9.0	9.3	10.5	7.0
D_{90}	21.1	18.0	16	10.5
Mean ± 1 SD	11.4±7.4	10.3±5.3	11.4±5.1	7.6±3.8

[1)] based on data from Burgherr & Ward (2000) and Burgherr et al. (2002)
[2)] excluding the main channel

7. VAL ROSEG – A NEAR NATURAL STUDY SYSTEM

The upper Val Roseg is a morphologically and hydrologically highly diverse Alpine catchment. It provides a unique opportunity to study a near natural glacial stream ecosystem. Relatively moderate human impacts include cattle grazing and a tourist facility (hotel) at the end of the flood plain plus enhanced atmospheric deposition that raise nitrate levels in precipitation (see Chapter 6). Cattle grazing is restricted to a few weeks in June and September and to those parts of the flood plain, where the access is not impeded by channels with high flow. Treated wastewater from the hotel is discharged into parafluvial gravels at the end of the flood plain but effects on water quality are small (within detection limits of chemical methods in the main channel).

8. REFERENCES

Barry, R. C. (1992). *Mountain weather and climate.* London: Routledge.

Burgherr, P., & Ward, J.V. (2000) Zoobenthos of kryal and lake outlet biotopes in a glacial flood plain. *Verhandlungen der Internationalen Vereinigung für Theoretische und Angewandte Limnologie, 27,,* 1587-1590.

Burgherr, P., & Ward, J. V. (2001). Longitudinal and seasonal distribution patterns of the benthic fauna of an alpine glacial stream (Val Roseg, Swiss Alps). *Freshwater Biology, 46,* 1705-1721.

Burgherr, P., Ward, J.V., & Robinson, C.T. (2002) Seasonal variation in zoobenthos accross habitat gradients in an alpine glacial flood plain (Val Roseg, Swiss Alps). *Journal of the North American Benthological Society, 21,*561-574.

Collins, D. N. (1998). Outburst and rainfall-induced peak runoff events in highly glacierized Alpine basins. *Hydrological Processes, 12,* 2369-2381.

Ellenberg, H. (1988). *Vegetation ecology of central Europe.* Cambridge: Cambridge University Press.

FOWG. (2000). *Hydrologisches Jahrbuch 1999.* Berne, Switzerland: Federal Office for Water and Geology.

Gensler, G. A. (1978). *Das Klima von Graubünden. Arbeitsbericht,* 77, pp. 122. Zürich: Schweizerische Meteorologische Zentralanstalt.

Haeberli, W., & Beniston, M. (1998). Climate change and its impacts on glaciers and permafrost in the Alps. *Ambio, 27,* 258-265.

Haeberli, W., Bösch, H., Scherler, K., Ostrem, G., & Wallén, C. C. (1989). *World glacier inventory - status 1988. A contribution to the Global Environment Monitoring System (GEMS) and the Internatioinal Hydrological Programme*: IAHS/GEMS/UNEP/UNESCO.

Haeberli, W., Maisch, M., & Paul, F. (2002). Mountain glaciers in global climate-related observation networks. *World Meteorological Organization Bulletin, 51*, 18-25.

Hermes, K. (1955). *Die Lage der oberen Waldgrenze in den Gebirgen der Erde und ihr Abstand zur Schneegrenze* (Vol. 5). Köln: Geographisches Institut der Universität Köln.

Hock, R. (1999). A distributed temperature-index ice- and snowmelt model including potential direct solar radiation. *Journal of Glaciology, 45*, 101-111.

Hoelzle, M. (1994). Permafrost und Gletscher im Oberengadin: Grundlagen und Anwendungsbeispiele für automatisierte Schätzverfahren. *Mitteilungen der Versuchsanstalt für Wasserbau, Hydrologie und Glaziologie*, No. 132, pp. 121. Zürich: Eidgenössisch Technische Hochschule.

Hoelzle, M., & Haeberli, W. (1995). Simulating the effects of mean annual air-temperature changes on permafrost distribution and glacier size: an example from the Upper Engadin, Swiss Alps. *Annals of Glaciology, 21*, 399-405.

Kilger, B. (1979). *Die Sedimente am Eisrand des Roseggletschers (Graubünden, Schweiz)*. PhD thesis, Eberhard-Karls-Univeristät, Tübingen.

Körner, C. (1999). *Alpine plant life*. Berlin: Springer.

Körner, C. (Ed.). (1995). *Alpine plant diversity: a global survey and functional interpretations* (Vol. 113). Berlin: Springer.

Maisch, M. (1992). *Die Gletscher Graubündens* (Vol. 33). Zürich: Geographisches Institut, Universität Zürich.

Maisch, M., Burga, C. A., & Fitze, P. (1999). *Lebendiges Gletschervorfeld. Von schwindenden Eisströmen, schuttreichen Moränenwällen und wagemutigen Pionierplflanzen im Vorfeld des Morteratschgletschers. Führer und Begleitbuch zum Gletscherlehrpfad Morteratsch* (2. Auflage). Pontresina: Gemeinde Pontresina.

Maisch, M., Wipf, A., Denneler, B., Battaglia, J., & Benz, C. (2000). *Die Gletscher der Schweizer Alpen. Gletscherhochstand 1850, aktuelle Vergletscherung, Glescherschwund-Szenarien.* Zürich: VdF-Hochschulverlag ETH Zürich.

Malard, F., Tockner, K., & Ward, J. V. (2000). Physico-chemical heterogeneity in a glacial riverscape. *Landscape Ecology, 15*, 679-695.

Müller, F., Caflisch, T., & Müller, G. (1976). *Firn und Eis der Schweizer Alpen, Gletscherinventar.* (No. 57-57a). Zürich: Geographischen Institut, Eidgenössich Technische Hochschule Zürich.

Richter, M. (1994). Die Pflanzensukzession im Vorfeld des Tschierva-Gletschers/Oberengadin. *Geoökodynamik, 15*, 55-88.

Rothenbühler, C. (2000). *Erfassung und Darstellung der Geomorphologie im Gebiet der Bernina (GR) mit Hilfe von GIS.* Diploma thesis, University of Zürich, Zürich.

Röthlisberger, H., & Lang, H. (1987). Glacial Hydrology. In A. M. Gurnell & M. J. Clark (Eds.), *Glacio-fluvial sediment transfer* (pp. 207-284). Chichester: Wiley.

Tockner, K., Malard, F., Burgherr, P., Robinson, C. T., Uehlinger, U., Zah, R., & Ward, J. V. (1997). Physico-chemical characterization of channel types in a glacial floodplain ecosystem (Val Roseg, Switzerland). *Archiv für Hydrobiologie, 140*, 433-463.

Töndury, G. A. (1954). Ursachen und Bekämpfungsmöglichkeiten der zunehmenden Hochwassergefahr im Engadin. *Wasser- und Energiewirtschaft, 46*, 308-323.

Trümpy, R. (1980). *Geology of Switzerland.* Basel: Wepf Publishers.

Uehlinger, U., Malard, F., & Ward, J. V. (2003). Thermal patterns in the surface waters of a glacial river corridor (Val Roseg, Switzerland). *Freshwater Biology, 48*, 284 -300.

Urfer, C., Gensler, G. A., Ambrosetti, E., & Zenone, E. (1979). *Regionale Klimabeschreibungen. 2. Teil. Alpennordhang, Graubünden und Alpensüdseite*, pp. 247-437. Zürich: Schweizerische Meteorologische Anstalt.

Zah, R., Niederöst, M., Rinderspacher, H., Uehlinger, U., & Ward, J. V. (2001). Long-term dynamics of the channel network in a glacial floodplain, Val Roseg, Switzerland. *Arctic, Antarctic, and Alpine Research, 33*, 440-446.

Zappa, M., Badoux, A., & Gurtz, J. (2000). The application of a complex distributed hydrological model in highly glaciated alpine river catchment. In J. Horvatic (Ed.), *Limnological Reports. 33rd Conference of the International Association for Danube Research. Osijek, Croatia, 3-9 September 2000.* (pp. 23-28).

CHAPTER 2

R. ZAH, M. MAISCH, U. UEHLINGER & C. ROTHENBÜHLER

GLACIAL HISTORY AND FLOODPLAIN EVOLUTION

1. INTRODUCTION

The Roseg valley was shaped by glaciers and glacial activity is still the dominant force altering floodplain habitats. Today, more than 40% of the catchment of the Roseg flood plain is glacierized. Glacial influence on floodplain dynamics is expected to increase in the future in response to climate change.

The impact of glaciers on floodplain habitats is manifold: the hydraulic regime of glacial flood plains is driven by diel and seasonal freeze-thaw cycles; most floodplain sediments originate from glacial moraines; and both glacial-driven sediment transport and flood events structure the fluviatile channel network. Consequently, today's floodplain habitats in the Val Roseg can be seen as a product of glacial history. To some extent, even the distribution of terrestrial and aquatic biocenoses reflects the history of glacial movements.

This chapter aims to highlight the glacial dynamics in the Roseg catchment and to link these environmental forces to the evolution of the Roseg flood plain. Because environmental dynamics are always scale-dependant, different scales of observation have been chosen to reveal different aspects of glacio-fluvial dynamics. On a long-term scale, this chapter outlines the glacial dynamics of Val Roseg since the last ice age, when the present flood plain was formed and developed. On a short-term scale, glacial movements and geomorphic response of the flood plain over the last 50 years are analyzed.

2. GLACIAL DYNAMICS

2.1. Glacial History

Glaciers or ice streams act like powerful transportation systems and can, therefore, in a closed catchment area, be considered as sediment evacuation media. They can produce, carry and redeposit masses of morainic material, building up new landforms, such as terminal and lateral moraines or ice margin terraces (i.e. kames), by glacial or fluvio-glacial action. Application of complex GIS-based techniques enables one to produce detailed glaciological reconstructions and to calculate equilibrium lines (ELA) and other paleo-climate parameters from contemporary

J.V. Ward & U. Uehlinger (eds.), Ecology of a Glacial Flood Plain, 17-36.
© 2003 Kluwer Academic Publishers. Printed in the Netherlands.

deposits (Kerschner et al., 2000). The interpretation of the morainic sequences allows one to draw a spatio-temporal system of the late-glacial ice retreat, which can be adapted and modified for different inneralpine regions, such as the Bernina (Fig. 1). The relative chronology of late-glacial stadials and the subsequent ice retreat phases (interstadials) inside the Alps is widely accepted today (Kerschner et al., 1999). However, the absolute chronology of these glacier retreat stadials is less well established, based only on a few, sometimes very questionable radiocarbon dates (base samples of peat bogs), some selected pollen profiles and the analysis of lake sediments located close to late-glacial or Holocene glacier activity.

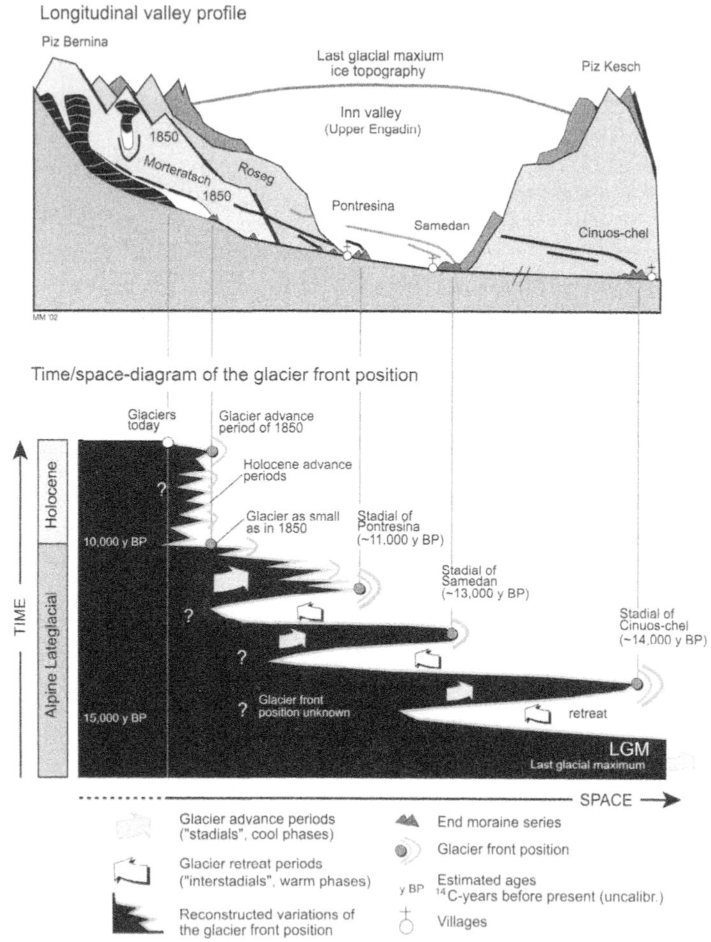

Figure 1. Schematic illustration of the regional system of late-Würmian glacier retreat in the Upper Engadin (Bernina-, Morteratsch-, Roseg- and Tschierva glaciers). The longitudinal valley profile (horizontal axis) is combined with their assumed chronological position (vertical axis). Modified from Maisch (1999), Fig. 87, p. 89.

Present knowledge holds that the Pleistocene ice stream network of the Alps retreated very quickly after the end of the LGM (Last glacial maximum) of the "Würmian ice-age". The overall recession of glaciers thereby followed the main trend of the warming process that began about 18,000 years ago and can be divided and subdivided into various steps and stages (stadials or interstadials). The late-glacial ice decay in and around the Bernina region formed remarkable morainic systems near Cinuos-chel (Clavadel-stadial), Samedan (Daun-stadial) and Pontresina (Egesen-stadial) (Gamper & Suter, 1982; Maisch et al., 2000). Equivalent morainic series, correlated by ELA-depressions and moraine geomorphology, can be traced also in other adjacent valleys of the Upper Engadin and serve as a geomorphic base for paleo-glaciological reconstructions and for explaining the late-Würmian landscape evolution (Fig. 2a-c).

Sediment cores taken in the lakes of Sils, Silvaplana, Champfèr and St. Moritz record detailed environmental changes due to distinct climatic variations (Ohlendorf, 1999). Most parts of the Upper Engadin seem to have been deglaciated after the Cinuos-chel (Clavadel-stadial) at around 14,000 yBP (uncalibr.). The Daun-stadial (13,000 yBP), except the readvance of the Bernina glacier into the Samedan basin, was restricted to smaller local valley glaciers. The Egesen stage in the eastern Alps or, for example, the regional Pontresina-stadial in the Bernina during the Younger Dryas period (10,800 to 10,000 years BP, uncalibr.), represent distinctive advance as well as rapid retreat phases (Maisch, 1995). A sharp crested morainic ridge at Julierpass, thought to be of Egesen-age (Suter, 1981), could be directly connected with the Younger Dryas by using exposure dating techniques (Ivy-Ochs et al., 1996). In the Val Roseg numerous stadial moraines can be found and the related former ice bodies reconstructed. According to Rothenbühler (2000) three distinct series could be mapped and correlated tentatively with phases of the Pontresina-stadial (Egesen, Younger Dryas).

With the beginning of the Holocene period about 10,000 years ago and the final melting back of the Alpine glaciers far into the upper reaches of the mountains, there appears evidence of an extraordinarily steady climatic and glacial development with only minimal fluctuations (Fig. 3). This general pattern can be revealed because glaciers of Holocene age passed over soils and organic layers and they overrode living trees that emerged during the ice-free periods. It is possible to trace the Holocene history of glacier variations far back to prehistoric times by analyzing these "silent witnesses", especially by radiocarbon dating (Gamper & Suter, 1982). Compared with the late-glacial epoch, it is now possible to document a much "higher", that is, warmer, climate level. Eight clearly substantiated advancing cycles can be compared with the same number of phases with minimal glacial extent similar to that found today. The maximum extent period of 1850 with its huge morainic systems thus represents a dimension of glacial history which is not only typical for that period of the Little Ice Age (1350 - 1850), but also for almost all older maximum extent periods of the Holocene (Gamper & Suter, 1982; Holzhauser, 1995; Patzelt, 1973). However, there is much more uncertainty in the knowledge about the "warm period" minimum extents than about the geomorphological-stratigraphic periods of advance, for which clear proof can usually be given.

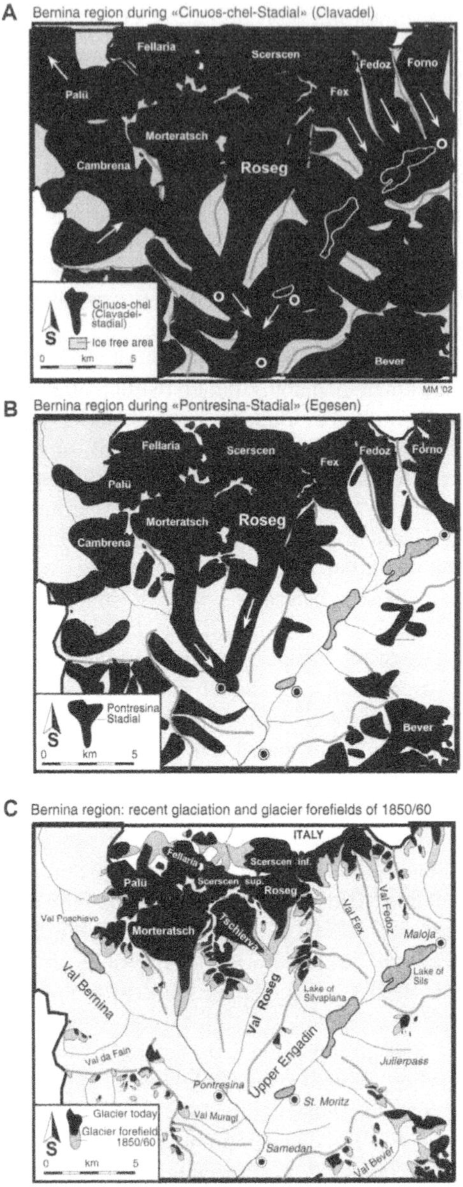

Figure 2. Maps of late-glacial ice extent during A) the Cinuos-chel- (appr. 14,500 y BP, oldest Dryas, Clavadel/Senders) and B) the Pontresina-stadial (Younger Dryas, Egesen) in comparison with C) the advance period of 1850 (Little Ice Age) and the contemporary glacierization. Modified from Maisch et al. (1999), Fig. 7, p. 11, Fig. 17, p. 21, and Fig. 89, p. 91.

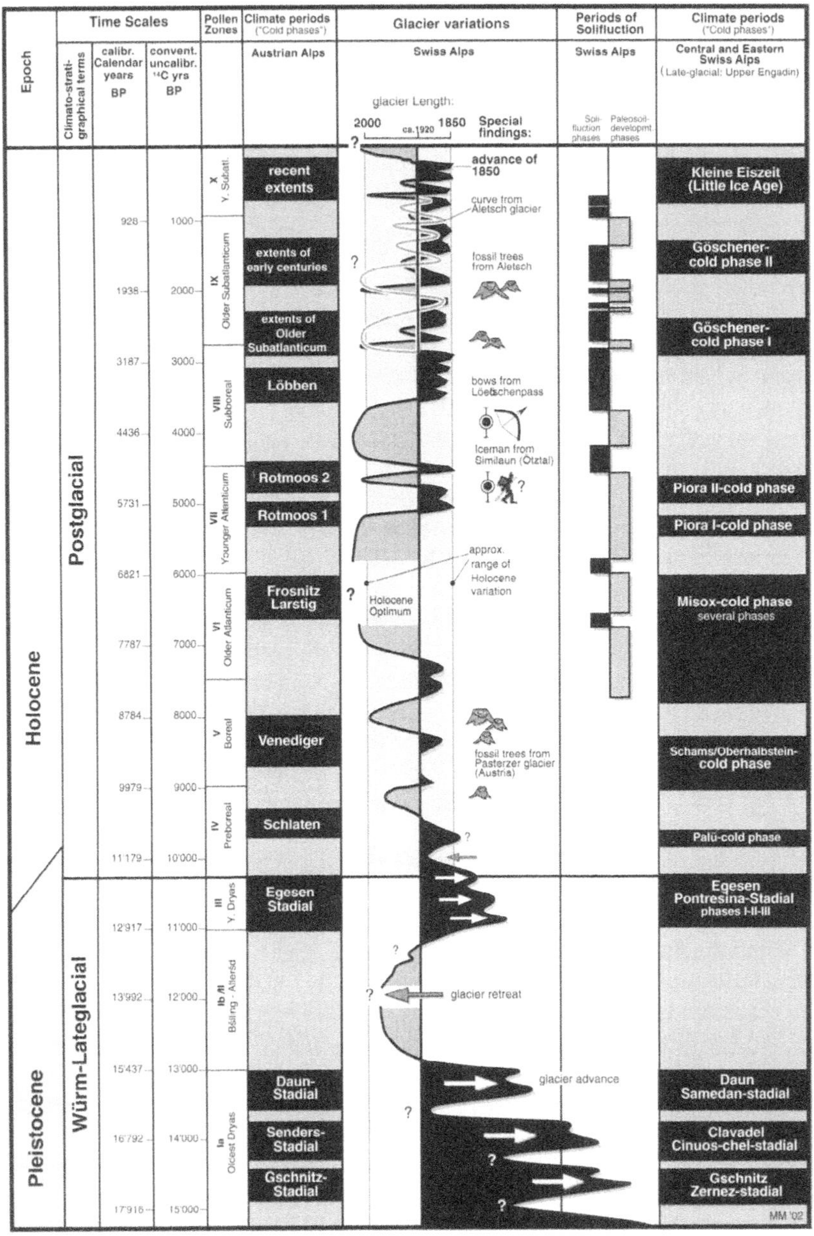

Figure 3. Chronology of late-glacial and Holocene climate and glacier fluctuations in the Eastern Alps, according to the results of different archives (glacier history, solifluction phases). Compiled from various sources after Maisch et al. (2000), Fig. 7, p. 11, Fig. 17, p. 21, and Fig. 89, p. 91.

Summarizing the main results of the Holocene climate and glacier fluctuations, which also apply to the Roseg and Tschierva areas, the last 10,000 years was a period of glacial and climatic variations of always similar size, and of comparable temperature amplitude (±1°C). Within the range of post-glacial activity, the increase of the glaciers around the middle of the 19th century marks a characteristic order of maximum extent which was reached repeatedly in the case of glacier advances, but was surpassed only rarely during cold phases in the past. The current situation is, thus, even after the strong melting period in the last 150 years, still within the "normal" Holocene variation, while being clearly in the "warm" transition zone of all reconstructable glacier and climate fluctuations. This range will probably be exceeded if the predicted temperature rise scenarios are realized (IPCC, 1990, 1995).

Roseg and Tschierva are among the 121 glaciers of the Swiss observation network, where annual measurements of the length changes have been taken for more than a century (Herren et al., 2001; Hoelzle et al., 1999). In this Swiss sample of predominantly large and long glaciers, Tschierva with a length (in 2000) of 4.7 km takes 40th place and Roseg with a length of 4.3 km 43rd place (Maisch, 2001). Observation of the length variations started in 1894 (Roseg) and in 1934 (Tschierva), when these glaciers separated from each other. The long-term signal of glacier and climate change can best be recognized and interpreted by drawing cumulative length change curves (Fig. 4). The selected glaciers show, depending on their inital size and length in 1850, an individual but clearly visible recession trend of first order quality. The overall retreat distances are highly variable and range from 500 m (Paradisin, a very small glacier) up to more than 2000 m (Roseg, Morteratsch, Tschierva, all being long glaciers). On this general picture of long-term ice retreat and therefore of long-term negative mass balances, three smaller periods of readvance activities – or at least of decreasing melting rates – are superimposed, which are very typical also in other glacierized regions of the Alps (advance periods of 1890, 1929 and 1965-1985).

During the observation period 1894-2001 Roseg glacier retreated totally –2261 m at an average rate of –24.3 m/year. The most positive year was 1917 (+56 m) and the most negative was recorded just recently in 2001 (–118 m), probably reflecting the extraordinary warm period of the 1990s. Since Tschierva was measured separately (from 1934-2001) this glacier lost totally –1045 m which accounts for a mean net retreat rate of –19 m/year. The reaction of this glacier – probably due to its steep bed topography – was quite sensitive to the slight cooling trend starting in the 1960s with a clear readvance signal (until 1985). The year with maximum advance rate was 1974 (+56 m). Despite the individual character on a year to year scale, the general retreat behavior of Roseg and Tschierva is in good correlation with the ongoing global warming trend and the worldwide observable acceleration in negative mass balances at the moment (Haeberli et al., 2002).

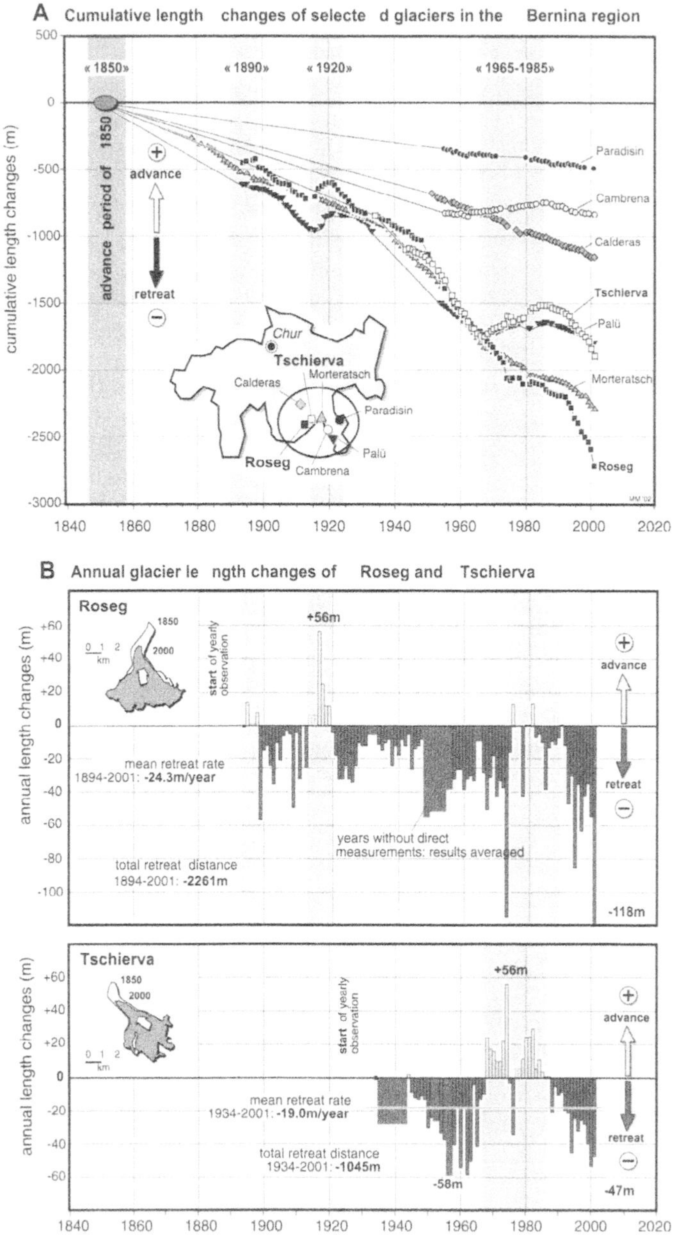

Figure 4. A) Cumulative glacier length changes of selected glaciers in the Bernina region and B) annual measurements of Roseg- and Tschierva glaciers since the start of their observation (data source: Glaciological commission of the SANW, Swiss Academy of Natural Sciences).

2.2 Future Scenarios

In order to model future glacier scenarios a simple geometrical approach was used and applied to the whole Swiss and also the whole Bernina glacier data set. The vertical interval between the 2:1-ELA (equilibrium line altitude) and the maximum glacier elevation (highest glacier point) were taken as the key parameters in estimating the sensitivity of each glacier under different new 2:1-ELA conditions. Those 2:1-ELA scenarios were calculated in 100 m steady state steps under the assumption that a glacier will disappear when its 2:1-ELA exceeds the upper glacier limit for a sufficient period of time. According to this hypothesis each glacier was classified into the categories *melting* or *remaining* by checking step by step its vertical extension compared to the virtual new 2:1-ELA position (2:1-ELA scenarios of +100 m, +200 m, +300 m, +400 m, +1000 m, etc.).

The results combined with given temperature rise scenarios (IPCC, 1990, 1995) were regionally summarized and plotted in Figure 5. The simulation and the statistics of future ice retreat scenarios result in a dense bundle of regionally different ice-decay curves. They indicate the percentage of remaining glacier bodies (y-axis) with respect to the assumed 2:1-ELA rise scenarios (x-axis). The steep slope of the outcoming curves leads to the general conclusion that a tendency of accelerated ice decay may even start to occur as early as the first third of this century. Regions with only weak glacierization and small glaciers in general will react much faster to the expected changes in climate (i.e. Lower Engadin), whereas the densely glacierized regions of Bergell and Bernina with still large glacier bodies will react a little bit slower but within the mean range of all tested Swiss glacier regions.

To visualize the consequences of the ongoing "global warming", a sequence of four block diagrams of Tschierva and Roseg glaciers is presented in Fig. 6a-d. These hypothetical scenarios may give us a better understanding of glacier dynamics but also of the increasing vulnerability of the glacial and periglacial environments in the future. It seems evident that accelerated disintegration of ice will change the discharge rates of the meltwater rivers and therefore probably affecting also the fluvial dynamics of the large flood plain in the Val Roseg. The process of continuing and probably accelerated glacier melting in high alpine environments could finally lead to a significant change in the risk potential of natural hazards (ice avalanches, debris flows, rock fall, melting of permafrost zones, etc.) under changed climate conditions (Haeberli et al., 1999; Zimmermann et al., 1997).

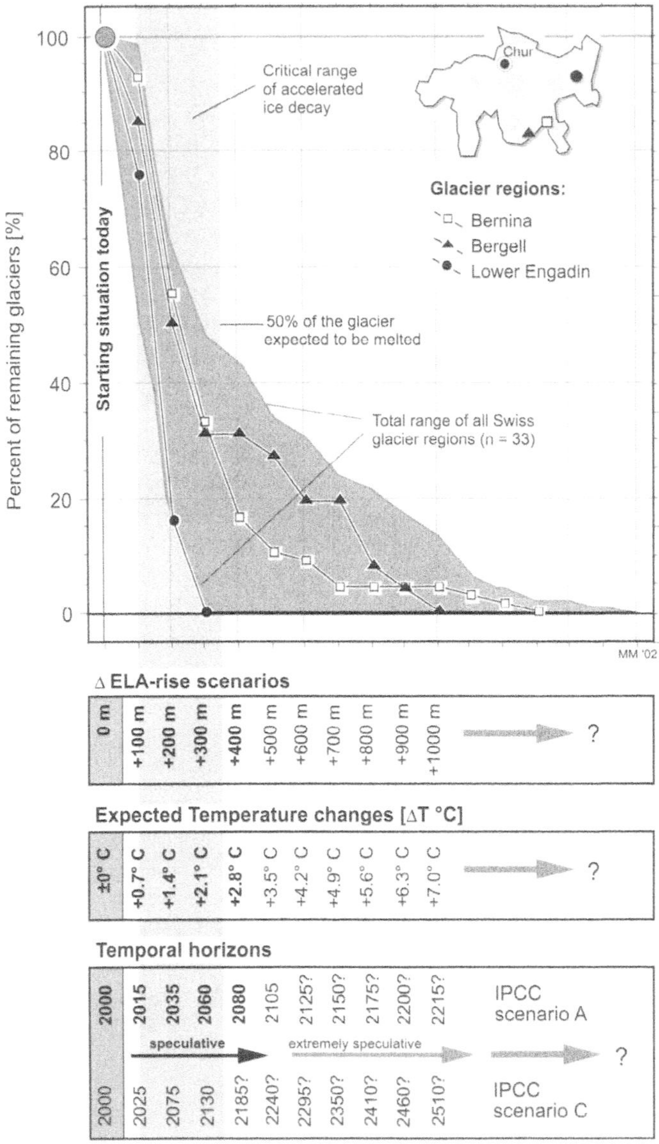

Figure 5. Summary of various glacier retreat scenarios for the Swiss Alps: the lines represent the relative numbers (percent) of the remaining ice bodies according to the different ELA-rise- and temperature change-scenarios (bottom). The shaded field displays the maximum range of the modeled decay curves for all 33 Swiss glacier regions. The mountain ranges of Bernina, Bergell and Lower Engadin are outlined. Modified from Maisch et al. (2000), Fig. 113, p. 284.

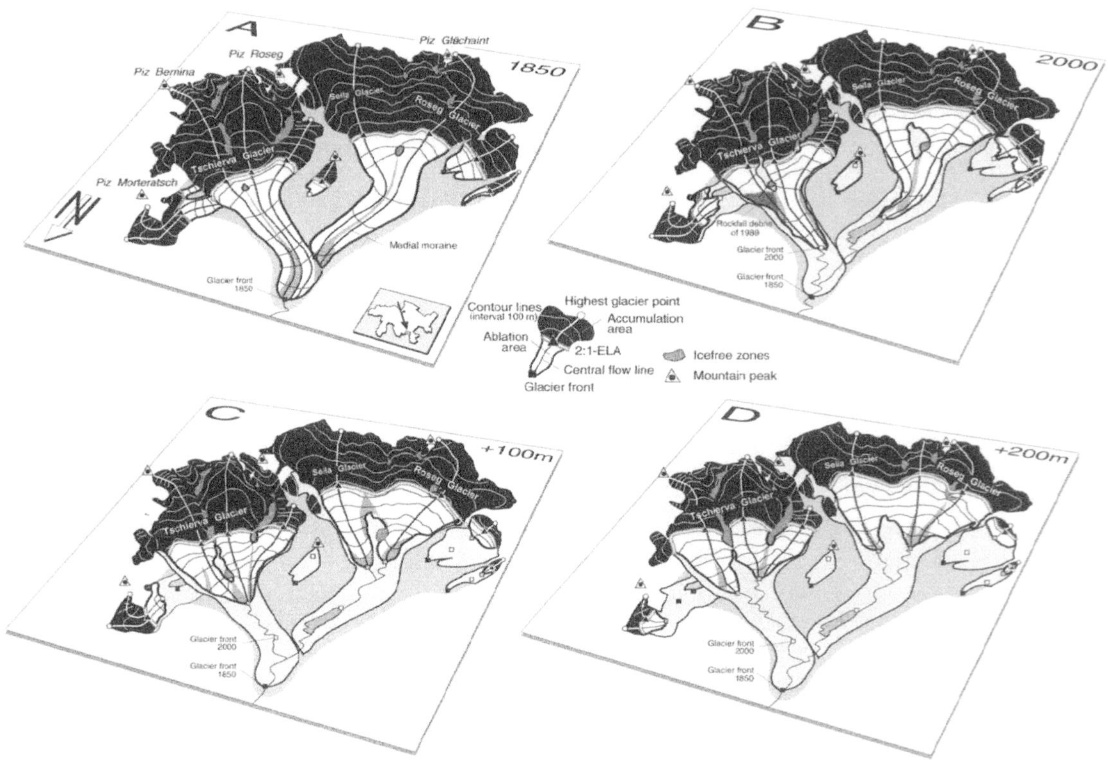

Figure 6. Block diagrams of the glacier advance period of 1850 (A), the actual situation in 2000 (B), and two glacier retreat scenarios based on an ELA-rise of +100 m (C) and +200 m (D), respectively. Modified from Maisch (1995), Fig. 9, p. 98, Fig. 10, p. 99.

3. FLOODPLAIN EVOLUTION

3.1. Geological time scale: landscape evolution

At the end of Pleistocene, following glacial retreat, the Val Roseg was a glacier formed, ice-free valley. The flood plain, however, was formed later by a huge landslide that originated from the left slope and filled the whole valley bottom (Fig. 7). The Roseg River was thereby dammed and the Roseg plain was filled with morainic deposits. Ground-penetrating radar measurements demonstrated a sediment depth of >30 m (P. Huggenberger, unpublished data). The dating of this landslide and therefore also the assumed age of the flood plain is uncertain and varies from 2,000 to 10,000 years ago. However, the morphology of this landslide implies that the event happened soon after the Younger Dryas,

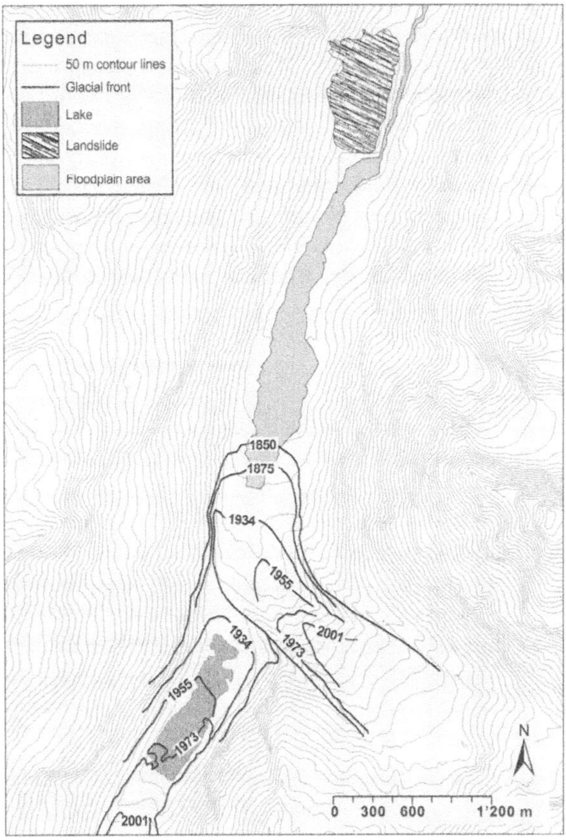

Figure 7. Floodplain evolution since the beginning of the Holocene: Variation of glacier margins as recorded in topographic maps after Gamper (1986); modified and updated to 2001.

probably 7,000 to 10,000 years ago (P. Huggenberger, personal communication). With reference to the stable climatic conditions during the Holocene (Fig. 3), the glacier extent over the last 10,000 years was never greater then during the Little Ice Age (1850). This implies that the lower part of Roseg flood plain was always ice-free (Fig. 7), while the upper flood plain reflects the last glacier retreat that started in 1850 (see also Chapter 7). The differing history of the upper and the lower parts of the Roseg flood plain is also reflected by the morphology: in the upper part of the valley, the Roseg River is constrained to a single channel and flows in a V-shape valley formed by morainic sediments. The lower part of the flood plain is characterized by a dynamic braided channel network that migrates laterally across the entire flood plain.

3.2. Ecological time scale: ecological succession

According to Gurnell et al. (1999), rapidly retreating glaciers at high altitudes, like the Roseg and Tschierva Glaciers, have proglacial zones that are dominated by the sediment regime of the proglacial river. The succession rate of terrestrial vegetation in the Val Roseg compared to other proglacial areas is shown in Figure 8. Riparian vegetation in floodplain areas not affected by shifting channels for more than 30 years consists of scattered shrubs (*Juniperus communis, Salix* spp.). Trees (*Larix decidua, Pinus cembra* and *Pinus montana*) are restricted to valley slopes not affected by glacier movements or floodplain dynamics. Because the Roseg flood plain is situated about 200 m below treeline, succession progresses relatively slowly. It might be expected that trees will not establish in the next decades unless fluviatile dynamics decline (see also Chapter 7).

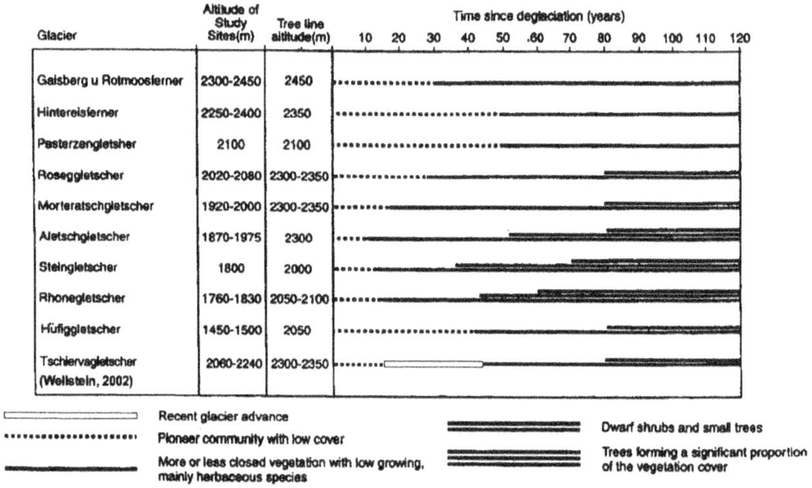

Figure 8. Postglacial succession of vegetation. Modified from Gurnell et al. (1999), Fig .3, p. 231.

In 1914, large parts of the upper flood plain, except the active channel network, were densely vegetated (Fig. 9). The continuous retreat of the glaciers in the following years obviously lead to a highly mobile glacier foreland: the vegetation density in the upper Roseg flood plain is lower today although the distance to the glacier front has increased.

Figure 9. The upper Roseg flood plain in 1914: the photograph shows the joint Roseg and Tschierva Glaciers, the moraine of the 1850-advance period below the glacial fronts, and the densely vegetated proglacial area (source and permission: "Sammlung Wehrli", Eidgenössisches Archiv für Denkmalpflege, Bern, Switzerland).

This trend continued over the last decades: Figure 10 shows that the active floodplain width continuously increased in the upper flood plain over the past 52 years. In the middle section, the active floodplain width remained more or less constant. A reduction in active floodplain width in the lower section was observed during the last 10 years, due to dams built from sediment to protect adjacent pastureland (Zah et al., 2001).

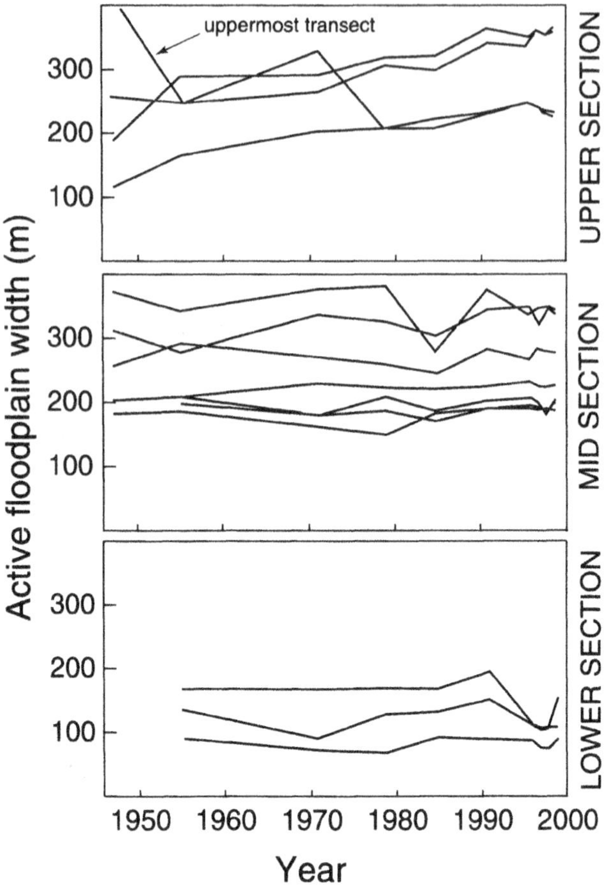

Figure 10. Changes in width of the active flood plain for different transects in the upper, middle, and lower sections.

4. FLOODPLAIN DYNAMICS

Aquatic habitats in flood plains are a product of short-term fluvial dynamics (Ward et al., 2002). The driving force of floodplain dynamics are floods. Erosion and sedimentation processes taking place during flooding change channel connections especially in the upper floodplain areas. This might lead to a new configuration of the whole channel network. Even in large flood plains, a single flood event is able to reconfigure a major portion of aquatic habitats (Arscott et al., 2002). Figure 11 shows the remarkable changes in the channel network of Val Roseg over the last 44 years (Zah et al., 2001).

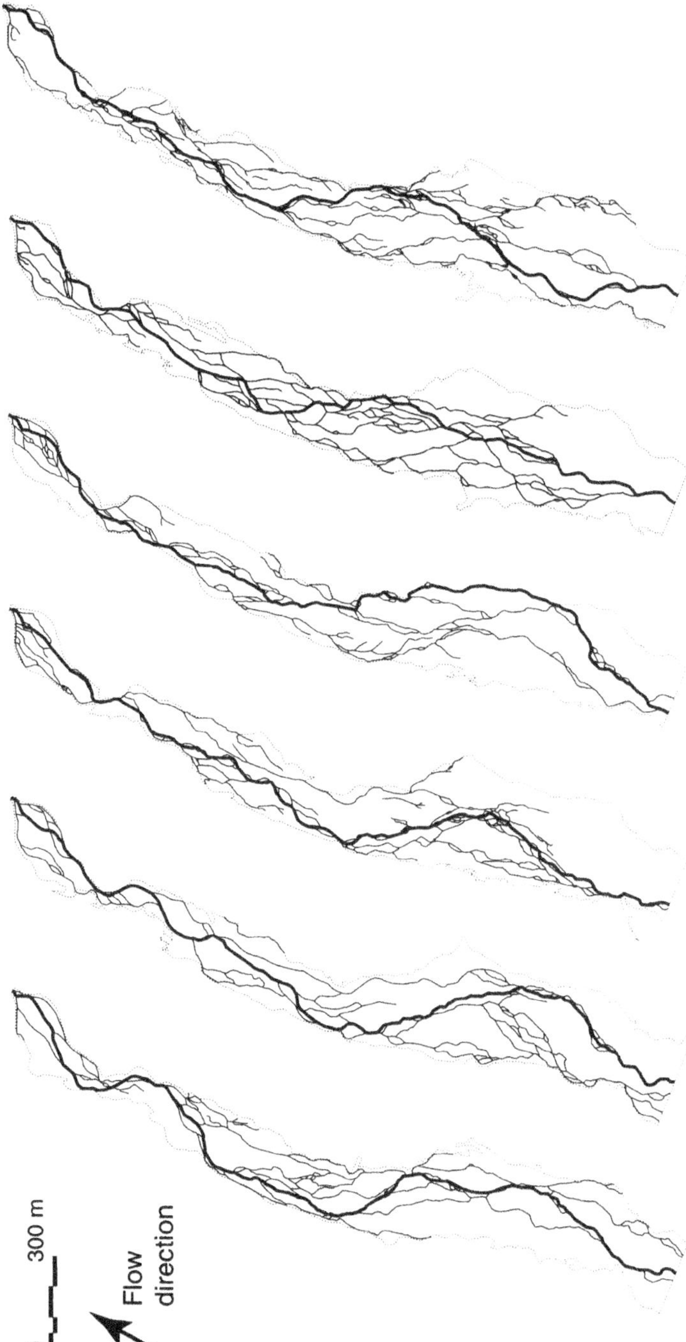

Figure 11. Short-term dynamics of the Roseg flood plain: Channel network and active floodplain area as derived from aerial images; bold lines indicate thalweg (main channel). Modified from Zah et al. (2001), Fig. 3, p. 443.

4.1. Dynamics of the channel network

Figure 12 shows the temporal development of three morphological indices. The sinuosity index of the Roseg River remained constant (1.15 ± 0.01). The braiding index varied only slightly (4.6 ± 1.2), being almost constant up to a discharge of 10 $m^3 s^{-1}$. The connectivity index, in contrast, showed high variation (6.6 ± 5.0), about 34% of which may be attributed to variation in discharge (regression analysis $P < 0.05$).

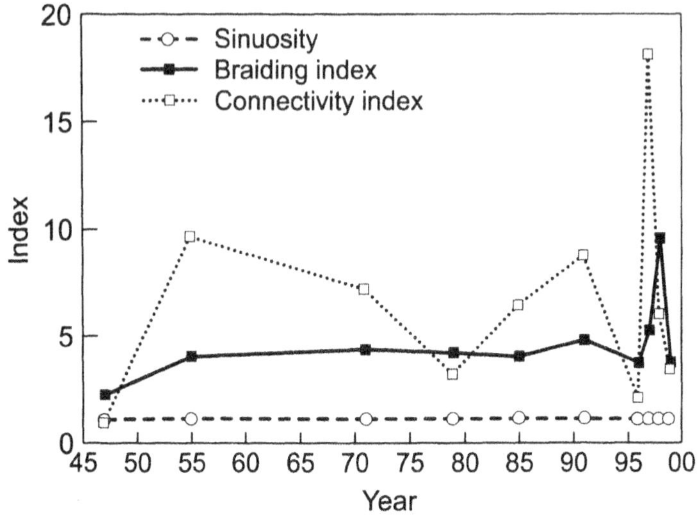

Figure 12. Changes in three morphological indices (sinuosity, braiding, connectivity) from 1947 – 2000. From Zah et al. (2001), Fig. 4, p. 444.

The main channel shifted laterally almost over the entire flood plain, but highest migration rates were measured in the upper section. This lateral movement varied distinctly between subsequent observation periods (20 to 100 m per period, see Fig. 12). Channel turnover calculated for various time periods increased from 30% (1-year interval) to 70% for time periods >20 years (Fig. 13). About 50% of the channel network was subject to change within a 10-year period but many former channels that persisted over time periods of >20 years were eventually reconnected. This may explain why channel turnover did not exceed 70%. Channel dynamics created a dynamic patch mosaic of terrestrial vegetation within the flood plain (see Chapter 7). Bare gravel or gravel with initial stages of pioneer plant communities prevailed in about 70% of the active floodplain area. In areas not affected by channel migration for more than 30 years, dense grassland with a few scattered shrubs can be found, while coniferous forests were restricted to valley slopes.

The active flood plain is relatively large in relation to the present flow regime. Disconnection and reconnection of former channels (many of the channels persist for more than 30 years) to the main channel in the upper floodplain section induce the major changes in the channel network. However, turnover rates were relatively low (30% mean annual turnover). For example, in the low elevation flood plain of the Tagliamento River with gravel as the dominating substratum, more than 50% of the channel network can be renewed within one year due to sheet flow (Kollmann et al., 1999). Sheet flow is lacking in the Roseg flood plain where channel erosion and aggradation mainly occur along the main channel (average lateral shift 2 to 10 m year^{-1}), leading to major channel reconnections every 15 to 25 years (Fig. 13). The relatively slow but continuous habitat renewal results in a persistent dynamic pattern of various habitats (Malard et al., 1999, 2000), that partly accounts for the diverse invertebrate community in the harsh proglacial environment (Burgherr & Ward, 2001).

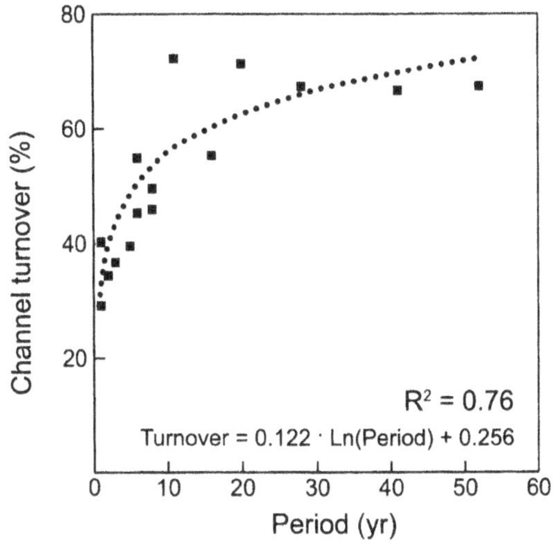

Figure 13. Turnover of the channel network of the Val Roseg flood plain. From Zah et al. (2001), Fig. 6, p. 444.

4.2. Linking glacial dynamics and floodplain dynamics

Regression analysis indicated no relationship between length change of the two glaciers and main channel turnover, main channel shift or other measured morphological parameters (Table 1). There also was no correlation between mean annual discharge or days with daily discharge exceeding 10, 12.5, and 15 m^3 s^{-1} and any of the morphological parameters. Only the annual maximum peak flow correlated significantly with the braiding and connectivity indices (Zah et al., 2001).

Substantial glacier retreat and a relatively persistent morphological configuration of the glacial flood plain characterize the last 50 years. The retreating Tschierva glacier exposed substantial amounts of sediments susceptible to fluvial erosion. Regression analyses showed that the number of days with daily discharge exceeding

10, 12.5, or 15 m^3 s^{-1} slightly increased during the past 45 years; we lack reliable information whether this increase enhanced the transport of sediments to the flood plain. The growing active floodplain area, particularly in the upper part, may be explained by higher transport capacity and sediment availability. Linking long-term floodplain evolution to glacial retreat requires consideration of sediment transport processes, which are determined by discharge and channel morphology (Lane & Richards, 1997; Lane et al., 1996). However, this requires data on sediment transport and information on morphological changes at higher temporal and spatial resolution than examined here. The conceptual model of Gurnell et al. (1999) that describes scenarios of alpine proglacial river evolution under changing climate conditions, predicts actively braiding proglacial river systems below rapidly retreating glaciers if proglacial zones are close to the treeline. Under these conditions, floodplain development is controlled by the sediment regime rather than by riparian vegetation. The Roseg flood plain appears to fit into this framework. In relation to the lateral shift of the main channel, the growth of woody vegetation was apparently too slow (because of high altitude) to stabilize channel margins. However, riparian succession also may be constrained by cattle grazing that takes place every year for about two weeks in late spring and early autumn.

Table 1. Regression analysis of flood peaks and glacial movements vs. indicators of floodplain morphology; asterisks indicate significant relationships.

	Flood peaks	*Glacial movement*	*Floods x Glacial mov.*
Sinuosity	0.06	0.85	0.05
Braiding index	0.003*	0.31	0.003*
Connectivity index	0.03*	0.77	0.03*
Channel turnover	0.09	0.29	0.08
Main channel shift	0.84	0.65	0.63
Active floodplain area	0.24	0.73	0.19

5. SUMMARY

At the end of the last Ice Age the Roseg- and Tschierva-Glaciers followed the overall trend of warming that began about 18,000 years ago and can be divided into various phases of glacier advance and retreat. This overall trend comprises a period of extraordinary steady climatic conditions over the last 10,000 years. During this period the joint Tschierva/Roseg Glacier never exceeded the extent of the 1850 advance. This implies that the lower part of the flood plain remained ice-free since its creation by a landslide about 7,000 – 10,000 years ago. Consequently, the aquatic ecosystem of Val Roseg is expected to be further developed, than in morphological similar flood plains in the Alps.

The glacial retreat over the last 150 years is reflected by the terrestrial vegetation succession that has now reached the forest phase (*Larici-Cembretum*) in old foreland areas, excluding the upper deglaciated flood plain (see Chapter 7). On a short-term perspective, structure and turnover rate of the channel network were surprisingly constant. The only parameter that showed a persistent change was the active floodplain width that continuously increased, particularly in the upper section, due to the high sediment mobility induced by the retreating glaciers. Factors limiting change in floodplain morphology are presumably the low magnitude and frequency of high flow periods and spates rather than sediment availability. Nonetheless, contemporary fluvial dynamics are sufficient to prevent succession beyond the shrub stage. Recent studies on the effects of climate change in the Alps predict an increase of hydrologic extremes such as floods (Imboden, 1998; Wanner et al., 1997). Because of the large sediment storage in the proglacial area of the Tschierva Glacier, we anticipate an increase in floodplain dynamics in the Val Roseg if current climate trends continue.

6. REFERENCES

Arscott, D. B., Tockner, K., Van der Nat, D., & Ward, J. V. (2002). Aquatic habitat dynamics along a braided alpine river ecosystem. *Ecosystems, 5*, 802-814.

Burgherr, P., & Ward, J. V. (2001). Longitudinal and seasonal distribution patterns of the benthic fauna of an alpine glacial stream (Val Roseg, Swiss Alps). *Freshwater Biology, 46*, 1705-1721.

Gamper, B. (1986). Bernina mit Tschierva- und Roseggletscher. Kommentar zum Schweizerischen Schulwandbilderwerk, Bild 212. Zürich.

Gamper, M., & Suter, J. (1982). Postglaziale Klimageschichte der Schweizer Alpen. *Geographica Helvetica, 2*, 105-114.

Gurnell, A. M., Edwards, P. J., Petts, G. E., & Ward, J. V. (1999). A conceptual model for alpine proglacial river channel evolution under changing climatic conditions. *Catena, 38*, 223-242.

Haeberli, W., Kääb, A., Hoelzle, M., Bösch, H., Funk, M., Von der Mühll, D., & Keller, F. (1999). *Eisschwund und Naturkatastrophen im Hochgebirge. Final Report NFP 31*. Zürich: v/d/f Hochschulverlag.

Haeberli, W., Maisch. M., & Paul, F. (2002). Mountain glaciers in global climate-related observation networks. *Bulletin of World Meteorological Organization, 51*, 18-25.

Herren, E. R., Hoelzle, M., & Maisch, M. (2001). *The Swiss Glaciers 1997/98 and 1998/99. Glaciological Report*. Zürich: Glaciol. Commission (GC) of the Swiss Academy of Sciences (SAS) and VAW ETH Zürich.

Hoelzle, M., Von der Mühll, D., & Maisch, M. (1999). Die Gletscher der Schweizer Alpen im Jahr 1997/98. *Die Alpen, 75*, 28-40.

Holzhauser, H. (1995). Gletscherschwankungen innerhalb der letzten 3200 Jahre am Beispiel des Grossen Aletsch- und des Gornergletschers. Neue Ergebnisse. In B. Salm (Ed.), *Gletscher im ständigen Wandel* (pp. 101-122). Mitteilungen der VAW Zürich: vdf-Hochschulverlag an der ETH Zürich.

Imboden, D. M. (1998). Views from the Alps - Regional perspectives on climate change - Introduction. In P. Cebon, U. Dahinden, H. Davies, D. M. Imboden, & C. C. Jaeger (Eds.), *Views From The Alps* (pp. 1-20). Cambridge, MA: MIT Press.

IPCC. (1990). *Climate Change - The IPCC Scientific Assessment of Climate Change*. Cambridge, UK: Intergovernmental Panel on Climate Change.

IPCC. (1995). *Climate Change 1995: The Science of Climate Change. Summary for Policymakers and Technical Summary of the Working Group I*. Cambridge, UK: Intergovernmental Panel on Climate Change.

Ivy-Ochs, S., Schlüchter, C., Kubik, P. W., Synal, H.-A., Beer, J., & Kerschner, H. (1996). The exposure age of an Egesen moraine at Julier Pass, Switzerland, measured with the cosmogenic radionuclides 10Be, 26Al and 36Cl. *Eclogae geologicae Helvetiae, 89/3*, 1049-1063.

Kerschner, H., Ivy-Ochs, S., & Schlüchter, C. (1999). Paleoclimatic interpretation of the early Lateglacial glacier in the Gschnitz valley, central Alps, Austria. *Annals of Glaciology, 28*, 135-140.

Kerschner, H., Kaser, G., & Sailer, R. (2000). Alpine Younger Dryas glaciers as palaeo-precipitation gauges. *Annals of Glaciology, 31*, 80-84.

Kollmann, J., Vieli, M., Edwards, P. J., Tockner, K., & Ward, J. V. (1999). Interactions between vegetation development and island formation in the Alpine river Tagliamento. *Applied Vegetation Science, 2*, 25-36.

Lane, S. N., & Richards, K. S. (1997). Linking river channel form and process: Time, space and causality revisited. *Earth Surface Processes and Landforms, 22*, 249-260.

Lane, S. N., Richards, K. S., & Chandler, J. H. (1996). Discharge and sediment supply controls erosion and deposition in a dynamic alluvial channel. *Geomorphology, 15*, 1-15.

Maisch, M. (1995). Gletscherschwundphasen im Zeitraum des ausgehenden Spätglazials (Egesen-Stadium) und seit dem Hochstand von 1850 sowie Prognosen zum künftigen Eisrückgang in den Alpen. In B. Salm (Ed.), *Gletscher im ständigen Wandel* (pp. 81-100). Zürich: v/d/f-Hochschulverlag.

Maisch, M. (2001). Characteristics and Retreat of the Swiss Measurement Network Glaciers since 1850. In E. R. Herren, M. Hoelzle & M. Maisch (Eds.), *The Swiss Glaciers 1997/98 and 1998/99* (Vol. 119/120, pp. 43-52). Zürich: Glaciol. Commission (GC) of the Swiss Academy of Sciences (SAS) and VAW.

Maisch, M., Burga, C. A., & Fitze, P. (1999). *Lebendiges Gletschervorfeld. Von schwindenden Eisströmen, schuttreichen Moränenwällen und wagemutigen Pionierpflanzen im Vorfeld des Morteratschgletschers. Führer und Begleitbuch zum Gletscherlehrpfad Morteratsch.* (2. Auflage). Pontresina: Gemeinde Pontresina.

Maisch, M., Wipf, A., Denneler, B., Battaglia, J., & Benz, C. (2000). *Die Gletscher der Schweizer Alpen. Gletscherhochstand 1850, Aktuelle Vergletscherung, Gletscherschwund-Szenarien. Final Report NFP 31* (No. 4031-033412). Zürich: ETH.

Malard, F., Tockner, K., & Ward, J. V. (1999). Shifting dominance of subcatchment water sources and flow paths in a glacial floodplain, Val Roseg, Switzerland. *Arctic, Antarctic, and Alpine Research, 31*, 135-150.

Malard, F., Tockner, K., & Ward, J. V. (2000). Physico-chemical heterogeneity in a glacial riverscape. *Landscape Ecology, 15*, 679-695.

Ohlendorf, C. (1999). *High Alpine lake sediments as chronicles for regional glacier and climate history in the Upper Engadine, southeastern Switzerland.* Berichte aus der Geowissenschaft, Shaker Verlag, Aachen.

Patzelt, G. (1973). Die postglazialen Gletscher- und Klimaschwankungen in der Venedigergruppe (Hohe Tauern, Ostalpen). *Z. Geomorph., Suppl.-Bd. 16*, 25-72.

Rothenbühler, C. (2000). *Erfassung und Darstellung der Geomorphologie im Gebiet Bernina (GR) mit Hilfe von GIS.* Diploma Thesis, Geographisches Institut, Universität Zürich.

Suter, J. (1981). Gletschergeschichte des Oberengadins: Untersuchungen von Gletscherschwankungen in der Err-Julier-Gruppe (pp. 47). Universität Zürich: Physische Geographie.

Wanner, H., Rickli, R., Salvisberg, E., Schmutz, C., & Schuepp, M. (1997). Global climate change and variability and its influence on Alpine climate - Concepts and observations. *Theoretical And Applied Climatology, 58*, 221-243.

Ward, J.V., Tockner, K., Arscott, D.B., & Claret, C. (2002) Riverine landscape diversity. *Freshwater Biology 47*, 517-539.

Wellstein, C. (2002). *Vegetationsdynamik auf Moränenflächen eines Gletschervorfeldes (Val Roseg, Schweiz) im Vergleich zur anschliessenden Schwemmebene.* Diploma Thesis, Albert-Ludwigs-Universität, Freiburg i.B., Germany.

Zah, R., Niederöst, M., Rinderspacher, H., Uehlinger, U., & Ward, J. V. (2001). Long-term dynamics of the channel network in a proglacial floodplain, Val Roseg, Switzerland. *Arctic, Antarctic, and Alpine Research, 33*, 440-446.

Zimmermann, M., Mani, P., Gamma, P., Gsteiger, P., Hunziker, G., & Heiniger, O. (1997). *Beurteilung der Murganggefährdung mit Hilfe eines Geographischen Informationssystems. Analyse der räumlichen Entwicklung infolge von Klimaänderungen - Final Report NFP 31.* Zürich: v/d/f-Hochschulverlag.

CHAPTER 3

F. MALARD

GROUNDWATER-SURFACE WATER INTERACTIONS

1. INTRODUCTION

Headwater streams have long been essentially considered as conduits for the downstream transport of water and solutes arising from different hydrological reservoirs of the catchment including ice, snow, and ground water. An alternative view is to look at streams as "part of the catchment's downvalley and downgradient transport of water and solutes" (Bencala, 1993). From this perspective, water is transported through the catchment by overland, surface channel (i.e. streams), shallow and deep subsurface flow pathways. As water moves downvalley, streams and ground water continuously exchange water, nutrients, organic matter and organisms (Jones & Mulholland, 2000). Stream water infiltrates into the sediment at multiple locations along the valley corridor, travels for some distance in unconfined aquifers, mixes with ground water and then returns to the stream. Thus, streams and subsurface compartments of the catchment are integrated components of a single hydrological network, the function of which is to transport, transform, retain, and connect (Fisher, 1997). The contribution of different hydrological reservoirs to the flow of water within the catchment as well as the location, direction, and strength of groundwater-surface water exchanges also change over time (Malard et al., 2002). The seasonal shift in sources and flow paths of water is an essential feature of glaciated catchments that has profound implications on biogeochemical processes and biodiversity in streams (Milner et al., 2001).

The aim of this chapter is to provide an holistic view of the different hydrologic reservoirs and flow pathways that contribute to the flow of surface water in the flood plain of the Val Roseg. Emphasis is put on subsurface flow pathways because groundwater inputs play a major role in structuring ecological conditions in surface water habitats of the flood plain, thereby sustaining biodiversity and productivity in this harsh glacial environment (see Chapters 10 & 11). First, the physico-chemical fingerprints of water sources and flow paths are identified. Second, spatial changes in hydrological linkages between the streams and underlying ground water are examined and areas of major groundwater upwelling along the river continuum are identified. Finally, based on measurements of hydrochemical tracers at multiple locations within the flood plain, a conceptual model is developed to portray the seasonal changes in water sources and flow paths which contribute to the flow of surface water in the Val Roseg flood plain.

J.V. Ward & U. Uehlinger (eds.), Ecology of a Glacial Flood Plain, 37-56.
© 2003 *Kluwer Academic Publishers. Printed in the Netherlands.*

2. PHYSICO-CHEMICAL FINGERPRINTS OF WATER SOURCES AND FLOW PATHS

2.1. Glacial water

The bulk of water from the Tschierva and Roseg valley glaciers enters the flood plain at a single location via a surface flow path (site 17, see Appendix I). Glacial water is then distributed throughout the flood plain by either surface connections between channels or hyporheic flow systems (i.e. lateral and longitudinal downwelling and upwelling of channel water). Water draining from the Tschierva and Roseg glaciers exhibits strong seasonal changes in physico-chemistry that reflect both the changing efficiency of glacier drainage and the relative contribution of different hydrological sub-systems to the flow of water at the outlet of the glaciers (Fig. 1). In the absence of significant ice and snow melting (i.e. from November to March), glacier discharge is sustained mainly by subglacial water that is stored in the distributed drainage system beneath the glacier (Röthlisberger & Lang, 1987; Tranter et al., 1997). Because of intimate contact with sediments that have accumulated beneath the glacier, subglacial water is enriched in calcium, magnesium, bicarbonate, and sulfate (Table 1). Because of low flow velocities, subglacial water has turbidity usually below 10 NTU. Atmospheric deposition alone cannot account for the sulfate content of glacial water in the Val Roseg. The weighted mean concentration of sulfate in high alpine snow packs (3000 to 4000 m a.s.l.) of the European Alps is usually below 15 μeq L^{-1} (Nickus et al., 1997), whereas sulfate concentration in subglacial water draining from the Tschierva Glacier averages 407 μeq L^{-1} (Table 1). Pyrite oxidation (equation 1) rather than simple dissolution of evaporites such as gypsum or anhydrite (equation 2) is considered as the most plausible source of sulfate in subglacial water of the Val Roseg catchment (Malard et al., 1999).

$$FeS_2 + 15/4\ O_2 + 7/2\ H_2O \rightarrow Fe(OH)_3 + 2\ SO_4^{2-} + 4\ H^+ \qquad (1)$$

$$CaSO_4 \,.\, 2\ H_2O \rightarrow Ca^{2+}\ SO_4^{2-} + 2\ H_2O \qquad (2)$$

Because the dissolution of evaporites occurs at a faster rate than the acid hydrolysis of nonevaporitic minerals, sources of evaporites, if present in the Val Roseg catchment, should result in elevated concentrations of sulfate in glacial water throughout the year. Pyrite oxidation which acts as a source of protons for weathering other minerals such as calcite is promoted by the slow transit of water in the distributed drainage system beneath glaciers (Tranter & Raiswell, 1991; Tranter et al., 1993, 1996).

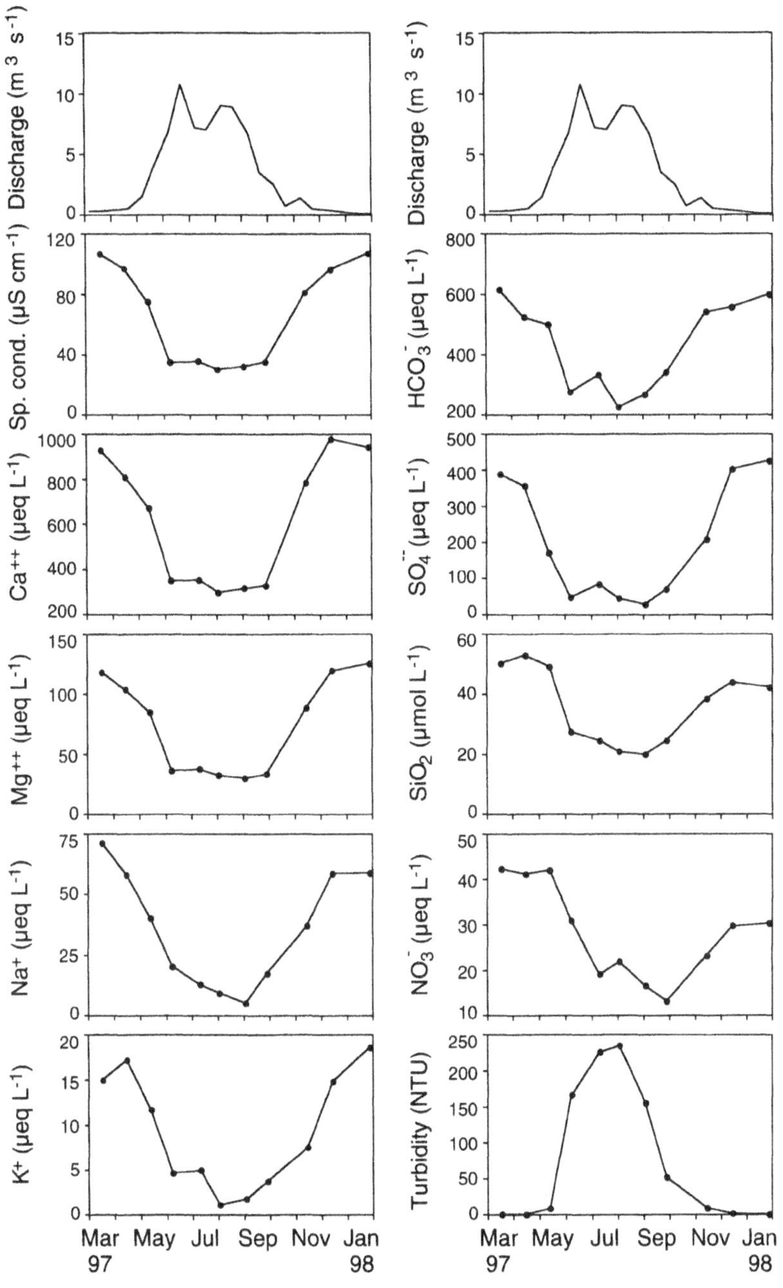

Figure 1. Temporal changes in river discharge (biweekly average) and solute composition of bulk meltwater draining the Tschierva Glacier. Samples were collected at site 7 located 400 m from the glacier terminus.

Melting of the snow pack on the glacier surface (May-June) and the subsequent melting of the ice as glacier ice is directly exposed to solar radiation by snowline recession (July, August), result in a marked decline in the specific conductance and concentrations of base cations, bicarbonates, and sulfates in bulk meltwater draining from the glacier (Fig. 1). The contact time between meltwater and rock materials is limited by the increasing drainage efficiency of the channelized system at the surface of the glacier (i.e. supraglacial water), within the glacier (i.e. englacial water) and beneath the glacier (subglacial water), thereby limiting the solute acquisition by meltwater. The bulk meltwater draining from the glacier also becomes increasingly turbid as subglacial sources of sediments are mobilized by the developing subglacial drainage network which consists of large open (free-surface) conduits (Clifford et al., 1995).

A striking feature of glacial water in the Val Roseg is its high relative concentration of Ca^{2+} over Na^+ and K^+ and its low relative content of SiO_2 over HCO_3^- (Table 1). The alteration of plagioclase, which is usually considered as the main source of Ca^{2+} and Na^+ in water from silicate rocks, releases Ca^{2+} and Na^+ to solution in approximately the same ratio as they occur in the rock (equation 3). The same phenomenon applies to the weathering of biotite, that provides K^+ and Mg^{2+} in solution (equation 4).

$$Na_{(1-x)}Ca_xAl_{(1+x)}Si_{(3-x)}O_8 + (1+x)\ CO_2 + 1/2\ (11-5x)\ H_2O \rightarrow (1+x)/2\ Al_2Si_2O_5(OH)_4$$

Plagioclase (x = mole fraction of anorthite) *Kaolinite*

$$+ (1-x)\ Na^+ + x\ Ca^{2+} + (1+x)\ HCO_3^- + 2\ (1-x)\ H_4SiO_4 \qquad (3)$$

$$KMg_xFe_{(3-x)}AlSi_3O_{10}(OH)_2 + (1+2x)\ CO_2 + 15/2\ H_2O + O_2 \rightarrow 1/2\ Al_2Si_2O_5(OH)_4$$

Biotite (x = mole fraction of magnesium) *Kaolinite*

$$+ K^+ + x\ Mg^{2+} + (1+2x)\ HCO_3^- + (3-x)\ Fe(OH)_2 + 2\ H_4SiO_4 \qquad (4)$$

A Na^+/Ca^{2+} ratio of 0.10 for the bulk glacial water would imply that the plagioclase contains only about 10% albite and 90% anorthite (Table 1). Although information on the mineralogical composition of the granite in the Val Roseg catchment is lacking, this ratio of albite to anorthite is highly improbable; rather, there must be an additional source of the Ca^{2+}. The most plausible source of Ca^{2+} is the dissolution of minor amount of calcite present in the bedrock (equation 5).

$$CaCO_3 + H_2O \rightarrow Ca^{2+} + 2HCO_3^- \qquad (5)$$

Table 1. Solute composition (mean ± standard deviation) of different water sources in the Val Roseg flood plain using examples of representative sites. All concentrations are in $\mu eq\ L^{-1}$ except SiO_2 and dissolved oxygen (DO) in $\mu mol\ L^{-1}$ and specific conductance (Sp. cond.) in $\mu S\ cm^{-1}$. Ion concentrations in snow and ice samples are reported as $\mu eq\ L^{-1}$ per liter of melt water. Snow samples were collected in March and April 1997 in the flood plain. Supraglacial water was obtained in August and September 1997 from several rivulets on the surface of the Tschierva Glacier. A total of 5 ice cores (3-m deep) were collected on the Tschierva Glacier

	Snow	Ice	Supraglacial water	Bulk glacial meltwater		Ground water		
				Winter (site 2)	Summer (site 2)	Shallow alluvial ground water (site 64)	Hillslope ground water (site 86)	Deep alluvial ground water (site 91)
	n = 6 [a]	n = 11	n = 8	n = 3	n = 3	n = 3	n = 3	n = 1
NH_4^+	3 ±1	1 ±2	nd	nd	1 ±0	nd	nd	36
NO_3^-	11 ±1	2 ±1	1 ±1	34 ±7	19 ±3	9 ±1	18 ±2	1
DON	13 ±6	7 ±10	2 ±1	4 ±1	2 ±2	2 ±3	2 ±2	nd
Ca^{++}	80 ±30	9 ±17	18 ±23	950 ±27	314 ±32	681 ±115	898 ±27	893
Mg^{++}	6 ±2	1 ±3	2 ±3	121 ±4	34 ±4	63 ±10	118 ±2	259
Na^+	nd	1 ±4	nd	63 ±7	13 ±3	24 ±6	79 ±1	135
K^+	nd	1 ±4	nd	16 ±2	6 ±7	10 ±1	12 ±2	30
HCO_3^-	99 ±31	37 ±17	45 ±26	591 ±30	264 ±60	583 ±58	766 ±52	1107
SO_4^{--}	14 ±15	27 ±13	34 ±24	407 ±19	35 ±17	147 ±66	279 ±25	110
SiO_2	nd	1 ±0	2 ±2	45 ±4	21 ±2	46 ±1	100 ±1	300
Sp. cond.	10.1 ±3.1	2.8 ±2.2	2.4 ±2.2	103.6 ±6.1	30.1 ±3.2	66.0 ±10.9	95.3 ±1.4	114.2
pH	-	-	5.8 ±0.3	6.8 ±0.0	6.7 ±0.2	6.8 ±0.2	6.8 ±0.2	-
DO	-	-	-	-	331 ±12	256 ±3	175 ±3	3

nd: non detectable
- : not measured
[a] 18 samples were taken but only these 6 relate to the end of the snow accumulation period

The rapid movement of water in freshly exposed rock materials deposited beneath the glaciers probably results in selective weathering that favors the preferential dissolution of calcite. Drever and Hurcomb (1986), Stauffer (1990), and Mast et al. (1990) also suggested that the selective weathering of trace amounts of carbonates in glaciated or recently deglaciated catchments occurred as a result of intensive mechanical erosion that exposed large amounts of fresh bedrock to dissolution.

2.2. Snow pack

Depth integrated snow samples collected at the end of the accumulation period (March, April) are enriched in Ca^{2+}, HCO_3^- and most nitrogen species (Table 1). Elevated concentrations of base cations and nitrogen (primarily dissolved organic nitrogen) are believed to originate from dry deposition, especially during the end of the snow accumulation period when the lower tropospheric convection extends upward to glacier altitudes (Nickus et al., 1997). As in many other mountain catchments (Charles, 1991), the spring snowmelt results in a pulse of nitrates in the Roseg River (Malard et al., 1999). However, the source of the elevated NO_3^- concentrations in surface water is not yet clear as it may either come from the snowpack or from catchment soils (see Chapter 6). Snow meltwater from non-glaciated areas of the catchment enters the flood plain at multiple locations mainly via subsurface flow paths. There is little overland flow in the Val Roseg catchment because the valley slopes are covered by thick permeable morainic and glacio-fluvial deposits. Snow meltwater from the surface of the glacier flows through the glacial drainage system and enters the flood plain at single location via a surface flow path (site 17). The end of the snowmelt period on the glacier surface / beginning of the ice ablation period is characterized by a strong decline of NO_3^- concentrations of glacial water (Fig. 1). At the Haut Glacier d'Arolla (Switzerland), NO_3^- concentrations were also higher in subglacial water fed by snowmelt than in englacial water fed by ice melt (Richards et al., 1996, Tranter et al., 1997).

2.3. Ground water

Ion concentrations (except NO_3^-) are distinctly higher in ground water than in bulk meltwater draining from the glaciers during the ice ablation period. However, there are great differences in the chemical composition and travel time of ground waters. We distinguish between three distinct types of ground water within the Roseg River flood plain (Table 1, Fig. 2). Shallow alluvial ground water is fed mainly by infiltration of stream water on the floodplain surface. Glacial water from the main channel infiltrates into the sediment at multiple locations within the flood plain, follows short subsurface flow paths (e.g. up to a few hundred m), and upwells in nearby surface upstream disconnected channels. Hillslope ground water is recharged by snow meltwater in spring and ice-melt water from hanging glaciers in summer.

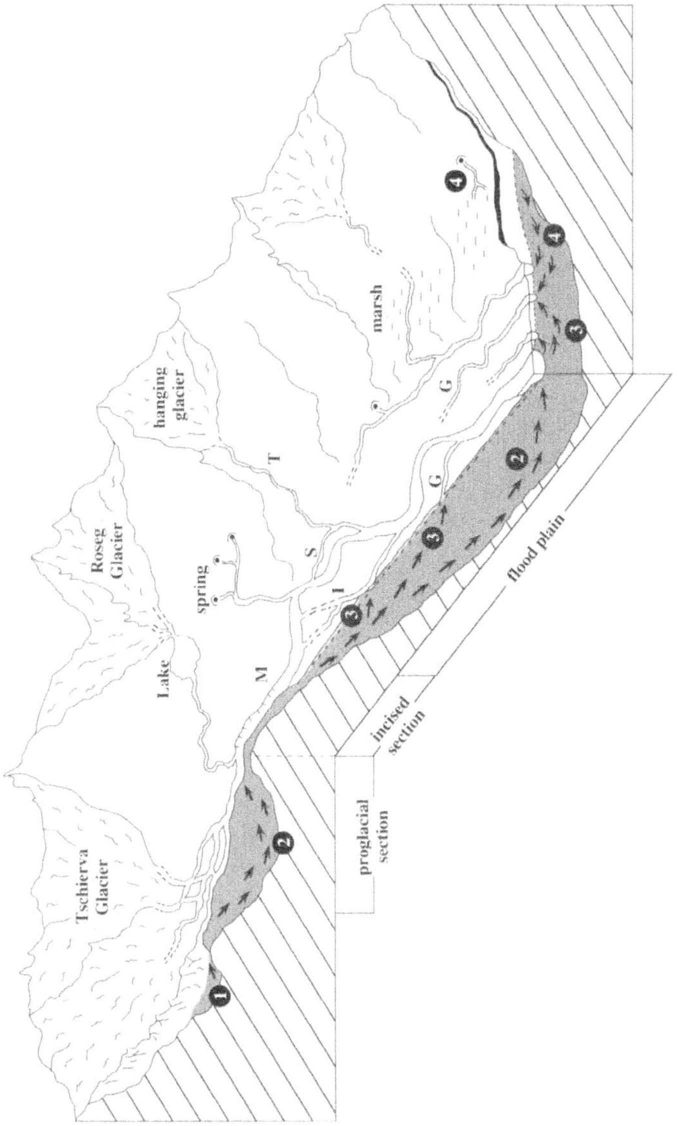

Figure 2. Schematic diagram showing the different reservoirs and pathways of ground water in the Val Roseg. 1 = subglacial ground water; 2 = large scale upwelling of ground water (deep alluvial ground water); 3 = infiltration and resurfacing of advected channel water (shallow alluvial ground water); 4 = inflow of hillslope ground water. M = main channel; S = side channel; I = intermittently-connected channel; G = groundwater-fed channel; T = tributary. Modified from Ward et al. (1999), Fig. 10, p. 291.

The discharge of hillslope ground water occurs by numerous rheocrene springs located on the catchment slopes and floodplain margins, extensive marshy areas on the floodplain terraces and direct groundwater seepage within the active flood plain. Deep alluvial ground water flows in glacio-fluvial deposits of the valley floor that extend to a depth of at least 30 m in the lower flood plain reach. Preliminary investigations with ground penetrating radar indicated unconsolidated sediment to a depth of at least 30 m at transects 4, 8, and 14 (P. Huggenberger, unpublished data). Moreover, two 30-m-deep wells which provide drinking water for the Val Roseg Hotel (site 91) do not reach bedrock.

Specific conductance and ion concentrations (except nitrogen species) increase in the following order: shallow alluvial ground water, hillslope ground water, and deep ground water (Table 1). Dissolved oxygen shows a reverse trend. Anoxic conditions in deep ground water may be caused by the oxidation of pyrite rather than by the oxidation of dissolved and particulate organic matter. Deep ground water is also enriched in dissolved iron (Fe = 960 μg L^{-1}). The Na^+/Ca^{2+} ratio is markedly higher in hillslope ground water and deep ground water than in shallow alluvial ground water, whereas the latter has a higher relative proportion of HCO_3^- to SiO_2. Channels located on the floodplain margins have higher relative concentrations of Na^+ over Ca^{2+} and SiO_2 over HCO_3^- than channels located in the active part of the flood plain (Fig. 3).

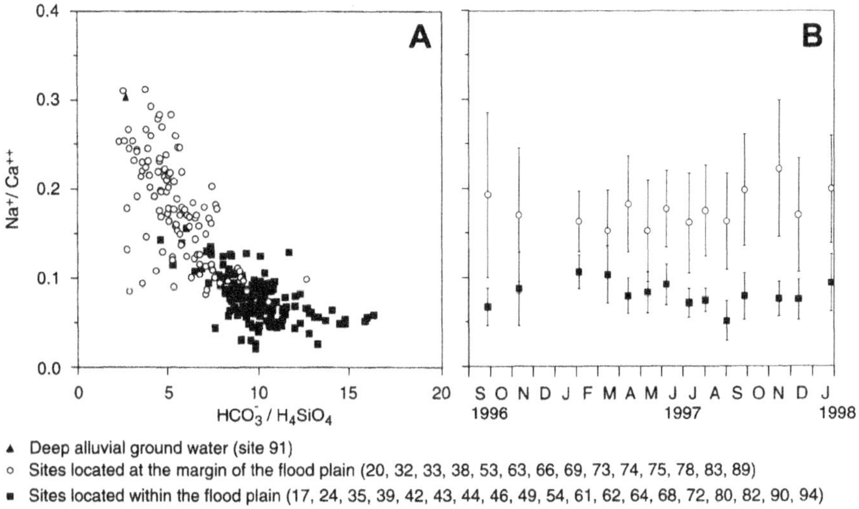

Figure 3. A) Molar ratio of Na^+ to Ca^{2+} plotted versus molar ratio of HCO_3^- to H_4SiO_4 for surface water samples collected from March 1997 to January 1998. B) Temporal changes in the average molar ratio of Na^+ to Ca^{2+} of surface water samples collected on the floodplain margins and within the active flood plain. Modified from Malard et al. (1999), Fig. 8, p. 147.

Elevated concentrations of Na^+ and SiO_2 in groundwater-fed channels showing high specific conductance is attributed to the dissolution of plagioclase that is enhanced during long flow paths in the more weathered rock materials of the valley slope and deep aquifers (equations 3 and 4). On the contrary, the dissolution of calcite is enhanced during short flow paths in shallow and freshly exposed sediments of the flood plain (equation 5). However, high Na^{+}/Ca^{2+} and low HCO_3^{-}/H_4SiO_4 ratios are also observed in a few floodplain margin channels showing low specific conductance. This suggests that soil contact may be a more important factor than the residence time of ground water in determining the solute content of hillslope ground water. Cation exchange in soils is a rapid process compared to chemical weathering that may also result in an increase of the Na^{+}/Ca^{2+} ratio (Drever 1988; Williams et al., 1993; Stumm & Morgan, 1996). Several studies have also demonstrated that water of low ionic strength can rapidly acquire substantial amounts of silica from soil (Kennedy, 1971; Zeman & Slaymaker, 1975; Brown & Lund, 1991).

Groundwater dating in the Val Roseg flood plain was carried out only once in February 1998 at three representative sites using the tritium-helium-3 (^{3}H-^{3}He) dating method (U. Beyerle, unpublished data). All floodplain channels at that time were fed by ground water because the main channel fed by subglacial water lost all surface water into the sediment in the upper part of the flood plain (transect 17). Shallow alluvial ground water was collected with an inertial pump at site 64 from a 1.5-m deep piezometer located in the middle of the flood plain (transect 10). Hillslope ground water was obtained from a 1.5-m deep piezometer installed in a meadow on the left side of the flood plain (site 86 in transect 5). Deep ground water was recovered from a 30-m deep well in the Val Roseg Hotel (site 91). Groundwater ages in shallow alluvial, hillslope and deep aquifers were 0.4 yr ± 0.5 (detection limit of the dating method), 3.7 yr ± 0.5, and 26.1 yr ± 1.4, respectively.

3. PATTERNS OF GROUNDWATER-SURFACE WATER INTERACTIONS ALONG THE RIVER CORRIDOR

3.1. Hydrologic linkages between the stream and underlying ground water

The groundwater table in the lower part of the Val Roseg flood plain (downstream of transect 10) is very shallow, intersecting the soil surface at many locations giving rise to numerous groundwater-fed channels. Channels are in direct saturated connection with the underlying ground water so that surface water-groundwater exchanges occur through a water-saturated layer of sediment (Fig. 4). In contrast, many channels in the upper flood plain appear to have their bed surface above the groundwater table. Holes dug near the shoreline of several channels reveal that an unsaturated layer of alluvium is interposed between the water-saturated bed-sediment and the deeper ground water (Fig. 4). Pumping with a hand piston pump in standpipes inserted at depths greater than 30 cm below the streambed provides a mixture of air and water or even no water at all. This situation is more commonly encountered in the uppermost part of the flood plain (upstream of transect 13) where channels have steeper slopes (i.e. 4-10 %, Zah et al., 2000). Some groundwater-fed tributaries entering the flood plain as surface water channels (e.g. site 20) may also lie above the groundwater table. Thus, it is important to distinguish between the source of water within a channel and the type of hydrologic linkages between the channel and the underlying ground water.

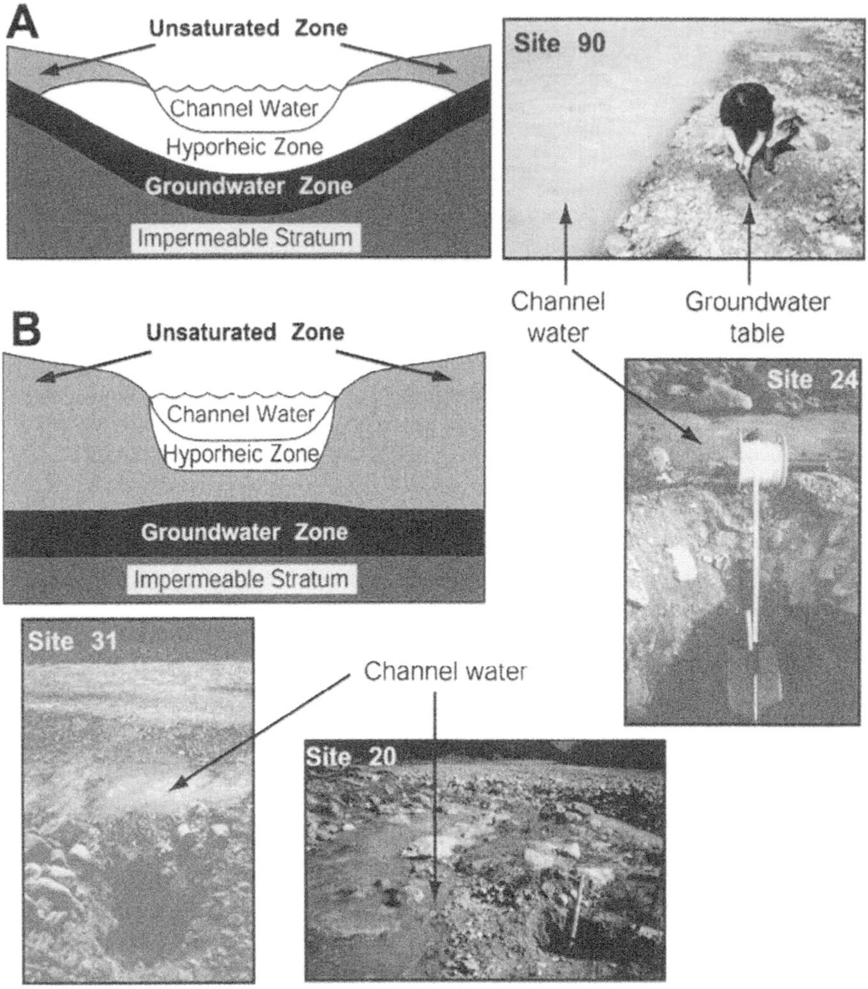

Figure 4. Conceptual cross-sectional models of surface channels and beds showing relationships of channel water to hyporheic and ground water in the lower flood plain (A) and the upper flood plain (B, upstream of transect 13). Holes dug near the shoreline show the absence / presence of ground water at a depth of about 80 cm.

During the end of the ice ablation period, the perched channels in the uppermost part of the flood plain start to dry at their downstream ends when the surface flow is too low to compensate for water infiltration through unsaturated and permeable sediments. This results in the surface disconnection between the channel networks of the upper and lower floodplains. Downstream changes in the type of hydrologic connections between streams and ground water are likely to occur in montane and piedmont flood plains where steep gradient rivers enter flat areas consisting of extensive alluvium. Saturated hydrologic disconnection between the surface channel network and ground water have been observed in other mountain river systems such as the Flathead River, Montana (Stanford & Ward, 1993) and the Tagliamento River, Italy (Malard et al., 2000). In the Val Roseg flood plain, variation in subsurface linkages between surface and ground water strongly influences the composition of the hyporheic invertebrate assemblages (see Chapter 11).

3.2. Distribution of groundwater upwelling

We used a two-end member mixing model, based on measured concentrations of sodium, to determine the relative contribution of hillslope ground water and glacial water (supraglacial, subglacial and englacial origin) to the flow of surface water along the river continuum (Malard et al., 2001a). Inputs of snowmelt water to the river are implicitly included in the model because they occur either via subsurface water flow paths through the catchment slopes (i.e. hillslope groundwater component) or via glacial flow paths (i.e. glacial water component). Sodium is considered as an effective hydrological tracer in the Roseg River because there is a large and consistent difference between concentrations in glacial water and hillslope and deep ground waters. Samples for the hillslope groundwater component were taken monthly from a piezometer installed at a depth of 80 cm into the bed of a groundwater-fed channel located on the margin of the flood plain (site 83). Na concentration at this site averaged 82 ±6 $\mu eq.L^{-1}$ (range = 72-94, n=12). Samples for the glacial water component were taken monthly at the lake outlet of the Roseg Glacier (site 10). Na concentration at this site averaged 15 ±6 $\mu eq.L^{-1}$ (range = 3-26, n=12). Surface water samples were collected monthly at 11 sampling sites located along the main channel of the Roseg River from 400 m below the terminus of the Tschierva Glacier (site 2) down to the end of the catchment (site 101). The relative contribution of hillslope ground water to river discharge is calculated at each site and date using the following formula:

$$f1=(Cs - C2)/(C1 - C2) \qquad (6)$$

where f1 is the fraction of river flow accounted for by hillslope ground water, and C1, C2, and Cs represent the Na concentrations of hillslope ground water (site 83), glacial water (site 10), and surface water at a particular site, respectively.

The relative contribution of hillslope ground water to river discharge varies seasonally from less than 10% during the ice ablation period in summer to more than 80% during winter (Fig. 5). The longitudinal pattern suggests that hillslope ground water already enters the river at the lower end of the proglacial reach. Decreasing groundwater contribution downstream of the proglacial reach is caused by the inflow of glacial water from the Roseg Glacier. The major input of hillslope ground water occurs from the downstream part of the flood plain. In April, the contribution of hillslope ground water more than doubles from the end of the flood plain to the end of the catchment.

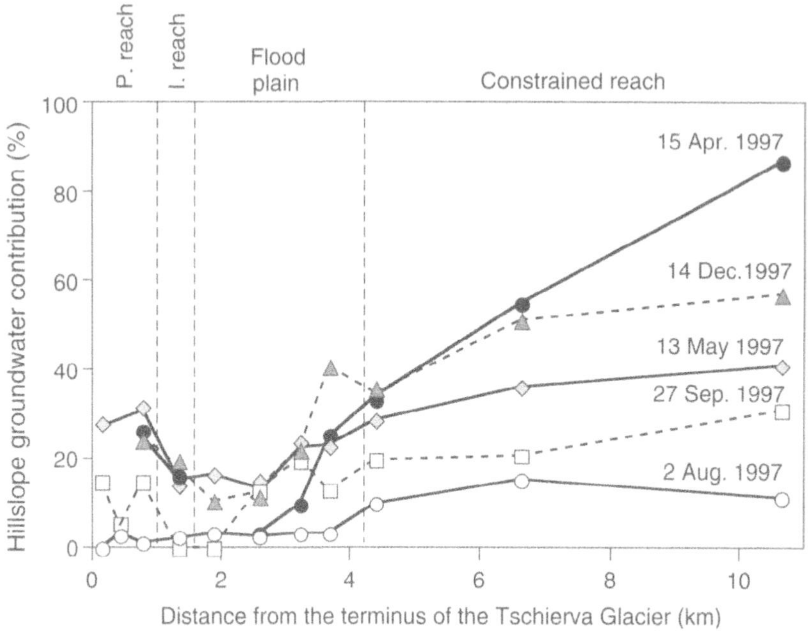

Figure 5. Longitudinal changes in the contribution of hillslope ground water to the flow of surface water on 5 different dates. From Malard et al. (2003), Fig. 3, p. 97.

In winter, the totality of the river discharge at the downstream end of the flood plain is from ground water because the main channel loses all its water into the sediment at the upstream end of the flood plain. However, river discharge is also sustained by the resurfacing of advected subglacial water from the main channel in the middle section of the flood plain. This downwelling–upwelling flow pattern results in a distinct longitudinal pattern of specific conductance and concentrations of SiO_2, HCO_3^-, Mg^{2+} and Na^+ in surface water of the flood plain (Fig. 6). Channels fed by resurfacing of advected main channel water in the upper flood plain have lower concentrations than those fed by upwelling of hillslope and deep ground waters in the lower flood plain. This upwelling flow pattern at the floodplain scale is probably caused by a combination of factors including a reduction in floodplain slope, a narrowing of the valley width, as well as a downstream decrease in sediment permeability. Upwelling of ground water ensures a permanent flow of surface water during winter in approximately 6 km of channels representing about one quarter of the maximum floodplain channel network length (i.e. 24 km).

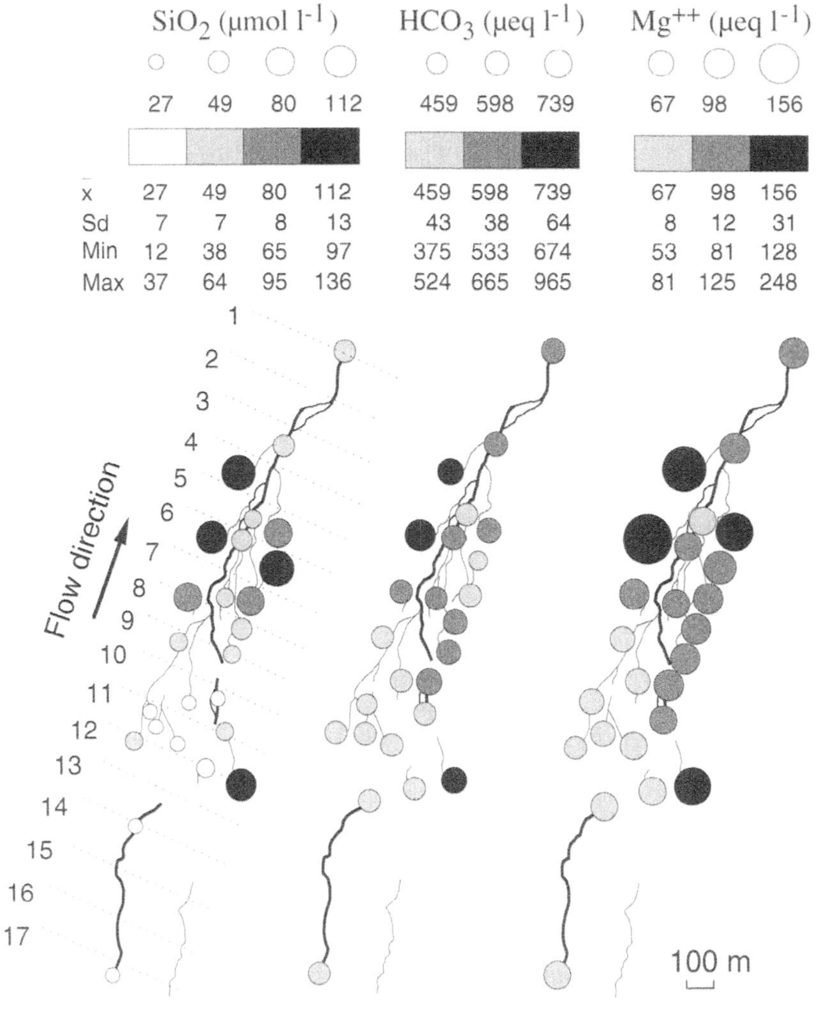

Figure 6. Spatial distribution in the concentrations of SiO_2, HCO_3^-, and Mg^{2+} of surface water during the period of surface disconnection between the upper and lower floodplain channel networks (14 December 1997).

4. SURFACE–SUBSURFACE WATER INTERACTIONS WITHIN THE FLOOD PLAIN

4.1. Diversity of groundwater-fed habitats

Groundwater-fed habitats share common features including low and constant discharge (< 100 $L.s^{-1}$), absence of bed movement, clear water (i.e. turbidity < 9 NTU), and warmer temperature. Clear water-fed habitats are expected to sustain much higher algal biomass and diversity and density of invertebrates than turbid water bodies (see Chapter 10). However, groundwater-fed channels encompass a range of habitats that have different sources of water, hydrological connections with underlying ground water, and life spans. Three distinct type of groundwater-fed habitats can be distinguished within the flood plain based on the source of water: (a) alluvial groundwater-fed channels that originate from the resurfacing of advected surface water from the main channel, (b) helocrenes, limnocrenes, and rheocrenes fed by seepage of hillslope ground water on the floodplain margins, and (c) hillslope groundwater-fed tributaries entering the flood plain as surface water channels (Fig. 2). Whereas the two former types of groundwater-fed channels have saturated hydrologic linkage with the underlying ground water, the latter may behave as perched streams.

These clear water habitats may disappear as a result of drying or by upstream connection with turbid glacial water. Thus, the flood plain is an ever changing mosaic of groundwater-fed and glacial-water fed patches that appear, disappear, and are relocated in response to the fluctuations of river discharge (Fig. 7). Only a few groundwater-fed channels located on the margin of the lower flood plain maintain clear water conditions throughout the year. These channels represent a length of only about 2 km, which corresponds to less than 10 % of the total channel network. These habitats constitute hotspots of biodiversity in this harsh glacial environment (see Chapters 10 and 11). All other clear water-fed habitats have a life span ranging from only 2 to 7 months, either because they establish an upstream surface connection with the main channel during the ice ablation period or because they dry up in autumn. Groundwater-fed tributaries at the margins of the upper flood plain have clear water throughout the summer but they become dry in autumn. All channels located in the lower flood plain carry clear water in winter but most of them are flooded with turbid glacial water in summer.

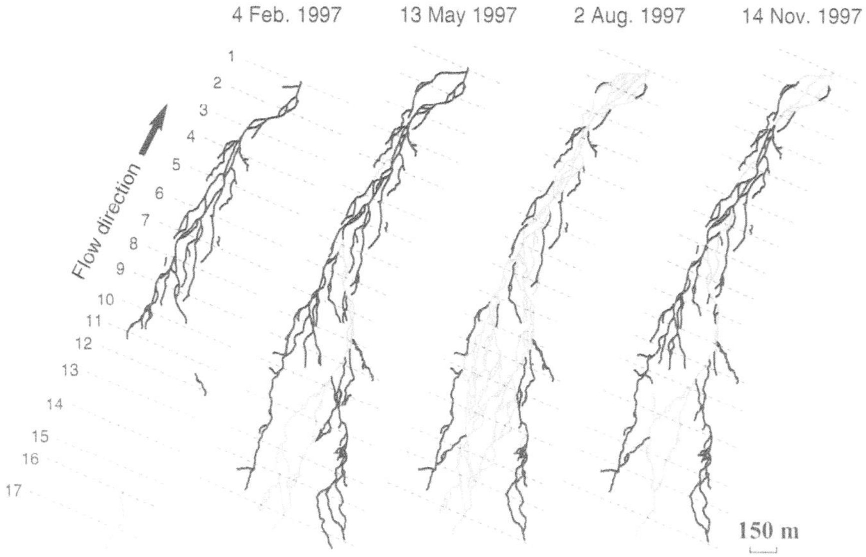

Figure 7. Distribution of groundwater-fed and glacial water-fed channels in the flood plain during a period of base flow (February, Q=0.2 m^3 s^{-1}), the expansion phase (May, Q=1.5 m^3 s^{-1}), the maximum extension of the floodplain channel network (August, Q=5.7 m^3 s^{-1}), and the contraction phase (November, Q=0.7 m^3 s^{-1}). Black and gray lines indicate channels fed by ground water (turbidity < 9 NTU) and channels fed by glacial water (turbidity > 9 NTU), respectively. From Malard et al. (2002), Fig. 9, p. 636.

4.2. Seasonal shift in water sources and flow paths

The block diagram in Figure 8 portrays the seasonal changes in the sources and flow paths of water in the flood plain based on monthly measurements of surface water chemistry at multiple sites distributed throughout the channel network (Fig. 9) (Malard et al., 1999). Elevated NO_3^- concentrations in spring (April–early June) indicates that the snowmelt is the main source of water in spring. The persistence of high NO_3^- concentrations for more than 2 months probably reflects the spatiotemporal heterogeneity of snowmelt within the catchment (altitude range > 2000 m) and the transient storage of snow meltwater in ground water and subglacial water. Snow meltwater from non-glaciated areas of the catchment enters the flood plain via subsurface flow pathways as indicated by the high specific conductance and concentrations of base cations in almost every channel of the flood plain. The progressive decline in specific conductance and concentrations of SiO_2, HCO_3^- and Ca^{2+} during the period of elevated NO_3^- concentrations indicates that there is an increase in the relative proportion of snow meltwater from snow on the surface of the glaciers, which enters the upper flood plain via surface flow paths.

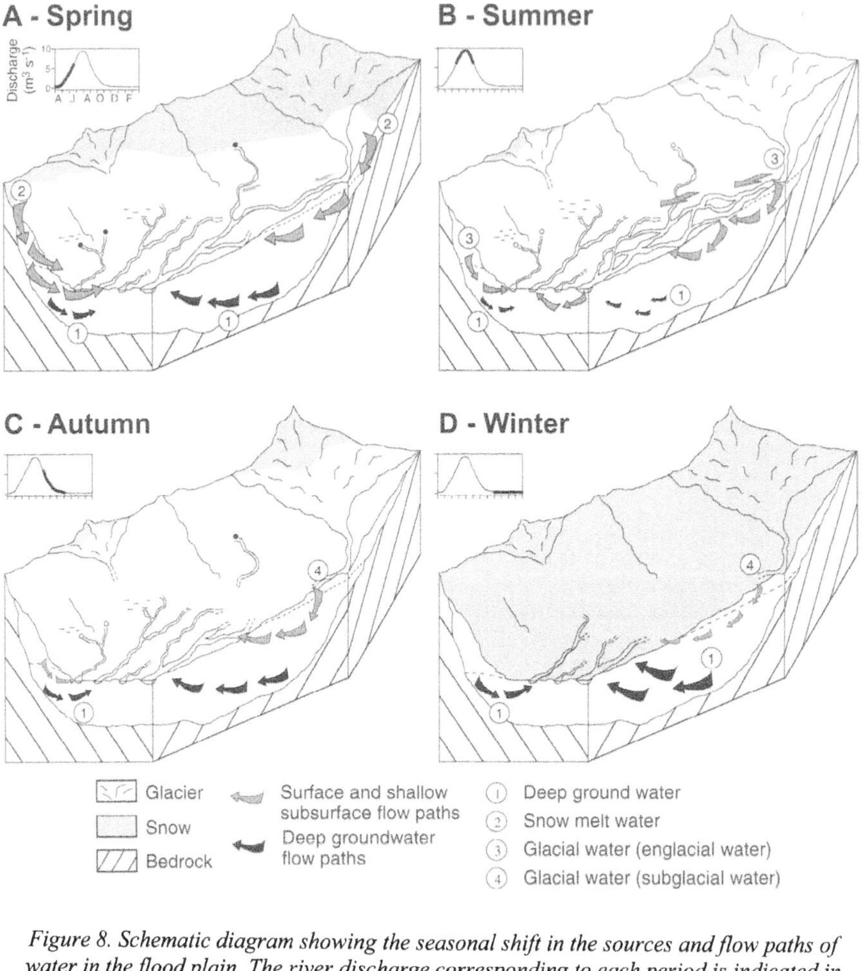

Figure 8. Schematic diagram showing the seasonal shift in the sources and flow paths of water in the flood plain. The river discharge corresponding to each period is indicated in bold. The relative size of the arrows show the relative importance of the different sources and flow paths of water. From Malard et al. (1999), Fig. 9, p. 148.

Ice meltwater is the main source of water in summer (July-mid-September), the bulk of which enters the flood plain via surface flow paths at site 17. Glacial water from the main channel expands laterally into the flood plain either via surface connections between channels or lateral downwelling–upwelling flow systems. This results in a pronounced decline in the specific conductance and ion concentrations in most of the floodplain channels, the chemical signature of hillslope aquifers (i.e. high silica concentration) being restricted to a few water bodies located on the margin of the flood plain (Fig. 9).

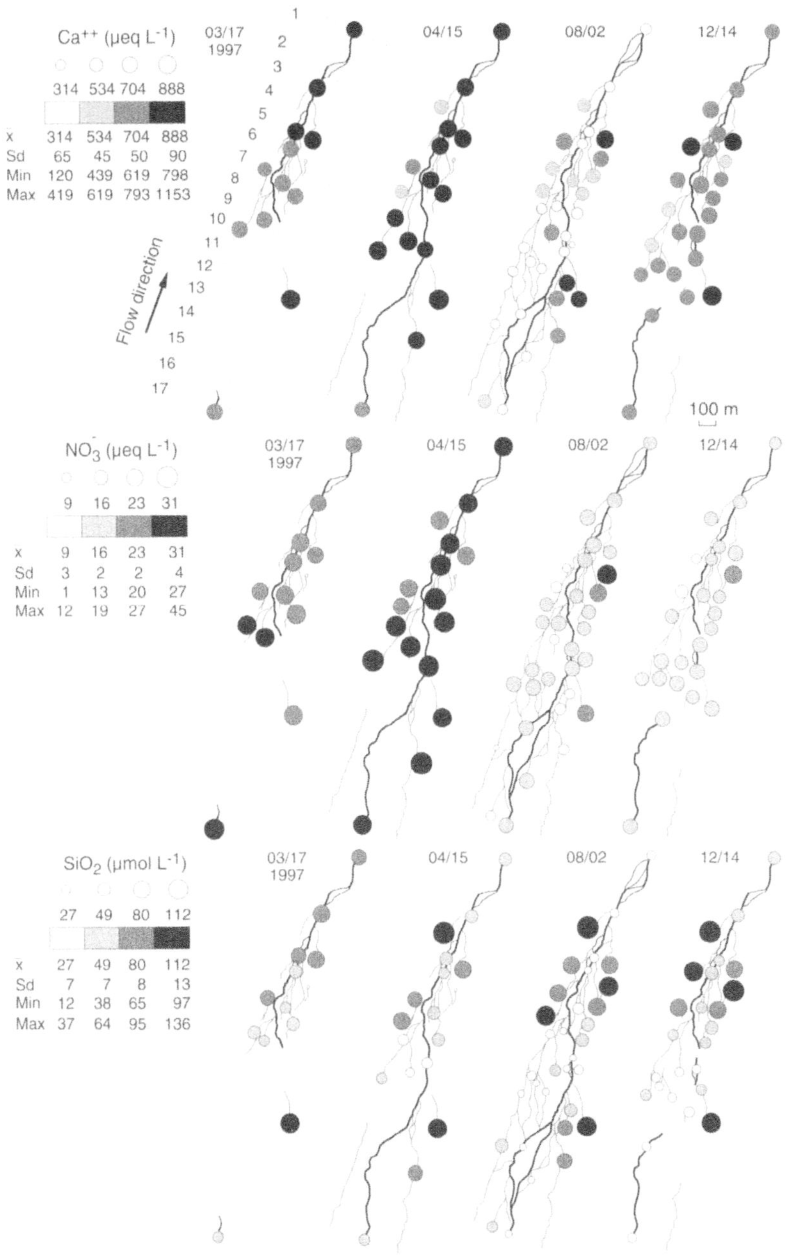

Figure 9. Temporal changes in the spatial distribution in Ca^{2+}, NO_3^- and SiO_2 concentrations of surface water in the flood plain. See Figure 6 for explanations.

The lateral expansion of glacial water restricts the drainage of hillslope ground water, which upwells on the floodplain terraces and generates extensive marshy areas.

Autumn (mid-September-December) is a period of declining discharge during which the relative contribution of subglacial water to glacier discharge increases. This results in a marked increase in specific conductance and concentrations of most ions. The upper and lower floodplain channel networks become disconnected when river discharge falls below 0.5 m^3s^{-1}. Then, resurfacing of main channel water is the primary source of water in the upper flood plain, whereas surface flow in the lower flood plain is also sustained by upwelling of hillslope and deep alluvial ground waters (Figs. 6 and 9).

In winter (January-March), the progressive increase in the concentrations of Na^+ and SiO_2 and Na^+/Ca^{2+} ratio suggests an increasing contribution of ground water from deep alluvial and hillslope aquifers. The vertical transport of heat induced by the upwelling of deep ground water prevents freezing of channel water. Whereas the main channel of the Roseg River is covered by snow along most of its length, the channel network in the lower flood plain is usually free of snow. Although daily average air temperature can be as low as –18 °C in winter, Malard et al. (2001b) found that hyporheic temperature at a depth of 80 cm into the sediment of 2 lower floodplain sites (sites 73 and 83) remained above 2.5 °C and 3.8 °C throughout the winter

5. SUMMARY

Seasonal variations in river discharge, extent of the channel network, and surface water chemistry are governed by changes in the relative contribution of different hydrologic reservoirs and flow paths of water operating within the Val Roseg catchment. Snow and ice meltwaters not only result in a distinct river flow pulse over the spring and summer ablation seasons but also replenish ground water and control the size of the subglacial water stores (Röthlisberger & Lang, 1987). In winter, surface flow is maintained essentially by groundwater upwelling and to a lesser extent by the discharge of subglacial water. Our model of shifting water sources and flow paths is essentially qualitative because natural chemical tracers alone do not allow quantification of each of the water sources to the river. Quantifying the spatiotemporal distribution of different water fluxes within the flood plain requires the use of alternative methods such as the measurements of isotopes (Sickman et al., 2001) or the modelling of subsurface flow based on the distribution of hydraulic heads and aquifer properties (Wondzell & Swanson, 1996). Interactive pathways between surface water and ground water, which are especially well developed in the flood plain, produce a shifting mosaic of groundwater-fed and glacial water-fed aquatic patches. Understanding the dynamics of this mosaic is of particular importance because the extent, spatial arrangement, and turnover of these contrasted habitat-patches are expected to control floodplain biodiversity and bioproduction (see Chapters 10 and 11).

6. REFERENCES

Bencala, K. E. (1993). A perspective on stream-catchment connections. *Journal of the North American Benthological Society, 12*, 44-47.

Brown, A. D., & Lund, L. J. (1991). Kinetics of weathering in soils from a subalpine watershed. *Journal Soil Science Society of America, 55*, 1767-1773.

Charles, D. F. (1991). *Acidic Deposition and Aquatic Ecosystems*. New York: Springer-Verlag.

Clifford, N. J., Richards K. S., Brown, R. A., & Lane, S. N. (1995). Scales of variation of suspended sediment concentration and turbidity in a glacial meltwater stream. *Geografiska Annaler, 77A (1-2)*, 45-65.

Drever, J. I. (1988). *The Geochemistry of Natural Waters*. 2nd ed. Englewood Ciffs. N.J.: Prentice Hall.

Drever, J. I., & Hurcomb, D. R. (1986). Neutralization of atmospheric acidity by chemical weathering in an alpine drainage basin in the North Cascade Mountains. *Geology, 14*, 221-224.

Fisher, S. G. (1997). Creativity, idea generation, and the functional morphology of streams. *Journal of the North American Benthological Society, 16*, 305-318.

Jones, J. B., & Mulholland, P. J. (Eds.) (2000). Streams and Ground Waters. San Diego: Academic Press.

Kennedy, V. C. (1971). Silica variation in stream water with time and discharge. In J. D. Hem (Ed.), *Nonequilibrium Systems in Natural Water Chemistry. Advances in Chemistry Series* (Vol. 106, pp. 94-130). Washington: American Chemical Society.

Malard, F., Tockner, K., & Ward, J. V. (1999). Shifting dominance of subcatchment water sources and flow paths in a glacial floodplain, Val Roseg, Switzerland. *Arctic Antarctic and Alpine Research*, 31, 135-150.

Malard, F., Ward, J. V., & Robinson, C. T. (2000). An expanded perspective of the hyporheic zone. *Verhandlungen der Internationalen Vereinigung für Theoretische und Angewandte Limnologie, 27*, 431-437.

Malard, F., Lafont, M., Burgherr, P., & Ward, J. V. (2001a). A comparison of longitudinal patterns in hyporheic and benthic oligochaete assemblages in a glacial river. *Arctic Antarctic and Alpine Research, 33*, 457-466.

Malard, F., Mangin, A., Uehlinger, U., & Ward, J. V. (2001b). Thermal heterogeneity in the hyporheic zone of a glacial floodplain. *Canadian Journal of Fisheries and Aquatic Sciences, 58*, 1319-1335.

Malard, F., Tockner, K., Dole-Olivier, M.-J., & Ward J. V. (2002). A landscape perspective of surface – subsurface hydrological exchanges in river corridors. *Freshwater Biology, 47*, 621-640.

Malard, F., Ferreira, D., Dolédec, S., & Ward, J.V. (2003). Influence of groundwater upwelling on the distribution of the hyporheos in a headwater river flood plain. *Archiv für Hydrobiologie, 153 80-116.*

Mast, M. A., Drever, J. I., & Baron, J. (1990). Chemical weathering in the Loch Vale watershed, Rocky Moutain National Park, Colorado. *Water Resources Research, 26*, 2971-2978.

Milner, A. M., Brittain, J. E., Castella, E., & Petts, G. E. (2001). Trends of macroinvertebrate community structure in glacier-fed rivers in relation to environmental conditions: a synthesis. *Freshwater Biology, 46*, 1833-1847.

Nickus, U., Kuhn, M., Baltensperger, U., Delmas, R., Gäggeler, H., Kasper, A., Kromp-Kolb, H., Maupetit, F., Novo, A., Pichlmayer, F., Seibert, P., Staudinger, M., Trockner, V., Wagenbach, D., & Winiwarter, W. (1997). SNOSP: ion deposition and concentration in high alpine snow packs. *Tellus, 49B*, 56-71.

Richards, K., Sharp, M., Arnold, N., Gurnell, A., Clark, M., Tranter, M., Nienow, P., Brown, G., Willis, I., & Lawson, W. (1996). An integrated approach to modelling hydrology and water quality in glacierized catchments. *Hydrological Processes, 10*, 479-508.

Röthlisberger, H., & Lang, H. (1987). Glacial hydrology. In A. M. Gurnell & M. J. Clark (Eds.), *Glacio-Fluvial Sediment Transfer* (pp. 207-284). Chichester: John Wiley and Sons.

Sickman, J. O., Leydecker, A., & Melack, J. M. (2001). Nitrogen mass balances and abiotic controls on N retention and yield in high-elevation catchments of the Sierra Nevada, California, United States. *Water Resources Research, 37*, 1445-1461.

Stanford, J. A., & Ward, J. V. (1993). An ecosystem perspective of alluvial rivers: connectivity and the hyporheic corridor. *Journal of the North American Benthological Society, 12*, 48-60.

Stauffer, R. E. (1990). Granite weathering and the sensitivity of alpine lakes to acid deposition. *Limnology and Oceanography, 35*, 1112-1134.

Stumm, W., & Morgan, J. J. (1996). *Aquatic Chemistry*. New York: Wiley Interscience (3rd ed).

Tranter, M., & Raiswell, R. (1991). The composition of the englacial and subglacial component in bulk meltwaters draining the Gornergletscher, Switzerland. *Journal of Glaciology, 37*, 59-66.

Tranter, M., Brown, G. H., Hodson, A. J., & Gurnell, A. M. (1996). Hydrochemistry as an indicator of subglacial drainage system structure: a comparison of alpine and sub-polar environments. *Hydrological Processes, 10*, 541-556.

Tranter, M., Brown, G., Raiswell, R., Sharp, M., & Gurnell, A. (1993). A conceptual model of solute acquisition by alpine glacial meltwaters. *Journal of Glaciology, 39*, 573-581.

Tranter, M., Sharp, M. J., Brown, G. H., Willis, I. C., Hubbard, B. P., Nielsen, M. K., Smart, C. C., Gordon, S., Tulley, M., & Lamb, H. R. (1997). Variability in the chemical composition of in situ subglacial meltwaters. *Hydrological Processes, 11*, 59-77.

Ward, J. V., Malard, F., Tockner, K., & Uehlinger, U. (1999). Influence of ground water on surface water conditions in a glacial flood plain of the Swiss Alps. *Hydrological Processes, 13*, 277-293.

Williams, M. W., Brown, A. D., & Melack, J. M. (1993). Geochemical and hydrologic controls on the composition of surface water in a high-elevation basin, Sierra Nevada, California. *Limnology and Oceanography, 38*, 775-797.

Wondzell, S. M., & Swanson, F. J. (1996). Seasonal and storm dynamics of the hyporheic zone of a 4th-order mountain stream. I: Hydrologic processes. *Journal of the North American Benthological Society, 15*, 3-19.

Zah, R., Niederöst, M., & Uehlinger, U. (2000). Application of photogrammetry in freshwater ecology: analysing the morphology of a high alpine floodplain. *International Archives of Photogrammetry and Remote Sensing, 33*, 1739-1746.

Zeman, L. J., & Slaymaker, H. O. (1975). Hydrochemical analysis to discriminate variable runoff source areas in an alpine basin. *Arctic and Alpine Research, 7*, 341-351.

CHAPTER 4

K. TOCKNER & F. MALARD

CHANNEL TYPOLOGY

1. INTRODUCTION

Riverine flood plains are dynamic systems characterized by a diverse array of aquatic habitats and a rich biota adapted to the high spatio-temporal heterogeneity (Tockner and Stanford, 2002). Hydrological processes interacting with geomorphic features play a major role in sustaining high levels of habitat heterogeneity within river corridors (Ward et al., 2002). Seasonal changes in discharge result in expansion/contraction of channel networks, accompanied by changes in hydrologic connectivity between floodplain waterbodies and between surface waters and groundwater aquifers, thereby forming a dynamic mosaic of shifting water sources and flow paths (Stanley et al., 1997; Malard et al. 1999; Tockner et al., 2002). Although the "flood pulse" (sensu Junk et al., 1989) is widely recognized as the major force structuring floodplain habitat conditions, even moderate increases in discharge ("flow pulses") may result in dramatic expansion of inundated area and can be important in sustaining habitat heterogeneity in river flood plains (Tockner et al., 2000).

The remarkably high level of aquatic habitat heterogeneity exhibited by the glacial flood plain of Val Roseg is largely attributable to its complex channel network consisting of six distinctive channel types. In this chapter we describe the environmental conditions characterizing each channel type and synthesize our current understanding of the factors responsible for sustaining a diversity of channel types, drawing on a vast data base established from frequent sampling of a suite of physicochemical parameters and morphological features over a 3-year period (Tockner et al., 1997; Malard et al., 1999, 2000). The channel network was mapped from digitized aerial photographs (color and IR images, resolution: 1.0 m) and ground surveys. Geochemical indicators, employed to identify spatio-temporal shifts in water sources during expansion/contraction cycles, included silica (ground water), nitrate (snowmelt water), and particulate phosphorus (glacial meltwater) (see Chapter 3). Channel heterogeneity (hereafter "riverscape" heterogeneity) was calculated using a diversity index with 8 turbidity classes representing "species" and the proportion of channel length in each class representing "abundance" (Malard et al., 2000). Based on the statistical relationship between discharge, channel network length, and riverscape heterogeneity (e.g., Benke et al., 2000; Van der Nat et al., 2002), a simple model was developed to predict habitat heterogeneity.

J.V. Ward & U. Uehlinger (eds.), Ecology of a Glacial Flood Plain, 57-73.
© 2003 *Kluwer Academic Publishers. Printed in the Netherlands.*

The chapter begins with an analysis of the major role of the annual expansion/contraction cycle in producing predictable spatio-temporal shifts in habitat heterogeneity within the glacial flood plain. Channel network heterogeneity is calculated using turbidity classes as a landscape-level indicator of surface connectivity patterns. Against this background, the aquatic habitat template of the flood plain is developed, based on characterization of individual floodplain channel types, their distribution and spatio-temporal dynamics.

2. FLOODPLAIN EXPANSION AND CONTRACTION

2.1. Expansion and contraction dynamics

The floodplain reach of the Val Roseg ("study area" in Fig. 1) is 2.6 km long, with width ranging from 130 to 510 m. Total channel length increased logarithmically with discharge, from about 5 km in winter to more than 20 km in summer. The number of nodes (points of convergence and divergence within the channel network) ranged between 83 in winter to more than 250 in summer (Table 1).

Expansion and contraction were much more pronounced in the upper than in the lower flood plain, which were separated by a geomorphic knickpoint. The upper and lower flood plains were characterized by distinct differences in slope (4.8 % in the upper floodplain vs 1.9 % in the lower flood plain), floodplain width (510 m vs up to 260 m) and average sediment size (12 vs 9 cm, average b-axis). As a consequence, surface water downwelled at the upstream end of the flood plain, flowed through the alluvial aquifer, and upwelled at this knickpoint half way along the flood plain. In winter the upper flood plain was completely disconnected from the surface waters of the lower section (Fig. 1; Tockner et al., 1997).

Table 1. Characterization of the four hydrological phases (see text for further explanation). Surface connectivity: Relative proportion (%) of the channel network length having an upstream connection with the main river channel. Turbidity, specific conductance and temperature are from the main channel at the downstream end of the flood plain (after Tockner et al., 2002).

	Low water	*Expansion*	*Glacial ablation*	*Contraction*
Season (months)	JFM	AMJ	JAS	OND
Discharge (bi-weekly average; $m^3 s^{-1}$)	0.13-0.23	0.30-7.0	3.1-10.4	0.3-2.0
Channel network length (km)	5.8-6.3	11.2-21.2	20.6-21.7	9.6-18.1
Maximum number of nodes*	83	120	262	141
Surface connectivity (%)	5-6	5-31	44-45	5-31
Turbidity (NTU)	0.7-4.5	3.0-48	85-164	2-54
Specific conductance ($\mu S\ cm^{-1}$)	84-88	61-91	35-47	40-77
Water temperature (°C)	0.0-1.2	0.3-6.2	4.0-7.4	0.1-4.8

*Points of convergence or divergence within the channel network

Floodplain inundation is a complex phenomenon that involves different water sources (lateral overflow, ground water, upland sources and direct precipitation) and flow paths (seepage flow, groundwater upwelling, active overbank flow). In the Val Roseg flood plain, several water sources contributed to the expansion of the channel network, including ground water, snowmelt, and glacial water. The relative proportion of these water sources changed during the annual cycle, with alluvial and hillslope ground water dominating in winter, snowmelt water in spring, and glacial meltwater in summer (Fig. 1; Malard et al., 1999; Tockner et al., 2002). Overall, the entire flood plain shifted from a relatively uniform groundwater dominated system in winter to a complex channel network in summer. The relative contribution of hillslope ground water to total floodplain discharge ranged from <10% in summer to >70% in winter (Fig. 2A). The shifting dominance of different water sources was reflected by a sequential change of physicochemical properties of floodplain waters. In winter, input of ground water led to a peak in dissolved organic carbon (DOC) and silica; transport of snowmelt water via subsurface pathways produced a pulse in nitrate in early spring; and runoff of glacial meltwater in summer resulted in a peak in particulate phosphorous (Fig. 2B). Silica, DOC, nitrate and PP are therefore good indicators for differentiating individual water sources (see Chapter 3).

Based on the relative contribution of individual water sources, channel network length, and surface connectivity over the expansion and contraction cycle, four seasonal periods were separated (Table 1). The dominance of ground water (low turbidity), maximum contraction of the channel network and a high specific conductance characterized the **low water period** in winter. The **expansion period** in spring was linked to the recharge of the aquifer with snowmelt water and a major change in channel network length and associated physicochemical attributes. The **glacial ablation period** in summer exhibited maximum channel network length and surface connectivity, thereby increasing the proportion of channels containing water of high turbidity and low specific conductance. The **contraction period** in autumn followed more or less a reversed pattern of the expansion cycle, although different hydrological processes caused the decrease of channel network complexity (Malard et al., 1999).

Connectivity, expressed as the relative proportion (%) of the channel network length having an upstream surface connection with the main channel, exhibited a clear 2-phase pattern during the expansion and contraction cycle (Fig. 3). During the expansion period connectivity remained low although total channel length more than doubled during the same period. During this period snow melt recharged hillslope and groundwater aquifers via subsurface flow pathways. The increase in channel network length was largely restricted to groundwater-fed channels and tributaries. When total channel network exceeded 15 km, there was a sharp increase in surface connectivity. During the contraction period the progressive decline in the water table of the floodplain aquifer created a connectivity trajectory that was similar (albeit reversed) to that occurring during the expansion period (see also Chapter 3).

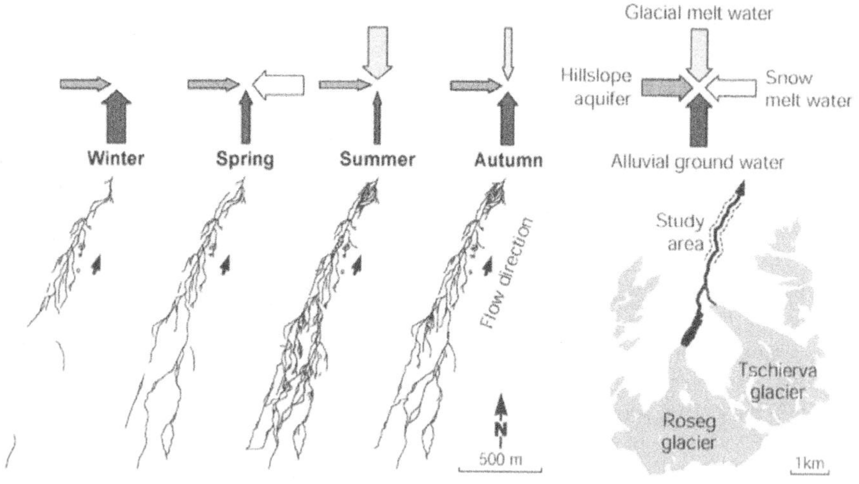

Figure 1. Expansion and contraction cycle of the channel network of the Val Roseg flood plain (four seasons) and the relative contribution of different water sources to total floodplain discharge. From Ward & Tockner (2001), Fig. 7, p. 813.

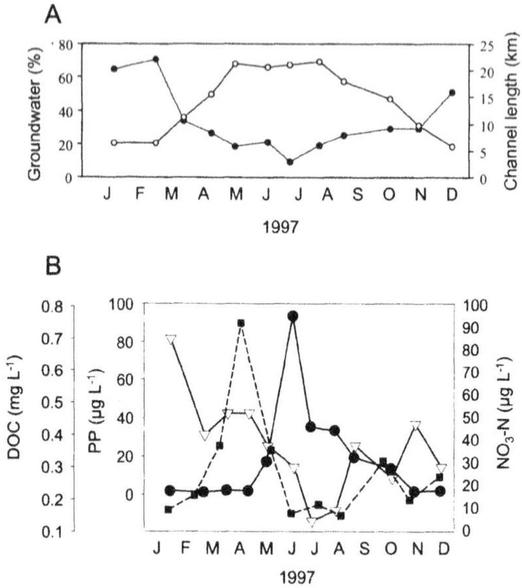

Figure 2. A) Total channel network length (open circles) and proportion of groundwater channels (filled circles) during an annual expansion/contraction cycle. B) Seasonal dynamics of dissolved organic carbon (DOC, triangle), nitrate-nitrogen (NO_3-N, square) and particulate phosphorous (PP, circle) at the downstream end of the flood plain.

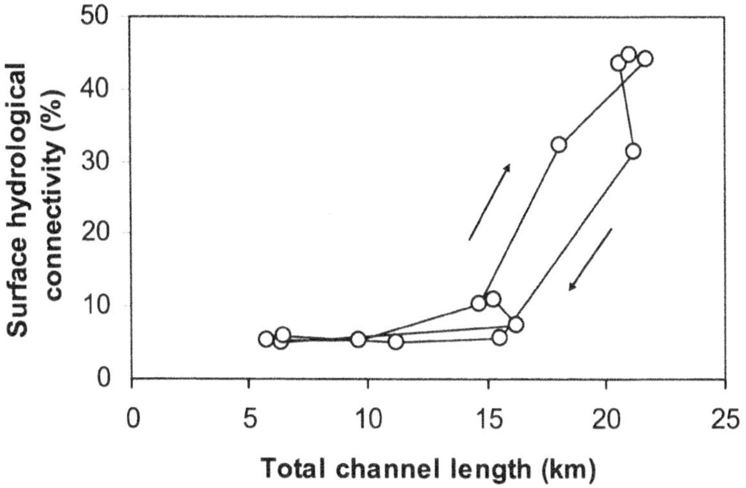

Figure 3. Relationship between surface hydrological connectivity (%) and channel expansion in the flood plain.

Models of floodplain inundation are often based on the flood pulse concept (Junk et al., 1989), which implies that river overspill floods a dry flood plain (Fig. 4A). Mertes (1997) examined the inundation patterns along several reaches of large rivers using remote sensing and showed that a variety of water sources led to flooded conditions prior to inundation from overflow of river water. Other components of inundation included direct precipitation, groundwater seepage to the surface, and local runoff (Fig. 4B). The inundation by mixed water sources was the more common although mostly neglected type of floodplain inundation with local waters as a major source (Mertes, 1997; Van der Nat et al., 2002). The Val Roseg is a high-gradient headwater flood plain that exhibits an additional type of inundation (Fig. 4C). All channels remain lotic throughout the annual cycle with floodplain water confined to stream channels during the seasonal flow pulse. Even a large unpredictable flood event ($Q > 20\ m^3s^{-1}$) caused by extreme rainfall did not increase the proportion of lentic habitats. Increases in discharge simply lead to an expansion of the channel network rather than to an inundation of the entire flood plain. Even at very high discharge ($>25\ m^3\ sec^{-1}$) the inundated area is small compared to the entire flood plain (Uehlinger et al., 2003). Further, the flow pulse in the Val Roseg is highly predictable and exhibits a monomodal pattern that is very similar to the flood pulse of very large tropical rivers. Smaller rivers in the tropics and the temperate zone typically show polymodal inundation patterns (e.g., Junk et al., 1989; Benke et al., 2000; Tockner et al., 2000; Van der Nat et al., 2002).

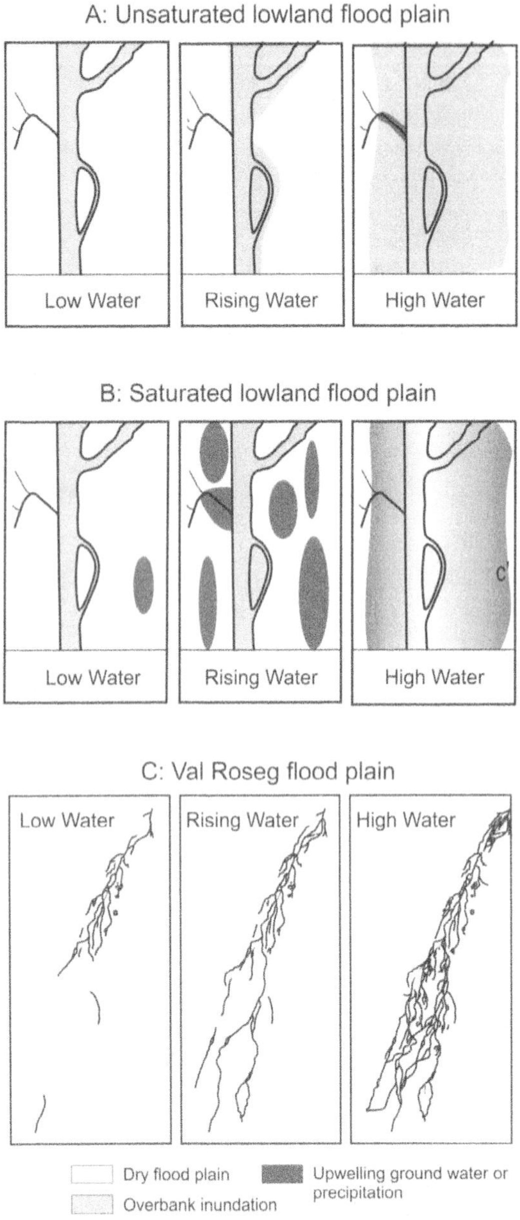

Figure 4. Inundation patterns for dry A) and saturated B) low gradient lowland flood plains compared to the high gradient glacial flood plain of Val Roseg C). A) and B) from Mertes (1977), Figs. 2a, b, p. 1751.

2.2. Spatiotemporal heterogeneity

Categorical maps of turbidity were employed to quantify seasonal changes in surface connectivity and riverscape heterogeneity. Turbidity was selected as an indicator of surface connectivity between the main channel and individual floodplain channels. The term "riverscape" refers to the heterogeneous spatial pattern that characterizes natural river corridors (e.g., Ward et al., 2002; Fausch et al., 2002).

In winter turbidity was very low in essentially all channels; this is a time when many floodplain channels were disconnected from the main channel at their upstream ends (Fig. 5). Most channels carried clear ground water with turbidity < 8 NTU (Nephelometric Turbidity Unit). With increasing discharge turbidity increased in those channels that experienced upstream surface connection with the main channel. Groundwater channels close to the main channel, fed by shallow hyporheic water emerging as seepage, transported moderately turbid water. Spring-brooks fed by deep alluvial and hillslope ground water carried clear water. In summer, glacial melting created a very turbid corridor (> 128 NTU) that was fringed by clear groundwater channels and a complex network of channels carrying a varying mixture of turbid glacial melt water and clear ground water.

Riverscape heterogeneity was significantly related to the flow pulse discharge, with maximum values during high summer flow conditions. A second order polynomial function described this positive relationship between discharge and heterogeneity ($y = -8.10\ x^2 + 0.10x + 0.35$; x = discharge in m^3s^{-1}). Based on this relationship we converted three years of discharge data to a time series of riverscape heterogeneity and total channel network length (Fig. 6). Results of the model indicated that heterogeneity remained high for more than eight months per year, starting with the expansion period in spring and lasting to the end of the contraction phase in autumn. Riverscape heterogeneity was already high at a discharge of 1 m^3s^{-1}, which led to high heterogeneity for periods longer than the duration of the "flow pulse". However, when discharge dropped below <1 m^3s^{-1}, heterogeneity rapidly decreased and stayed at a low level during the base flow period in winter. Increasing heterogeneity in spring and summer resulted from a rapid expansion in total channel network length owing to additional water sources such as snowmelt and glacial melt water. In summer, the contribution of ground water creates a very high heterogeneity although the relative proportion of ground water to total floodplain discharge is only about 10%. That means that small changes in the relative proportion of individual water sources can substantially modify floodplain patterns and processes. By comparing individual years, the model showed a very similar seasonal pattern over the 3-year period (Fig. 6).

In the Val Roseg flood plain, the shifting dominance of different water sources combined with a complex channel network created a remarkably high habitat heterogeneity, in particular during high flow conditions. However, overall floodplain heterogeneity (expressed as the coefficient of variation) was much lower in the Val Roseg compared to dynamic lowland rivers (Tockner et al., 2000). In most flood plains in temperate and tropical regions flooding creates very homogenous physicochemical conditions in the inundated plain. In such cases river water is the

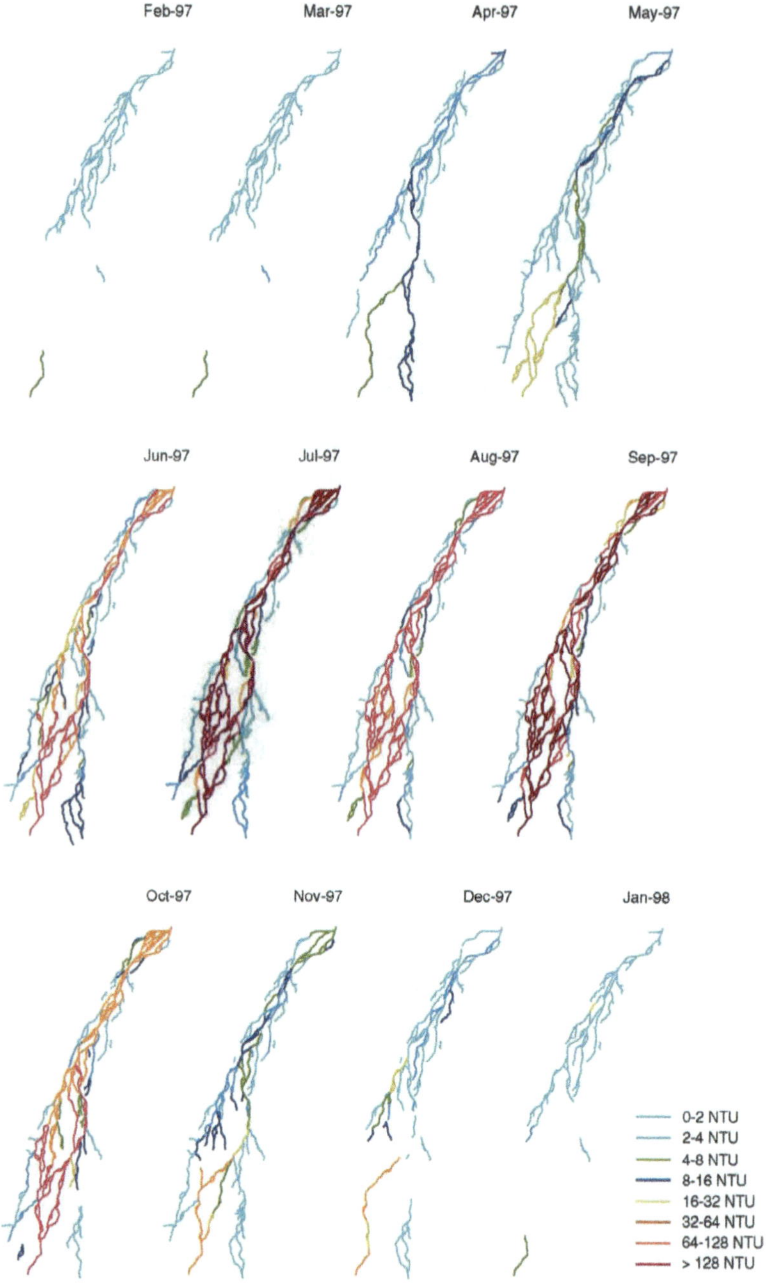

Figure 5. Map showing the distribution of turbidity in the Val Roseg flood plain during an annual expansion and contraction cycle. Modified from Malard et al. (2000), Fig. 3, p. 685.

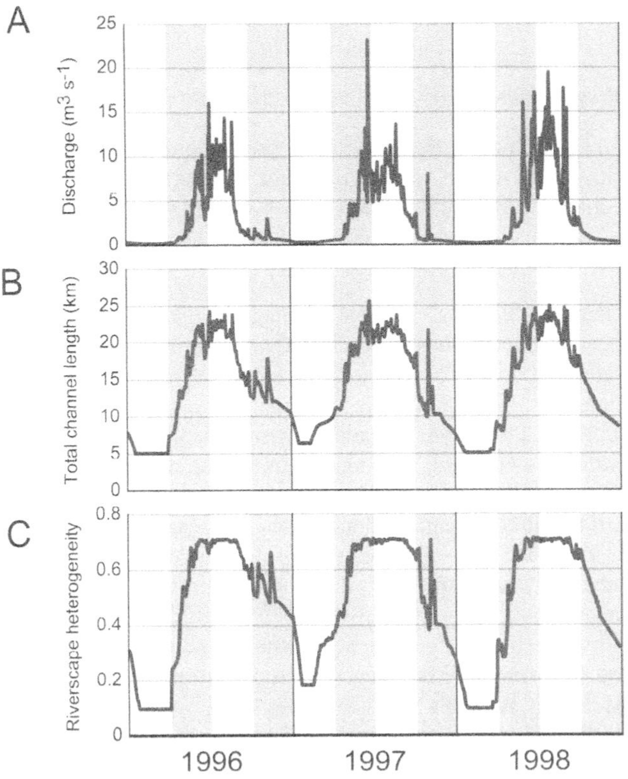

Figure 6. Seasonal dynamics of discharge A), total channel length B) and riverscape heterogeneity C) predicted from discharge-channel length and discharge-heterogeneity relationships. Riverscape heterogeneity (RH1 in Malard et al., 2000) combines three landscape indices of diversity, namely relative richness of turbidity classes, relative evenness of turbidity classes and relative contrast (based on the number of boundaries between adjacent patches of different turbidity classes).

primary source of flooding, despite minor contributions of other water sources (see Fig. 4B). In lowland rivers, local autogenic processes (e.g. nutrient uptake and transformation processes) lead to a high diversification of individual water bodies at low water conditions (e.g., Robertson et al., 1999). The Val Roseg flood plain primarily serves as a mixing zone for different water sources with autogenic processes playing an insignificant role (see Tockner et al., 2002). The glacial flood plain lacks the "flood pulse advantage" (sensu Bayley, 1991) typical for lowland flood plains. The "flood pulse advantage" reflects the amount by which productivity of a system is increased by a natural predictable flood pulse. However, in the Val Roseg flood plain the high channel heterogeneity ameliorates to some extent the negative effects of diurnal flow peaks associated with high sediment loads, channel instability, resource limitation and low temperature.

3. CHANNEL CHARACTERIZATION AND DISTRIBUTION

3.1. Physicochemical characterization of channel types

The channel network within the Val Roseg flood plain by no means consisted of uniformly cold, highly turbid waters. Based on the correspondence between water source, hydrological connectivity and physicochemical attributes six different channel types were distinguished (Figs. 7 and 8; Tockner et al., 1997; Malard et al., 1999):

(M) Main channel: Thalweg eurhithral channel that carries the majority (>70%) of surface water at all flow levels and is mainly fed by melt water from the valley glaciers.

(S) Side channels: Eurhithral side channels connected by surface flow to the main-channel at the upstream and downstream ends.

(I) Intermittently-connected channels: Pararhithral channels having permanent surface connections at their downstream ends, but an intermittent surface connection with the main channel at their upstream ends.

(X) Mixed channels: This channel type is fed by two or more water sources: main-channel, a side tributary, the alluvial aquifer, a tributary aquifer, or by marshes.

(G) Groundwater channels: Groundwater channels are fed by the alluvial aquifer or by a tributary aquifer. They have no upstream surface connection with type M, S, I, X channels.

(T) Tributaries: Channels that enter the flood plain as a surface water channel originating in the uplands. Upland sources range from ground water to hanging glaciers.

Groundwater channels can be further separated into three subtypes that are either fed by shallow alluvial ground water, deep alluvial ground water or hillslope ground water (Chapter 3). Individual channel types are primarily discernible by their spatiotemporal variability and not by average values of environmental variables (Table 2). Channels influenced by glacial meltwater showed pronounced diel and seasonal variations in their physicochemical properties. In intermittently-connected channels, for example, turbidity increased from <30 NTU's at low flow to >2000 NTU's at high flow. In the main channel concentrations of particulate phosphorous (PP) may vary from ca. 70 $\mu g\ L^{-1}$ to > 800 $\mu g\ L^{-1}$ over a 24-hr cycle during the main ablation period. In contrast to connected channels, PP concentrations remained low (<10 $\mu g\ L^{-1}$) and very constant in groundwater channels. Overall, temporal (diel and seasonal) variation of physicochemical variables is highest in intermittently-connected channels, followed by side channels, tributaries, the main channel, mixed channels, and finally groundwater channels. Groundwater channels, however, although exhibiting low temporal variability, may be very different from each other (high spatial heterogeneity).

Table 2. Idealized characterization of the six channel types in the Val Roseg flood plain (modified from Tockner et al., 1997).

Criteria	*Main channel (M)*	*Side channel (S)*	*Intermittently-connected channel (I)*	*Mixed channel (X)*	*Groundwater channel (G)**	*Tributary (T)*
Water source	Summer: glacial water Winter: subglacial water and alluvial ground water	Summer: glacial water Winter: mostly dry	Summer: water from main channel or shallow alluvial ground water Winter: dry	Summer: variable sources Winter: alluvial ground water	Alluvial and hillslope tributary ground water	Hanging glaciers
Surface connectivity	Complete in summer, upper and lower main channel disconnected in winter	Upstream and downstream with the main channel	Downstream with main channel	Variable connectivity with other channel types	Downstream with main channel	Only downstream connection with main channel
Flow variability	Distinct seasonal and diel fluctuations	Distinct seasonal and diel fluctu-ations	Distinct seasonal and diel fluctuations	Variable	Low fluctuations	Variable
Surface flow	50 % dry in winter	>80 % dry in winter	100 % dry in winter	>60 % dry in winter	Permanent flow	>95 % dry in winter
Bed stability	Low	Medium	Medium	Medium to high	High	Medium to high
Turbidity	Highly variable	Highly variable	Highly variable	Variable	Low	Low to moderate
Summer temperature (°C)	2-4	2-4	2-5	3-5	4-8	4-8
Within channel type spatial heterogeneity	Medium	Low	Very low	Medium	High	Low
Within channel type temporal variability	High	Very high	Very high	Medium	Low	Medium
Expected biodiversity	Medium/low	Low	Low	Medium/high	High	Low/medium

*Malard (Chapter 3) distinguishes three types of groundwater channels

Figure 7. Examples of different channel types in the upper section of the Val Roseg flood plain (M = Main channel, S = Side channel, I = Intermittently-connected channel, X = Mixed channel, G = Groundwater channel, T = Tributary (Photo: R. Zah).

For alpine landscapes, three major stream types have been identified: krenal streams fed by ground water, rhithral streams fed by snowmelt water, and kryal streams that carry mainly glacial melt water (Ward, 1994; Hieber et al., 2002). Although glacial flood plains were not considered in this stream classification, within the Val Roseg flood plain individual channel types are functionally comparable to these alpine stream types. The main channel and side channels are equivalent to kryal streams, groundwater-fed channels correspond to krenal streams and many tributaries are similar to rhithral streams. However, a krenal-like floodplain channel can be transformed into a "rhithral" and finally into a "kryal"-channel type during a single hydrological cycle. This means that at a relatively small scale – total floodplain area is 0.67 km^2 - a similar range of lotic systems occur as in an entire alpine catchment. In large lowland flood plains water bodies tend to reflect lateral successional and hydrological connectivity gradients, (eu-, para-, plesio- and palaeopotamal; Amoros & Petts, 1993). The parapotamal channels are analogous to some extent to intermittently-connected and groundwater channel types in the Val Roseg. The eupotamal channel type in lowland rivers includes the main and side channels in the Roseg flood plain, both of which are characterized by permanent surface connectivity upstream and downstream. Neither the palaeopotamal nor the plesiopotamal have clear analogues in the Val Roseg.

Figure 8. Close-up views of the six channel types. See Figure 7 legend for abbreviations of channel types (Photos: R. Zah & K. Tockner).

3.2. Spatiotemporal distribution

The distribution of channel types differed between the upper and the lower flood plain. Groundwater channels dominated the lower sections whereas intermittently-connected channels were only present in the upper plain (Fig. 9). The relative proportion of individual channel types also changed with season (Table 3). During glacial ablation in summer all channel types were well represented. During the low water period in winter, however, tributaries and intermittently-connected channels ceased to flow. Most tributaries started to flow with the beginning of the expansion period. Intermittently-connected channels started flowing later and therefore carried water only during the main ablation season. At low water level, the remaining mixed and main-channel segments were transformed into groundwater channels lacking upstream surface connectivity. Based on total length, groundwater channel types were the most abundant channel type during all seasons (Table 3).

Table 3. The relative proportion (%) of individual channel types during the four hydrological phases.

Channel Type	*Low water ($0.1\ m^3\ s^{-1}$)*	*Expansion ($1.5\ m^3\ s^{-1}$)*	*Glacial ablation ($6.0\ m^3\ s^{-1}$)*	*Contraction ($0.8\ m^3\ s^{-1}$)*
Main channel (M)	27.3	25.9	16.0	27.7
Side channel (S)	3.4	2.3	17.0	7.0
Intermittently-connected channel (I)	0.7	0.0	14.2	3.9
Mixed channel (X)	24.1	19.4	14.7	15.9
Groundwater channel (G)	43.0	34.7	25.5	35.0
Tributary (T)	1.5	17.6	12.7	10.5

4. SUMMARY

Based on the correspondence between hydrological connectivity and physico-chemical attributes six distinct channel types have been identified within the floodplain ecosystem. Distinct seasonal and daily runoff patterns, caused by ice melt, change the hydrological connectivity between the individual channel types. This functional characterization of channel types serves as the habitat template to structure future ecological research in the Val Roseg flood plain.

Major insights derived from the intensive research on habitat heterogeneity of the Val Roseg flood plain include: (1) recognition of the high level of spatiotemporal heterogeneity that characterizes this glacial flood plain, (2) an elucidation of the importance of expansion/contraction cycles for creating this heterogeneity, (3) recognition of the major differences compared to flood plains at lower elevation, and (4) appreciation of the important role that regional processes (e.g., hillslope groundwater influence) play in this glacial floodplain complex.

Compared to single-thread high alpine streams, flood plains are presumably more resistant to expected changes in flow regime and land use (e.g., Tockner & Stanford, 2002). They provide regional ecosystem stability to otherwise very sensitive and rapidly changing high alpine systems (Körner & Spehn, 2002). The Val Roseg flood plain provides a complex shifting mosaic of channel types, thereby contributing significantly to the regional diversity of an otherwise harsh alpine environment.

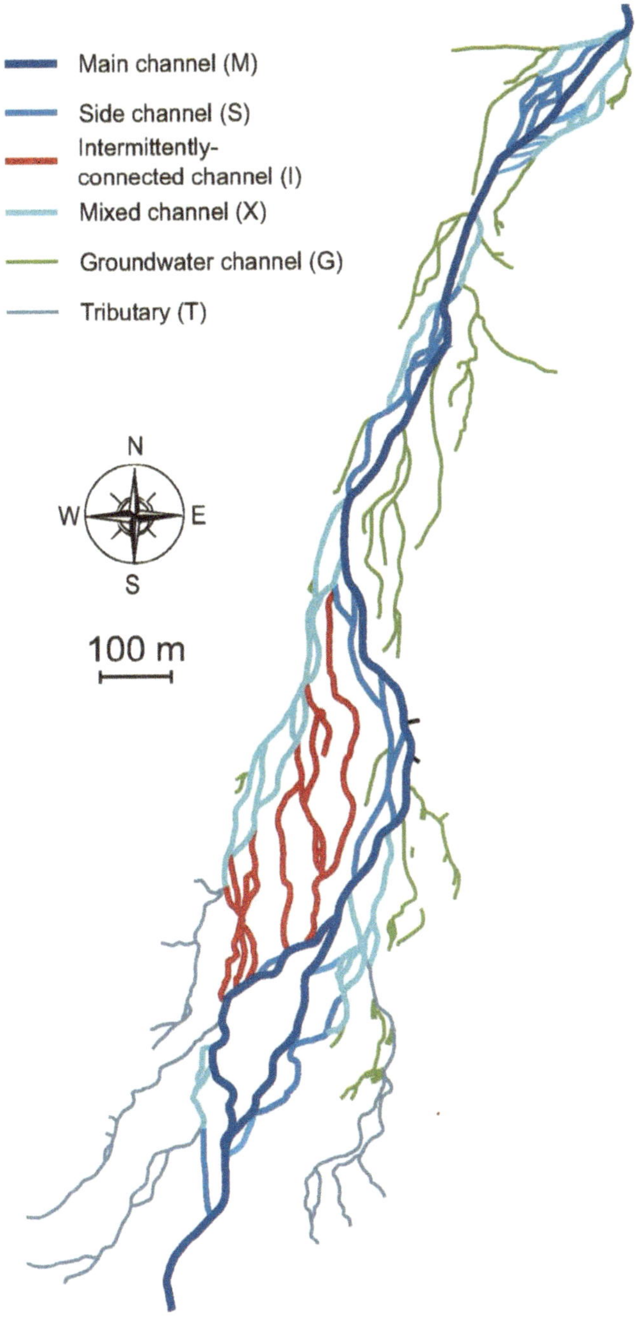

Figure 9. Spatial distribution of channel types in the Val Roseg flood plain.

5. REFERENCES

Amoros, C., & Petts, G. E. eds. (1993). *Hydrosystemes Fluviaux*. Paris: Masson.

Bayley, P. B. (1991). The floodpulse advantage and the restoration of river-floodplain systems. *Regulated Rivers: Research & Management, 6*, 75-86.

Benke, A. C., Chaubey, I., Ward, G. M., & Dunn, E. I. (2000). Flood pulse dynamics of an unregulated river floodplain in the southern U.S. coastal plain. *Ecology, 81*, 2730-2741.

Burgherr, P., Ward, J. V., & Robinson, C. T. (2002). Seasonal variation in zoobenthos across habitat gradients in an alpine glacial flood plain (Val Roseg, Swiss Alps). *Journal of the North American Benthological Society*, 21, 561-575.

Fausch, K. D., Torgersen, C.E., Baxter, C. V., & Li, H. W. (2002). Landscapes to riverscapes: bridging the gap between research and conservation of stream fishes. *BioScience, 52*, 483-498.

Füreder, L. (1999). High alpine streams: cold habitats for insect larvae. In R. Margesin & F. Schinner (Eds.), *Cold adapted organisms: Ecology, physiology, enzymology and molecular biology* (pp. 181-196). Springer: Berlin.

Hamilton, S. K., Sippel, S. J., & Melack, J. M. (1996). Inundation patterns in the Pantanal wetland of South America determined from passive microwave sensing. *Archiv für Hydrobiologie, 137*, 1-23.

Hieber, M., Robinson, C. T., Uehlinger, U., & Ward, J. V. (2002). Are alpine lake outlets less harsh than other alpine streams. *Archiv für Hydrobiologie, 154*, 199-223.

Junk, W. J., Bayley, P. B., & Sparks, R. E. (1989). The flood pulse concept in river-floodplain systems. *Canadian Special Publication of Fisheries and Aquatic Sciences, 106*, 110-127.

Klein, B., & Tockner, K. (2000). Biodiversity in springbrooks of a glacial floodplain (Val Roseg, Switzerland). *Verhandlungen der Internationalen Vereinigung für Theoretische und Angewandte Limnologie, 27*, 704-710.

Körner, Ch., & Spehn, E. M. eds. (2002). *Mountain biodiversity: a global assessment*. London: Parthenon Publishing.

Malard, F., Tockner, K., & Ward J. V. (1999). Shifting dominance of subcatchment water sources and flow paths in a glacial floodplain (Val Roseg, Switzerland). *Arctic, Antarctic and Alpine Research, 31*, 135-150.

Malard, F., Tockner, K., & Ward, J. V. (2000). A landscape-level analysis of phsyico-chemical heterogeneity in a glacial floodplain. *Landscape Ecology, 15*, 679-695.

Mertes, L. A. K. (1997). Documentation and significance of the perirheic zone on inundated floodplains. *Water Resources Research, 33*, 1749-1762.

Robertson, A. I., Bunn, S. E., Boon, P. I., & Walker, K. F. (1999). Sources, sinks and transformations of organic carbon in Australian floodplain rivers. *Marine and Freshwater Research*, 50, 813-29.

Sippel, S. J., Hamilton, S. K., Melack, J. M., & Novo, E. M. M. (1998). Passive microwave observations of inundation area and the area/stage relation in the Amazon River floodplain. *International Journal of Remote Sensing, 19*, 3055-3074.

Sporka, F. (1998). The typology of floodplain water bodies of the Middle Danube (Slovakia) on the basis of the superficial polychaete and oligochaete faunas. *Hydrobiologia, 386*, 55-62.

Stanford, J. A., & Ward, J. V. (1993). An ecosystem perspective of alluvial rivers: connectivity and the hyporheic corridor. *Journal of the North American Benthological Society, 12*, 48-60.

Stanley, E. H., Fisher, S. G., & Grimm, N. B. (1997). Ecosystem expansion and contraction in streams. *BioScience, 47*, 427-435.

Tockner, K., & Stanford, J. A. (2002). Riverine floodplains: present state and future trends. *Environmental Conservation, 29*, 308-330.

Tockner, K., Malard, F., Burgherr, P., Robinson, C. T., Uehlinger, U., Zah, R., & Ward, J. V. (1997). Physico-chemical characterization of channel types in a glacial floodplain ecosystem (Val Roseg, Switzerland). *Archiv für Hydrobiologie, 140*, 433-463.

Tockner, K., Malard, F., & Ward, J. V. (2000). An extension of the flood pulse concept. *Hydrological Processes, 14*, 2861-2883.

Tockner, K., Malard, F., Uehlinger, U., & Ward, J. V. (2002). Nutrients and organic matter in a glacial river-floodplain. *Limnology & Oceanography, 47*, 266-277.

Uehlinger, U., Malard, F., & Ward, J. V. (2003). Thermal patters in the surface waters of a glacial river corridor (Val Roseg, Switzerland). *Freshwater Biology, 48*, 284 -300.

Van der Nat, D., Schmidt, A., Tockner, K., Edwards, P .J., & Ward, J. V. (2002). Inundation dynamics in braided floodplains. *Ecosystems, 5*, 636-647.

Ward, J. V. (1994). Ecology of alpine streams. *Freshwater Biology, 32*, 277-294.

Ward, J. V., & Tockner, K. (2001). Biodiversity: toward a unifying theme in river ecology. *Freshwater Biology, 46*, 807-819.

Ward, J. V., Tockner, K., & Schiemer, F. (1999). Biodiversity of floodplain river ecosystems: ecotones and connectivity. *Regulated Rivers: Research & Management, 15*, 125-139.

Ward, J. V., Tockner, K., Arscott, D. B., & Claret, C. (2002). Riverine landscape diversity. *Freshwater Biology, 47*, 517-539.

CHAPTER 5

U. UEHLINGER & F. MALARD

THERMAL HETEROGENEITY

1. INTRODUCTION

Temperature is a primary factor structuring biotic communities and regulating ecosystem functions in streams and rivers. Temperature affects surface-subsurface interactions, the solubility of gases, chemical and biochemical reactions, ecosystem processes, and the structure of invertebrate communities (Constanz & Murphy, 1991; Vannote et al., 1980; Vannote & Sweeney, 1980; Ward, 1992; Ward & Stanford, 1982). Constant temperatures prevail only in special aquatic habitats such as some spring sources, whereas thermal heterogeneity is a characteristic feature of most freshwater ecosystems. Stream temperature is determined by the fluxes of heat comprising net radiation, sensible heat exchange, heat input by tributaries and groundwater, friction, incoming stream flow and losses by radiation or latent heat exchange (Webb & Zhang, 1997). Heat budgets depend on the regional climate and local characteristics such as riparian vegetation, aspect, channel morphology, stream water-groundwater interactions and hydrology (Constanz, 1998; Gu et al., 1998; Hawkins et al., 1997; Ward, 1985).

Latitude, altitude, and continentality determine the regional climate that in turn influences the thermal regimes of lotic ecosystems (Vannote & Sweeney, 1980; Ward, 1985). Air temperature is an important climatic factor that directly effects stream temperature. Groundwater influence, ice cover, spates or snow and ice melt may confound the relation between air and water temperatures, but in general stream temperatures tend to track air temperature. The average air temperature lapse rate varies between –0.4 and –0.6 °C/100m; from the foothills of the Alps (600 m a.s.l.) to the upper boundary of the alpine zone air temperatures decline by about 12°C. Low temperatures are a characteristic feature of alpine streams and in particular of glacial-fed streams. Low summer temperatures (<2°C) have been considered to be the primary factor responsible for the reduced diversity of zoobenthos in the uppermost reaches of glacial streams (Milner & Petts, 1994; Ward, 1994). However, despite the fact that temperature is considered to play a dominant role for the longitudinal zonation of invertebrate taxa downstream of glacier termini (Castella et al., 2001; Milner et al., 2001), thermal characteristics are poorly documented.

In this chapter we focus on thermal patterns in surface waters along the corridor of the Roseg River. In addition we examine hyporheic temperature at few sites. The complexity of surface and subsurface flow paths and different water sources in Val Roseg provide a potential for high thermal heterogeneity, within the constraints of

J.V. Ward & U. Uehlinger (eds.), Ecology of a Glacial Flood Plain, 75-90.
© 2003 *Kluwer Academic Publishers. Printed in the Netherlands.*

the regional climate. Floodplain systems are generally characterized by high thermal variation in space and over both short and long time periods (Arscott et al., 2001; Malard et al., 2001; Mosley, 1983; Uehlinger et al., 2003; Ward et al., 2002). Surface water temperatures were recorded at hourly intervals with temperature loggers at 27 sites, mainly from 1997 to 1998 (Fig. 1, Malard et al., 2001; Uehlinger et al., 2003). In 1997 and 1999 temperature was also measured with a temperature probe in each floodplain channel along 17 transects at monthly intervals (Uehlinger et al., 2003). To assess hyporheic temperatures loggers were deployed in piezometers at 30 and 80 cm below the stream bed at 8 sites (Malard et al., 2001).

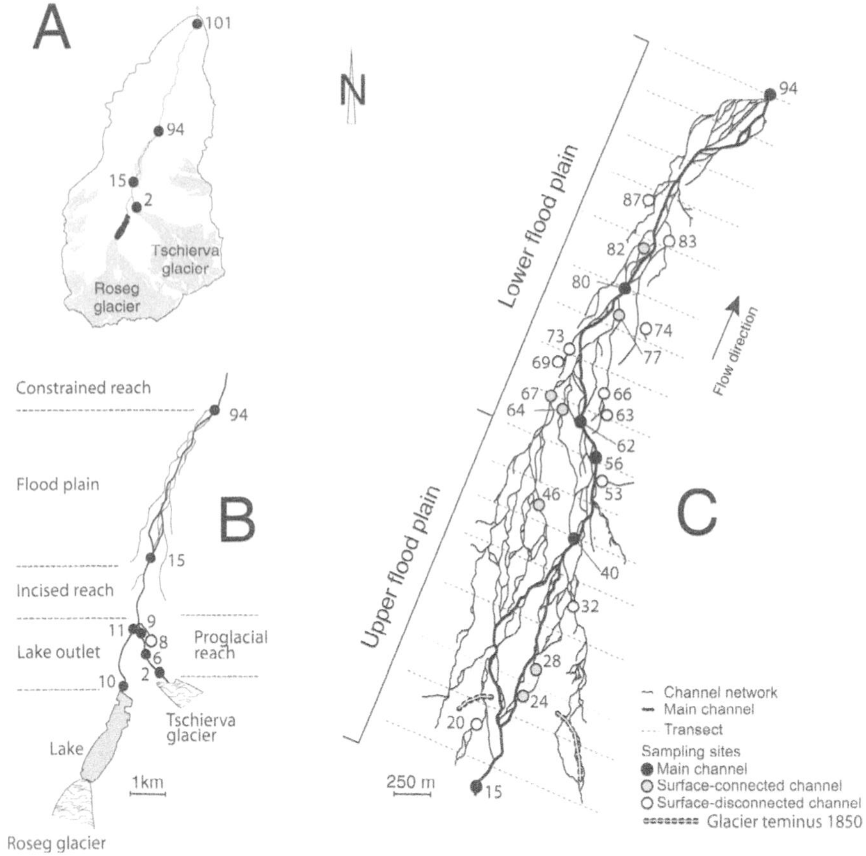

Figure 1. Val Roseg temperature monitoring: Location of the sites (circles) where temperature loggers had been deployed. A) Val Roseg catchment. B) Upper part of the catchment. C) Detail map of the flood plain. Modified from Uehlinger et al. (2003), Fig. 1, p. 286.

2. LONGITUDINAL PATTERNS ALONG THE MAIN STEM

2.1. The glacial temperature regime

The mean daily temperatures in the main stem of the Roseg River are characterized by a relatively rapid temperature increase in April and May, a subsequent moderate decline between June and September, and a rapid decrease until mid November (Fig. 2). This temporal pattern has also been observed in other glacial streams (Schütz et al., 2001). Rising air temperature and solar radiation increase the release of cold melt water into the lake below the Roseg Glacier and the proglacial stream below the Tschierva Glacier. Cold melt water and high discharge break the vernal increase of stream temperature and even result in a persistent temperature reduction over the summer. The autumnal decline of air temperature and solar radiation terminate glacier melt and give rise to a rapid decline of stream temperatures. The "glacial" temperature signature is evident at all main channel sites within the Roseg River corridor but weakens from the proglacial reach to the end of the catchment; about 33 km downstream from the Tschierva and Roseg Glaciers the temperature curve approaches that of Alpine streams carrying no or minor amounts of glacial melt water (site IR in Fig. 2).

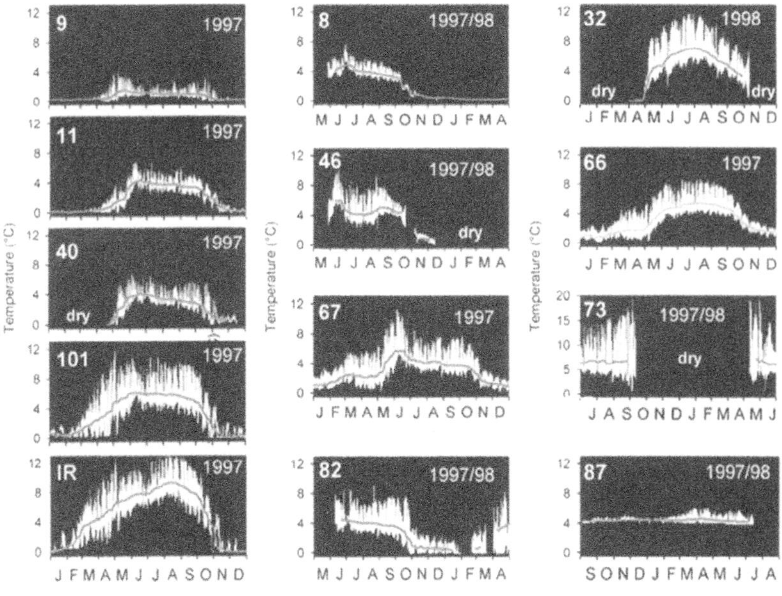

Figure 2. Surface temperature curves in the Roseg River corridor and at a site in the Inn River (IR). Hourly values (white) and 30-day moving median (bold gray line). Left panel: Main channel sites and Inn River. Middle panel = surface-connected floodplain channels. Right panel = surface-disconnected floodplain channels. Site number in the top left corner of each graph. Modified from Uehlinger et al. (2003), Fig. 3, p. 290.

2.2. Link between flow and temperature

Annual patterns of river temperature are in general highly predictable and largely independent of catchment size, reflecting the dominant influence of regional climate (e.g., Johnson, 1971; Smith & Lavis, 1975; Uehlinger, 2000; Webb & Walling, 1986). In small catchments, however, discharge patterns are less predictable unless glaciers or snow become major sources of water. A characteristic feature of glacial streams is the coincidence of the annual flow and temperature pulses (Fig. 3), resulting from the fact that air temperature and solar radiation directly control both parameters. The generally favorable conditions for aquatic biota that characterize most streams during summer are offset in glacial streams by high flow, bed load transport, and high turbidity. The positive link between discharge and temperature declines with decreasing glacierization and altitude (Uehlinger et al., 2003).

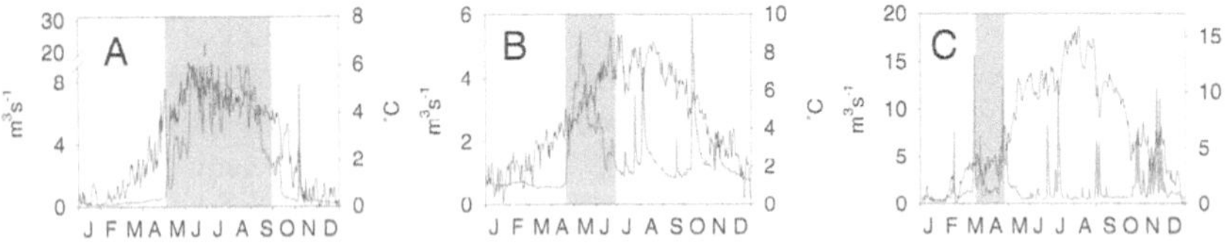

Figure 3. Daily discharge (gray line) and temperature (black line). Gray areas delimit the period of snow or ice melt. A) Roseg River at site 94. B) Subalpine river (1707 m a.s.l., catchment <0.1% glacierized). C) River in the northern foothills of the Alps (800 m a.s.l.).

2.3. Longitudinal patterns

Figure 4 depicts temperature patterns along the main channel of the Roseg River and at one Inn River site (IR) for selected periods in the annual cycle. In the Roseg River, the mean daily temperature peaked in June with 0.8°C at site 2 to 7.2°C at site 101. At the Inn River site (33 km downstream of the Tschierva glacier), mean daily temperatures reached the annual maximum (9.9°C) in August. Temperatures in the Roseg River were minimum in January, when daily temperatures at main channel sites ranged from 0 – 1.4°C and monthly averages from 0 to 1.3°C. Between March and October maximum temperatures within the main stem of the Roseg River were recorded at the end of the catchment (site 101). From December to February maximum temperatures occurred in the major upwelling zone in the middle of the flood plain (site 62).

The diel amplitude (or diel temperature pulse) is considered to be an ecologically significant variable for aquatic insects (e.g., Kamler, 1965; Vannote & Sweeney, 1980; Ward & Stanford, 1982). Headwaters at low elevation show little diel variation if they are fed largely by ground water and well shaded by tree canopy. In the open proglacial reach of the Roseg River, regional climate, high discharge, and the source of water (water temperature close to 0°C throughout the year) keep diel temperature fluctuation within a relatively narrow range. Temporal and spatial patterns of the diel amplitude more or less parallel those of mean daily temperatures

(Fig. 4). Amplitudes peaked in May at all sites except sites 10 and 11 (lake outlet), where the maximum occurred in June. The maximum diel pulse of 8.8°C was observed at the end of the catchment in April 1998.

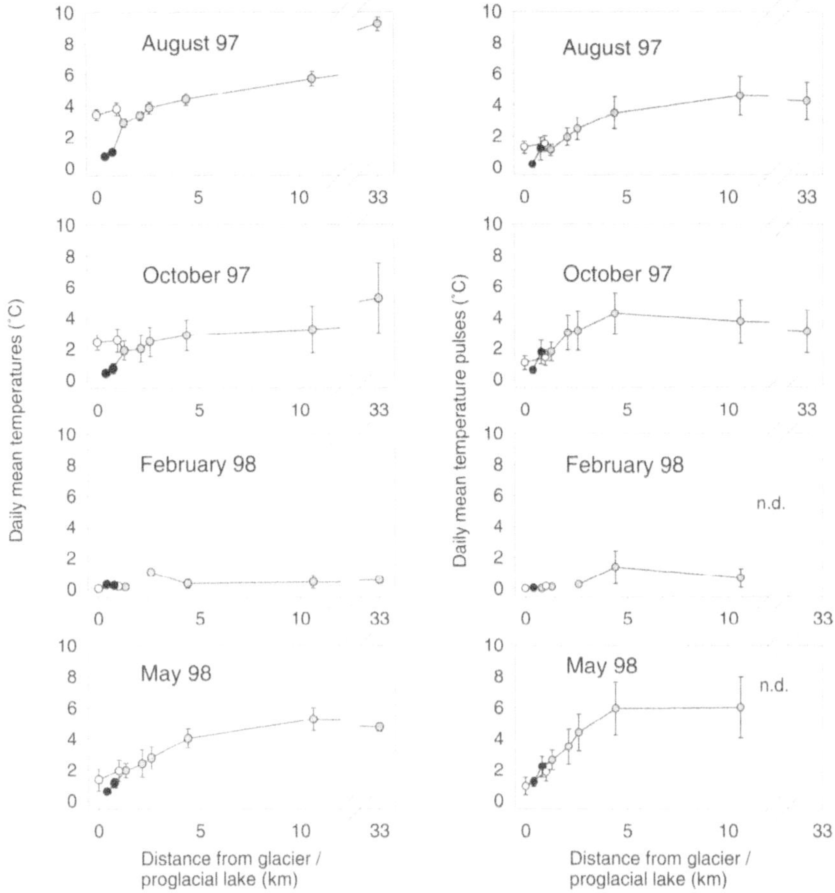

Figure 4. Average (circles) and standard deviations (error bars) of daily temperatures (left panel) and diel temperature pulses (right panel) in the main channel of the Roseg River during four seasons. Open circles = lake outlet sites, black circles = sites in the proglacial reach, gray circles = main channel sites and Inn River site (33 km downstream of the Tschierva Glacier). n.d. = no data. Modified from Uehlinger et al. (2003), Fig. 4, p. 291).

2.4. *Influence of geomorphic features*

2.4.1. *The proglacial lake*

The occurrence of a proglacial lake, which acts as a heat trap, results in a distinct amelioration of the temperature conditions in the outlet stream. The average annual temperature in the lake outlet stream was about 1.2 °C higher than in the adjacent proglacial reach below the Tschierva Glacier. During summer, average daily temperature increased by 3 to 5.5 °C between the Roseg Glacier terminus and the lake outlet (a distance of about 1.2 km). Stream temperatures in the lake outlet declined after commencement of the main icemelt period, but not as much as in the proglacial stream despite the short residence time of water. The autumnal temperature decrease in the lake outlet stream distinctly lags behind that of the proglacial reach.

The presence of a lake within the river corridor has an important influence on the thermal characteristics of the glacial stream and as a consequence on the structure of zoobenthic community (Burgherr & Ward, 2000). The lake at the terminus of the Roseg Glacier, through its influence on the thermal regime, changes the outlet stream to a rhithral reach, rather than a kryal reach that would occur there without a lake.

2.4.2. *The flood plain*

The presence of a braided flood plain serves as a local modifier of the temperature regime. Splitting of the parent channel into distributary channels reduces flow velocity and water depth (Richards, 1982), both of which are inversely related to the warming and cooling of stream water (Gu et al., 1998). Most main channel water diverted in the network of small channels in the upper left part of the Roseg flood plain flows through site 67 (Fig. 1). In July and August differences in daily temperatures between sites 67 and 62 averaged 0.33 ± 0.15°C, reflecting additional warming in the network of smaller channels compared to the main channel. However, stream temperature at the end of the flood plain increased by <0.2°C because the main channel carried more than 80% of the floodplain discharge. In summer, the influence of the relatively warm groundwater channels and tributaries on the main channel temperature is presumably small since their contribution to the main channel discharge is <10%. Average temperature of shallow alluvial groundwater at 80cm depth, measured in a well about 150 m downstream of site 62 (Malard et al., 2001), was only 0.9°C higher (mid May to mid September) than in the main channel. Assuming that 5% to the main channel discharge is of hyporheic origin, main channel temperature increases by <0.1°C. The flood plain has a minor influence on the main channel temperature pulse during summer. This concurs with results of a study by (Tockner et al., 2002) who found a negligible flood plain influence on concentrations of dissolved nutrients.

3. THE LATERAL DIMENSION OF THERMAL HETEROGENEITY

Although the Roseg flood plain had only a minor influence on main channel temperatures, it contributed markedly to the thermal heterogeneity of the glacial river corridor. The annual flow pulse results in a distinct expansion/contraction cycle of the floodplain channel network that is paralleled by corresponding alterations of surface hydrological connectivity between different channel types and exchange pathways between surface and hyporheic waters (Malard et al., 1999, 2000, 2001). These dynamics create a three-dimensional shifting mosaic of thermally distinct habitats that characterizes the thermal heterogeneity of this glacial flood plain.

3.1. Surface water

The glacial temperature regime prevails in the main channel and in surface connected flood plain channels (Fig. 2). Thermal patterns were less uniform in

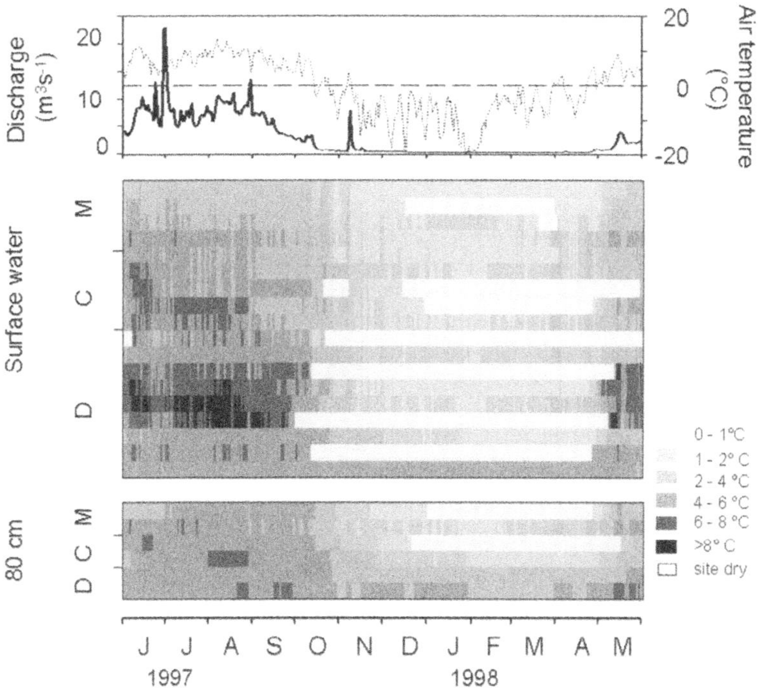

Figure 5. Upper panel: daily discharge (solid line) and daily air temperature (dotted line). Middle panel: daily temperatures in the surface waters of the flood plain. M = 4 main channel sites, C = 5 sites in surface-connected channels, D = 9 sites in surface- disconnected channels. Lower panel: daily temperatures in the hyporheic zone (depth = 80 cm). M = 2 main channel sites, C = 2 sites in surface-connected channels, D = 2 sites in surface-disconnected channels.

surface-disconnected channels, where daily mean temperatures reach the annual maximum between July and September or remain stable as for example in alluvial springs (site 87) or exfiltration zones of side slope aquifers (site 73).

Figure 5 illustrates thermal heterogeneity within the flood plain based on daily temperatures recorded at 18 sites during the annual expansion and contraction cycle of the channel network. From June to mid September daily temperatures ranged from 2.3°C (site 15, September) to 10°C (site 32, August). Between December and March daily temperatures varied between 0 and 4.5°C. Diel temperature maxima ranged from 7.6 to 19.4°C (May-September) and from 0.0 to 4.5°C (December – February) and the corresponding diel minima were 0.0 - 3.7°C and 0.0 - 0.2°C (Fig. 6). At site 87 (spring brook) daily winter temperatures varied between 4.5 and 5.1°C. Spot measurements along floodplain transects showed that "winter-warm" habitats (>2°C) are restricted to a few spring brooks at the floodplain border.

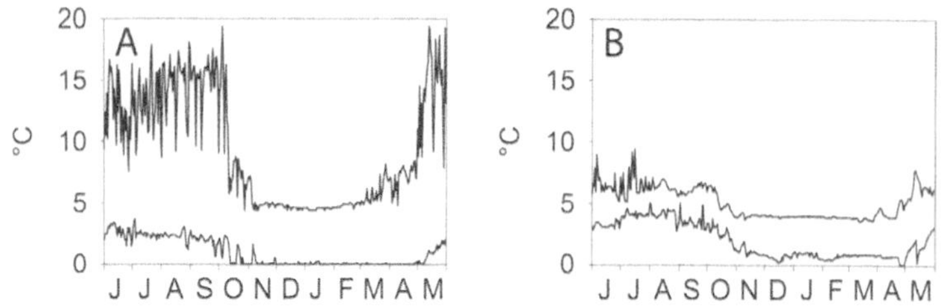

Figure 6. Minimum and maximum temperatures in the Roseg flood plain (9-18 sites). A) Surface water. B) Hyporheic water at a depth of 80 cm (6-sites).

During summer, the main channel formed a relatively cold corridor within the flood plain, where annual degree days ranged from 543 to 943 (Figs. 7, 8).

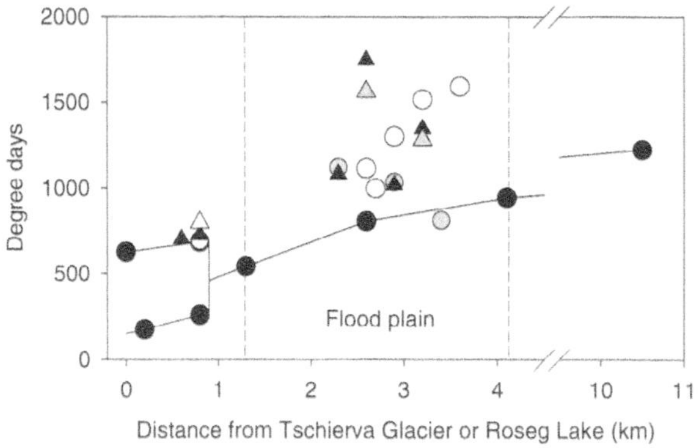

Figure 7. Annual degree days at various sites along the Roseg River corridor. Surface sites = circle: black filled circles = main channel, gray filled circles = surface-connected channels, open circles = surface- disconnected channels. Hyporheic sites = triangles: black filled = 80 cm, gray filled = 30 cm). Modified from Uehlinger et al. (2003), Fig. 5, p. 291.

Temperatures in surface-connected channels parallel the pattern of the main channel (annual maximum in June) but are significantly higher (Uehlinger et al., 2003).

Surface-disconnected channels included the warmest floodplain habitats; three such sites accumulated between 1302 and 1595 degree days (Fig. 7). Diel temperature pulses parallel more or less daily temperatures (Fig. 8). Maximum pulses (> 14°C) were observed at the surface disconnected channel sites 20 and 73 immediately after these sites became wet (May) or shortly before they fell dry (end of September/beginning of October), i.e. when low discharge coincides with relatively high diel air temperature amplitudes. Spot measurements performed along the 17 transects (Fig. 1) within the flood plain at 33 to 165 sites (depending on the number of channels carrying water) provide a spatially more comprehensive spatial picture of thermal heterogeneity than the logger records. These measurements confirm the thermal patterns identified by logger records (Uehlinger et al., 2003).

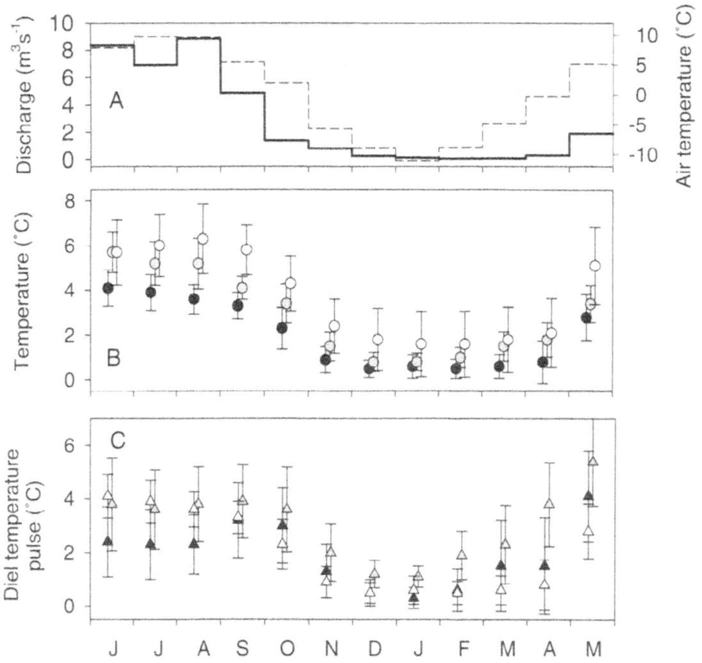

Figure 8. A) Monthly discharge (solid line) and air temperature (dashed line), B) Monthly averages of daily temperatures and standard deviations, and C) monthly averages and standard deviations of diel temperature pulses. Black symbols = main channel, gray symbols = surface-connected channels, open symbols = surface-disconnected channels. Modified from Uehlinger et al. (2003), Fig. 9, p. 292.

In the Roseg flood plain thermal patterns are quite stable throughout the summer (Fig. 5). Even during the distinct flow pulse, water is confined to stream channels (i.e. the floodplain surface is not inundated) because of the relatively steep slope (1.9

– 4.8%). Only at discharge exceeding 25 m^3s^{-1} (e.g. by rainfall induced spates) does water spill over the banks of the main channel to form a 30 - 60m wide inundated corridor, which is a relatively small portion of the entire floodplain area. In contrast, bank overflow in low gradient systems may inundate most of a flood plain and level out thermal differences between floodplain habitats (Ward et al., 2002). Moreover, low gradient flood plains contain different standing water bodies, such as isolated ponds, scour pools, or oxbow lakes. Depending on subsurface connectivity and canopy cover these habitats may heat up rapidly and substantially increase thermal heterogeneity between floods (Arscott et al., 2001). However, lentic water bodies are rare in the relatively steep flood plains of headwater reaches (Arscott et al., 2001).

4. THE VERTICAL DIMENSION OF THERMAL HETEROGENEITY

Temperatures in the hyporheic zone are influenced by the infiltration of suface water and ground water, heat transfer across the sediment/water and sediment/air interfaces and heat transfer in the sediment. However, correspondence between vertical patterns of temperature and SiO_2 indicates a dominating influence of surface water-groundwater exchange particularly in the Roseg flood plain (Malard et al., 2001). Malard et al. (2001) studying vertical temperature patterns at site 73 identified six periods based on the predominant direction of water exchanges between the stream channel and hyporheic zone (Fig. 9). In summer (Period A), the surface stream is flowing and gaining water from the hyporheic zone. Early autumn (Period B) is a period of decreasing discharge and declining groundwater table during which the surface stream is losing water into the bed sediment. By the middle of October (Period C), the streambed falls dry and becomes completely exposed to atmospheric conditions until snow blankets the flood plain (Period D). Snow cover disappears from the flood plain by the beginning of May, but the channel at site 73 does not regain surface water for a short time (Period E). Although the end of spring (Period F) is a period of rising groundwater table, surface water downwells into the sediment until the beginning of June when the streams becomes fed by ground water (upwelling flow pattern as in Period A). During periods of upwelling (Period A) there is no correspondence between temperature series of surface and hyporheic water. Daily patterns of surface and hyporheic temperatures are strongly correlated when stream water downwells into the bed sediment (Periods B and F). However, only in autumn (Period B) did downwelling surface water result in distinct diel fluctuations of hyporheic temperatures at a depth of 80 cm. Temperature of hyporheic water at a depth of 30 cm shows diel variation in the absence of surface flow and snow cover (Periods C and E).

4.1. A warm hyporheic corridor in the proglacial reach

Hyporheic temperatures (80 cm depth) at the main channel site 6 of the proglacial reach were characterized by a steep vernal increase (up to 7.3 °C) and a subsequent

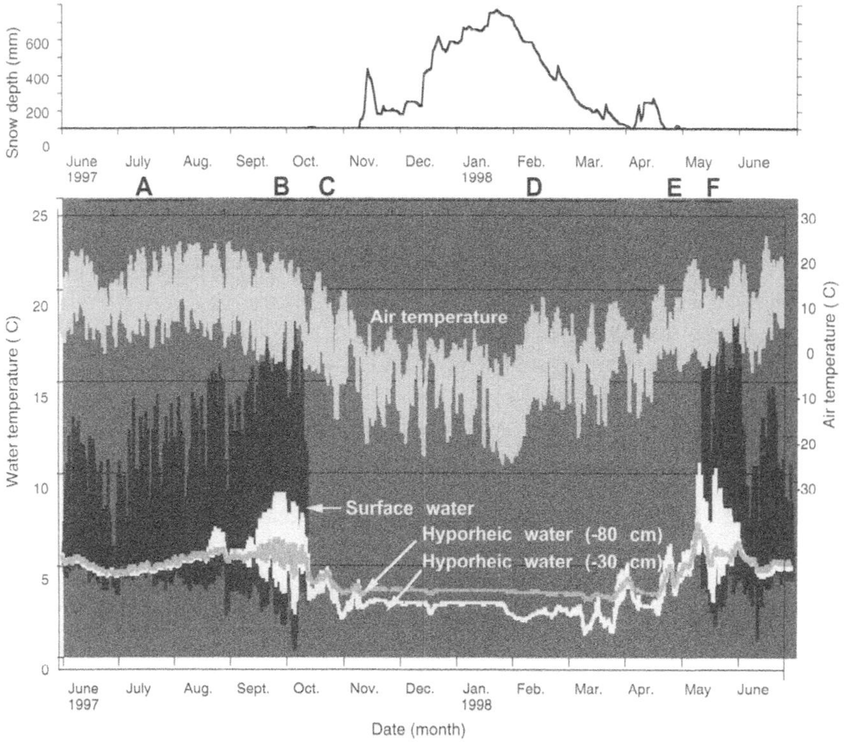

Figure 9. Snow depth, air temperature, and surface and hyporheic water temperatures recorded at site 73 (surface-disconnected channel). Snow depth and air temperature were measured at the nearest station of The Swiss Meteorological Survey (Samedan 1705 m a.s.l.). Temperature was corrected for difference in altitude (Uehlinger et al., 2003) The six time periods (A-F) are discussed in the text. From Malard et al. (2001), Fig. 6, p. 1330.

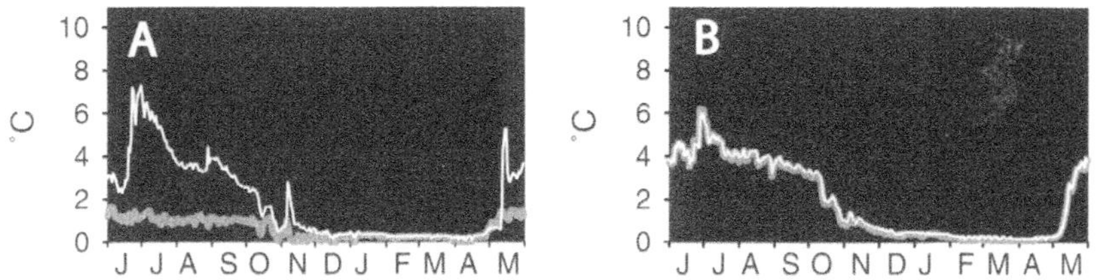

Figure 10. Daily temperature records for the surface stream water and the hyporheic zone (depth 80 cm) in the proglacial reach. A) Main channel: gray line = surface temperature at site 9, white line = hyporheic temperature at site 6. B) Site 8 (surface-disconnected channel): gray line = surface temperature, white line = hyporheic temperature.

decline starting in June and continuing until October (Fig. 10A). This remarkable pattern was attributed to the infiltration of snowmelt water displacing warm ground water stored in the alluvial sediments during winter (see Malard et al., 2001, for details). The similarity of the hyporheic temperature regimes of sites 6 and 8 (Fig. 10, white lines) indicates a general influence of deep ground water on shallow subsurface temperature in the proglacial reach. At site 6 the annual amplitude (difference between maximum and minimum daily temperature at 80 cm) was 7.2 °C, far exceeding the annual amplitude recorded in the surface waters of the proglacial stream. About 630 m downstream of the glacier terminus the hyporheic water accumulated 703 annual degree days compared to 248 degree days in the nearby surface water of the main channel. This suggests the presence of a relatively warm hyporheic corridor that underlies the cold metakryal reach. Whether the occurrence of such a warm hyporheic corridor is a common phenomenon of glacial streams needs further investigation.

4.2. Lateral patterns of hyporheic temperatures

Figures 5 and 11 depict temperatures in the hyporheic zone at 6 floodplain sites. At a depth of 80 cm instantaneous hyporheic temperatures ranged from 0.0°C to 9.4°C

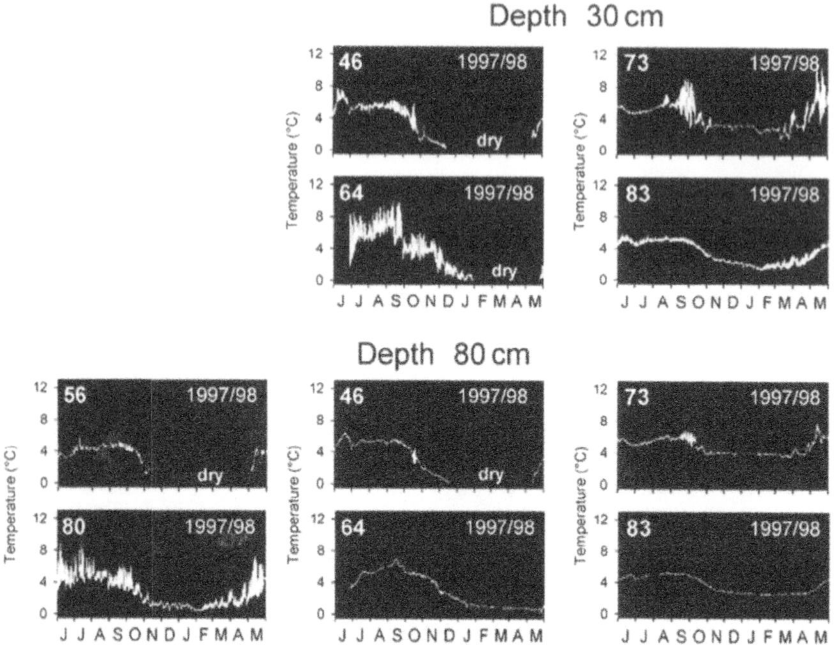

Figure 11. Temperature records (hourly values) for the hyporheic water at 30 and 80 cm depths in the flood plain. Sites 56, 80 = main channel, sites 46, 64, = surface-connected channels, sites 73, 83 = surface-disconnected channels.

and mean daily temperatures from 0.0 and 7.4 °C. Diel variations are small compared to surface water (Figs. 7, 10). At site 80, a main channel site is located in the center of the lower part of the flood plain, diel thermal fluctuations are relatively high and reflect the percolation of incoming water through sun-exposed bare gravel of the river bank (Malard et al., 2001). The influence of surface conditions decreases with increasing depth within the sediment: diel variations at 30 cm are distinctly higher than at 80 cm depth. Annual degree days in the hyporheic zone varied between 1084 and 1749 at various floodplain sites (Fig. 7). Differences between thermal patterns of surface and subsurface waters and between different hyporheic sites within the flood plain can be attributed to the inflow from three groundwater reservoirs (see chapter 4, Malard et al., 1999, 2001; Ward et al., 1999): (1) shallow alluvial ground water mainly derived from infiltrating stream water, (2) hill slope ground water recharged by snowmelt water in spring and ice-melt water from small hanging glaciers, and (3) deep ground water.

Temperature records for the Val Roseg demonstrated that thermal heterogeneity along the vertical dimension of a glacial flood plain could be as high as along lateral and longitudinal dimensions. Maximum differences between instantaneous temperatures of different surficial water bodies in the flood plain reached 15.1 °C in summer and 5 °C in winter. Instantaneous measurements of surface and hyporheic temperatures at a given site could differ by as much as 12.1°C in summer and 2.7°C in winter over a distance of only 0.8 m.

5. SPATIAL AND TEMPORAL STABILITY OF THERMAL PATTERNS

High flow can alter surface connectivity and as a consequence thermal regimes. For example, temperatures dropped distinctly in previously surface-disconnected channels after spates established an upstream connection to the main channel (Fig. 12). The spatial configuration of different habitats reflects the continuous turnover of floodplain channels under the discharge and sediment regime of the past 20 – 30 years (Zah et al., 2001). Channel turnover is not uniform throughout the flood plain; groundwater channels at the floodplain margin may persist for years, whereas the lifetime of channels within the braided corridor can be less than one year. Nevertheless, the overall spatial pattern is assumed to be rather stable unless flow or sediment supply is substantially altered. The rapidly retreating Tschierva Glacier exposes large amounts of sediments, which are susceptible to fluvial transport. This may enhance the channel turnover in the flood plain (Gurnell et al., 1999; Zah et al., 2001) and as a consequence affect thermal heterogeneity. A reduction of the life span of thermally benign habitats such as groundwater channels at the floodplain margins would be expected to influence the diversity of benthic communities.

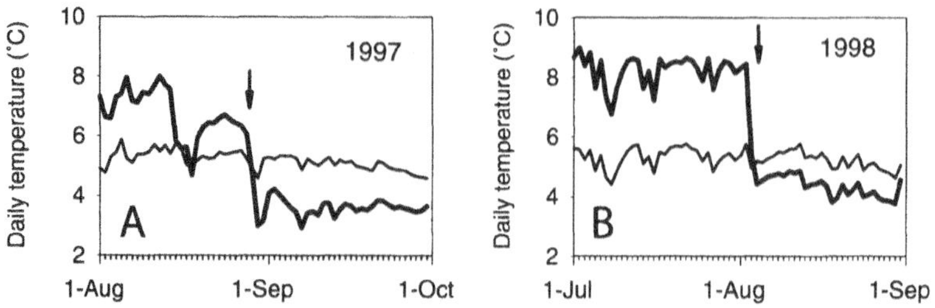

Figure 12. Change in mean daily temperature (bold line) in originally surface-disconnected channels after spate induced surface connection to the main channel. A) site 64. B) site 83. Fine line = temperature of a reference site (66) not affected by spate. Arrows indicate the occurrence of spates. Modified from Uehlinger et al. (2003), Fig. 11, p. 294.

Temperature records from stations in the Alps demonstrate an increase of minimum daily air temperature of 2°C since 1901 (Beniston et al., 1994). The response of stream temperatures to this warming is presumably small but local settings such as a glacier may magnify the effect of warming. Since the end of the Little Ice Age the Roseg Glacier receded by about 2.2 km and in 1998 the Tschierva Glacier ended 1.5 km from the 1850 terminus of the Roseg Glacier. The proglacial Roseg Lake raises annual temperatures in the outlet stream by 1.2°C. Assuming that stream temperatures at the snout of glaciers are close to 0°C, water temperature in the main channel at the upper end of the flood plain must have been 1.3°C lower in 1850 than today; this corresponds to a gain of 475 annual degree days. One may infer that during the past 150 years the main channel and surface connected floodplain channels became more favorable habitats, at least with respect to temperature.

6. SUMMARY AND CONCLUSIONS

The river continuum concept (Vannote et al., 1980) and the universal river zonation scheme (Illies & Botosaneau, 1963) both considered temperature to play a major role in structuring biotic communities along the course of rivers. However, these concepts did not consider lateral and vertical aspects of thermal heterogeneity. Comprehensive studies across flood plains or along the vertical dimension were lacking until recently (Arscott et al., 2001; Malard et al., 2001). Therefore, the high thermal heterogeneity along the three dimensions of the Roseg River corridor was not anticipated.

Without the proglacial lake and especially without the flood plain, the headwater reach of the Roseg River would be a thermally harsh environment. The glacial flood plain provides thermally benign habitats, namely surface-disconnected channels and channels with minor surface connection to the main channel, in close proximity to the glacier. Moreover, thermal heterogeneity within the flood plain is higher than along the main channel between the glacier and the end of the catchment.

The warm hyporheic corridor in the proglacial reach is a remarkable thermal feature of the glacial river corridor. Relatively high annual temperature maxima also characterize the hyporheic zone in the flood plain; lateral differences of hyporheic temperatures are relatively small compared to vertical temperature gradients. Neglecting the vertical dimension may result in an inappropriate picture of the thermal environment in glacial streams, and as a consequence, lead to untenable concepts of benthic communities in glacial streams (Malard et al., 2001).

Temperature effects the ecology of aquatic ectotherms through influences on growth and development, life history, behavior, and competitive hierarchies, all of which determine community composition, species diversity, and abundance levels (Ward & Stanford, 1982). In the Val Roseg, thermal heterogeneity within the river corridor may explain the unexpectedly abundant and diverse invertebrate communities in the flood plain, even close to the glaciers (Burgherr & Ward, 2000; Burgherr & Ward, 2001; Burgherr et al., 2001; Malard et al., 2001; Tockner & Klein, 2000).

7. REFERENCES

Arscott, D. B., Tockner, K., & Ward, J. V. (2001). Thermal heterogeneity along a braided floodplain river in the Alps. *Canadian Journal of Fisheries and Aquatic Sciences, 58*, 2358-2373.

Beniston, M., Rebetez, M., Giorgi, f., & Marinucci, M. R. (1994). An analysis of regional climate change in Switzerland. *Theoretical and Applied Climatology, 49*, 135-159.

Burgherr, P., & Ward, J. V. (2000). Zoobenthos of kryal and lake outlet biotopes in a glacial floodplain. *Verhandlungen der Internationalen Vereinigung für Theoretische und Angewandte Limnologie, 27*, 1587-1590.

Burgherr, P., & Ward, J. V. (2001). Longitudinal and seasonal distribution patterns of the benthic fauna of an alpine glacial stream (Val Roseg, Swiss Alps). *Freshwater Biology, 46*, 1705-1721.

Burgherr, P., Ward, J. V., & Glatthaar, R. (2001). Diversity, distribution and seasonality of the Simuliidae fauna in a glacial stream system in the Swiss Alps. *Archiv für Hydrobiologie, 152*, 19-37.

Castella, E., Adalsteinsson, H., Brittain, J. E., Gislason, G. M., Lehmann, A., Lencioni, V., Lods-Crozet, B., Maiolini, B., Milner, A. M., Olafsson, J. S., Saltveit, S. J., & Snook, D. L. (2001). Macrobenthic invertebrate richness and composition along a latitudinal gradient of European glacier-fed streams. *Freshwater Biology, 46*, 1811-1831.

Constanz, J. (1998). Interaction between stream temperature, streamflow, and groundwater exchanges in alpine streams. *Water Resources Research, 34*, 1609-1615.

Constanz, J., & Murphy, F. (1991). The temperature dependence of ponded infiltration under isothermal conditions. *Journal of Hydrology, 122*, 119-128.

Gu, R., Montgomery, S., & Austin, T. A. (1998). Quantifying the effects of stream discharge on summer river temperature. *Hydrological Sciences, 43*, 885-904.

Gurnell, A. M., Edwards, P. E., Petts, G. E., & Ward, J. V. (1999). A conceptual model for alpine proglacial river channel evolution under changing climatic conditions. *Catena, 38*, 223-242.

Hawkins, C. P., Hogue, J. N., Decker, L. M., & Feminella, J. W. (1997). Channel morphology, water temperature, and assemblage structure of stream insects. *Journal of the North American Benthological Society, 16*, 728-749.

Illies, J., & Botosaneau, L. (1963). Problèmes et méthodes de la classification et de la zonation écologiques des eaux courantes considerées surtout du point de vue faunistique. *Mitteilungen der Internationalen Vereinigung für Theoretische und Angewandte Limnologie, 12*, 1-57.

Johnson, F. A. (1971). Stream temperatures in an alpine area. *Journal of Hydrology, 14*, 322-336.

Kamler, E. (1965). Thermal conditions in mountain waters and their influence on the distribution of Plecoptera and Ephemoptera larvae. *Ekologia Polska Seria A, 13*, 377-413.

Malard, F., Lafont, M., Burgherr, P., & Ward, J. V. (2001). Longitudinal patterns of hyporheic versus benthic oligochaetes in a glacial river. *Arctic, Antarctic, and Alpine Research, 457-466*, 33.

Malard, F., Mangin, A., Uehlinger, U., & Ward, J. V. (2001). Thermal heterogeneity in the hyporheic zone of a glacial floodplain. *Canadian Journal of Fisheries and Aquatic Sciences, 58*, 1319-1335.

Malard, F., Tockner, K., & Ward, J. V. (1999). Shifting dominance of subcatchment water sources and flow paths in a glacial floodplain, Val Roseg, Switzerland. *Arctic, Antarctic, and Alpine Research, 31*, 135-150.

Malard, F., Tockner, K., & Ward, J. V. (2000). Physico-chemical heterogeneity in a glacial riverscape. *Landscape Ecology, 15*, 679-695.

Milner, A. M., Brittain, J. E., Castella, E., & Petts, G. E. (2001). Trends of macroinvertebrate community structure in glacier-fed rivers in relation to environmental conditions: a synthesis. *Freshwater Biology, 46*, 1833-1847.

Milner, A. M., & Petts, G. E. (1994). Glacial rivers: physical habitat and ecology. *Freshwater Biology, 32*, 295-307.

Mosley, M. P. (1983). Variability of water temperatures in the braided Ashley and Rakaia rivers. *New Zealand Journal of Marine and Freshwater Research, 17*, 331-342.

Richards, K. (1982). *Rivers. Form and process in alluvial channels.* London: Methuen.

Schütz, C., Wallinger, M., Burger, R., & Füreder, L. (2001). Effects of snow cover on the benthic fauna in a glacier-fed stream. *Freshwater Biology, 46*, 1691-1704.

Smith, K., & Lavis, M. E. (1975). Environmental influences on the temperature of a small upland stream. *Oikos, 26*, 228-236.

Tockner, K., & Klein, B. (2000). Biodiversity in alluvial springs of a glacial flood plain (Val Roseg, Switzerland). *Verhandlungen der Internationalen Vereinigung für Theoretische und Angewandte Limnologie, 27*, 704-710.

Tockner, K., Malard, F., Uehlinger, U., & Ward, J. V. (2002). Nutrients and organic matter in a glacial river floodplain system (Val Roseg, Switzerland). *Limnology and Oceanography, 47*, 521-535.

Uehlinger, U. (2000). Resistance and resilience of ecosystem metabolism in a flood-prone river system. *Freshwater Biology, 45*, 319-332.

Uehlinger, U., Malard, F., & Ward, J. V. (2003). Thermal patterns in the surface waters of a glacial river corridor (Val Roseg, Switzerland). *Freshwater Biology, 48*, 284 -300.

Vannote, R. L., Minshall, G. W., Cummins, K. W., Sedell, J. R., & Cushing, C. E. (1980). The river continuum concept. *Canadian Journal of Fisheries and Aquatic Sciences, 37*, 130-137.

Vannote, R. L., & Sweeney, B. W. (1980). Geographic analysis of thermal equilibria: A conceptual model for evaluating the effect of natural and modified thermal regimes on aquatic insect communities. *The American Naturalist, 115*, 667-695.

Ward, J. V. (1985). Thermal characteristics of running waters. *Hydrobiologia, 125*, 31-46.

Ward, J. V. (1992). *Aquatic insect ecology. Biology and Habitat.* New York: Wiley.

Ward, J. V. (1994). Ecology of alpine streams. *Freshwater Biology, 32*, 277-294.

Ward, J. V., Malard, F., Tockner, K., & Uehlinger, U. (1999). Influence of ground water on surface water conditions in a glacial floodplain of the Swiss Alps. *Hydrological Processes, 13*, 277-293.

Ward, J. V., & Stanford, J. A. (1982). Thermal responses in the evolutionary ecology of aquatic insects. *Annual Review of Entomology, 27*, 97-117.

Ward, J. V., Tockner, K., Arscott, D. B., & Claret, C. (2002). Riverine landscape diversity. *Freshwater Biology, 47*, 517-539.

Webb, B. W., & Walling, D. E. (1986). Spatial variation of water temperature characteristics and behaviour in a Devon River system. *Freshwater Biology, 16*, 586-608.

Webb, B. W., & Zhang, Y. (1997). Spatial and seasonal variability in the components of the river heat budget. *Hydrological Processes, 11*, 79-101.

Zah, R., Niedröst, M., Rinderspacher, H., Uehlinger, U., & Ward, J. V. (2001). Long-term dynamics of the channel network in a glacial floodplain, Val Roseg, Switzerland. *Arctic, Antarctic, and Alpine Research, 33*, 440-446.

CHAPTER 6

K. TOCKNER, R. ILLI, F. MALARD & U. UEHLINGER

NUTRIENT DYNAMICS

1. INTRODUCTION

Nutrients are pivotal in the structure and function of river ecosystems. As essential resources they control primary production and decomposition processes, and influence metabolic pathways in the riparian, parafluvial and hyporheic zones (e.g. Duff & Triska, 2000; Wetzel, 2001). Surprisingly few year-round studies, however, have quantified concentrations and fluxes of all major nutrient fractions; this is particularly true for remote alpine and arctic areas (Tockner et al., 2002, and references therein). The fractionation of nutrients into dissolved and particulate forms, and into inorganic and organic components, is essential for a comprehensive understanding of nutrient dynamics.

Alpine and arctic streams appear resource limited in terms of both nutrients and organic matter. For example, Lock et al. (1990) showed that the addition of phosphorous to an arctic river resulted in a substantial stimulation of both heterotrophic and autotrophic processes. Peterson et al. (1993) documented bottom-up control of tundra rivers subjected to long-term phosphorous addition. Robinson and Gessner (2000) demonstrated that nutrient addition accelerated leaf breakdown in a glacial stream. Further, there is strong evidence that zoobenthic communities in glacial streams are not solely structured by temperature and channel stability (cf. Milner & Petts, 1994), but are strongly influenced by seasonal shifts in water sources and corresponding availability of nutrient and organic matter resources (Ward, 1994; Füreder, 1999).

This chapter synthesizes the results of an intensive year-round investigation of nutrient dynamics in the Val Roseg river-floodplain ecosystem. Following a chemical characterization of the different water sources (snowmelt, glacial ablation, ground water), spatial gradients of nutrients are analyzed along the river's course, as a function of connectivity in the floodplain reach, and between surface water and hyporheic water. The strong seasonal patterns are first described and then cast in the context of temporal coherence to identify spatiotemporal differences in the behavior of individual nutrient fractions. Nutrient fluxes are examined at catchment and floodplain scales to identify source-sink phenomena. The N:P ratio was calculated to detect potential nutrient limitations and the C:N ratio was used to assess organic matter quality.

J.V. Ward & U. Uehlinger (eds.), Ecology of a Glacial Flood Plain, 91-107.
© 2003 *Kluwer Academic Publishers. Printed in the Netherlands.*

2. CHARACTERIZATION OF NUTRIENT SOURCES

The Val Roseg river-floodplain system is fed by three main water sources: ground water, snowmelt, and glacial melt water (see Chapter 3 for a geochemical characterization of water sources). Each of these water sources is characterized by a distinct nutrient composition (Table 1). Snowmelt water is rich in dissolved organic nitrogen (DON) and particulate and dissolved organic carbon (POC, DOC), glacial meltwater in particulate phosphorous (PP) and nitrate. Concentrations of silica (SiO_2), sodium (Na), and DOC were distinctly higher in hillslope ground water than in glacial meltwater.

Table 1. Nutrient and organic matter composition (mean ± standard deviation) of different water sources in the Val Roseg flood plain using examples from representative sites (Snow: 17 April 1997; Glacial melt water: site 3, July to September 1997; Hillslope ground water: site H2, July to September 1997). Nd: not detectable. n = 3 per water source (from Tockner et al., 2002).

Variable	*Snow*	*Glacial melt water*	*Hillslope ground water*
NH_4-N(μg L^{-1})	55 ± 10	15 ± 3	2 ± 2
NO_3-N(μg L^{-1})	148 ± 17	227 ± 24	180 ± 25
DON (μg L^{-1})	230 ± 64	32 ± 23	81 ± 54
PN (μg L^{-1})	71 ± 16	20 ± 8	22 ± 12
SRP (μg L^{-1})	10 ± 5	2 ± 0	0.3 ± 0.5
NRP (μg L^{-1})	8 ± 2	5 ± 1	3 ± 2
PP (μg L^{-1})	12 ± 5	97 ± 40	4 ± 2
DOC (mg L^{-1})	1.1 ± 0.1	0.3 ± 0.2	0.9 ± 0.6
POC (mg L^{-1})	1.2 ± 0.1	0.3 ± 0.2	0.2 ± 0
SiO_2 (mg L-1)	0.1 ± 0.1	1.2 ± 0.2	5.1 ± 0.3
Na (mg L^{-1})	Nd	0.2 ± 0.1	1.6 ± 0.1

Various processes contributed to the release of nutrients from the individual water sources. Melting of the snowpack results in a pulse of nitrate in the river, a common phenomenon for arctic and alpine rivers (e.g. Williams & Melack, 1989; Sickman et al., 2001). High concentrations of nitrogen in snow samples likely originate from dry and wet deposition, especially during the end of the snow accumulation period when the lower tropospheric convection extends upward to glacial altitude (Nickus et al., 1997). According to maps of atmospheric deposition for Switzerland, the deposition of nitrogen in the Val Roseg catchment is 10-15 kg N ha^{-1} yr^{-1} (Rihm, 1996). DON and NH_4-N represent a high proportion of TDN (48% DON) in the snowpack; NO_3-N, however, was the dominant fraction (87%) in stream water (Tockner et al., 2002). The source of high NO_3-N concentrations during spring runoff is attributable to (1) oxidation of NH_4-N released from the snowpack and (2) mineralisation processes of DON in soils and subsequent transport by meltwater percolating through the soil (see Williams & Melack, 1989). Recent isotopic analyses suggest that the majority of the nitrate pulse during snowmelt is not from the snowpack, but rather from catchment soils (Sickman et al., 2001).

Labile N pools in soils are increasingly built up during years of high snowfall. Therefore, the duration of snow cover and the timing of snowmelt runoff are expected to play a predominant role on the variability of N cycling and N export.

The **glacier** was the primary source for PP as shown by peaks during the period of maximum ablation in mid-summer. Concentrations of PP closely tracked suspended solid (SS) dynamics, reflecting the resuspension of glacial flour enriched in apatite during high-flow periods. During the main ablation period in summer with its distinct diel flow fluctuations, PP exhibited strong diel changes, with highest concentrations generally before peak flow ("hysteresis" effect, Fig. 1). In summer, an up to ten-fold increase in PP was observed during a diel freeze-thaw cycle. PP could also be an important source for SRP, as demonstrated by peaks in SRP during the main ablation period (see Bretschko, 1966). Nevertheless, SRP concentrations were very low in the Val Roseg flood plain, with concentrations often close to the detection limit (Tockner et al., 1997, Table 2).

Hillslope **ground water** was a source of silicate, nitrate and DOC. Silicate was always highest in ground water: however, nitrate was lower during the snowmelt period, when nitrate entered the flood plain primarily via surface pathways. In contrast to glacial and snow melt water, DOC was the dominant carbon fraction in hillslope ground water (Table 1).

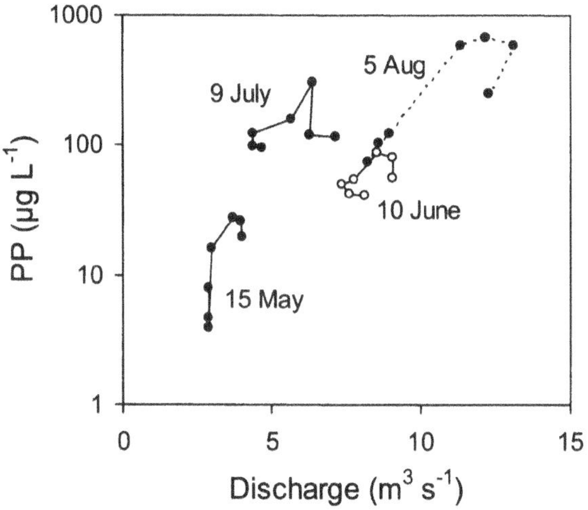

Figure 1. Diel variation of particulate phosphorous (PP; log-scale) on 4 dates. Samples were collected every 3 hours between 0600 and 2400h at the downstream end of the flood plain (7 samples per date).

Table 2: Nutrient concentrations (µg L^{-1} for NO_3, NH_4, DON, SRP and PP; mg L^{-1} for all other fractions) in glacial and non-glacial alpine, arctic and Antarctic streams, compared to the word average of rivers. Sampling intensity differs between individual investigations. Average values or ranges are presented. Nd = Not determined.

River/catchment	*Sampling intensity*	*NO_3-N*	*NH_4-N*	*DON*	*SRP*	*PP*	*DOC*	*POC*	*SiO_2*	*Reference*
Val Roseg(main channel, range)	Year-round	190-353	Jan-14	1-219	1–3	2-102	0.2-0.4	0.13-0.26	1.6-3.7	Tockner et al., 2002
Various glacial streams in Switzerland	Year-round	139-257	13-27	Nd	0.8-2.4	31-112	0.3-0.5	0.2-1.8	Nd	Robinson et al., 2001
Rotmoosache, Austria (glacial stream close to snout)	Year-round	180	37	Nd	3.3	326	0.46	Nd	Nd	Füreder et al., 2001
Königsbach, Austria (spring-fed alpine stream)	Year-round	191	2.6	Nd	1.1	4	1.5	Nd	Nd	Füreder et al., 2001
Alpine lake outlets, Switzerland (rhithral streams)	Year-round	170	7.6	62	1.3	3.3	0.5	0.2	Nd	Hieber et al., 2002
Alpine lake outlets, Switzerland (kryal streams)	Year-round	166	26	41	1.4	28	0.4	0.3	Nd	Hieber et al., 2002
Glacial streams in Western Greenland	July	41	39	Nd	16	Nd	Nd	Nd	Nd	Friberg et al., 2001
Groundwater-fed streams in Greenland	July	4	23	Nd	22	Nd	Nd	Nd	Nd	Friberg et al., 2001
Prince of Wales Island , Alaska	July-August	100-140	14-41	Nd	6-9	Nd	Nd	Nd	0.74-1.54	Wissmar et al., 1997
Rock Creek, Denali NP, Alaska	June-October	100-280	<10-40	Nd	Nd	Nd	0.3-1.2	Nd	Nd	Stottlemyer, 1992
Schirmacher Oasis, Antarctica	Antarctic summer	9.5-15.5	2.4-8.5	Nd	0.6-1.6	Nd	Nd	Nd	Nd	Pandey et al., 1995
World average		100	15	260	12.5	530	5.75	4.8	Nd	Meybeck, 1982

3. SPATIAL GRADIENTS

The average DOC, SRP and nitrogen concentration for world rivers in the pristine state has been given as 0.5 mg L^{-1} DOC, 2 µg L^{-1} SRP, 15 µg L^{-1} NH_4-N, 50 µg L^{-1} DON and 100 µg L^{-1} NO_3-N (Meybeck, 1982; Meybeck et al., 1989). In the Val Roseg River, concentrations of DON, NH_4-N and SRP were within the range reported for relatively unpolluted rivers (Meybeck et al. 1989); DOC concentrations were among the lowest reported (see Tockner et al., 2002). However, nitrate and particulate phosphorous concentrations were much higher than in most unpolluted alpine, arctic and Antarctic rivers; nitrate is even high compared to the world average. (Table 2). Compared to selected alpine, arctic and Antarctic rivers, the Val Roseg exhibits low NH_4-N and SRP but relatively high nitrate concentrations. For most other variables, comparable information among alpine and arctic rivers is scarce (Table 2).

3.1. Longitudinal patterns

Patterns of nutrients and organic matter were disrupted at two locations along the course of the Roseg River: at the confluence of the proglacial reach with the glacial lake outlet ("L" in Fig. 2) and at a knickpoint at the downstream end of the flood plain ("F" in Fig. 2) characterized by upwelling of alluvial ground water. The flood plain itself had a minor influence on the longitudinal patterns, except for ISS and PP (Fig. 2). The lake outlet led to either an increase or a decrease of nutrient concentrations, depending on the individual variable and the season. For example, NO_3-N concentrations decreased downstream of the lake outlet, whereas DOC concentrations increased, at least in July and October. In the constrained section downstream of the flood plain, particulate parameters and NO_3-N remained relatively constant, whereas SiO_2 and DOC increased in March and October (Fig. 2). However, as the water traversed the 8 km long constrained section only small changes in nutrient concentrations were observed.

3.2. Connectivity gradients at the floodplain scale

Based on connectivity, we distinguished three different channel types within the flood plain: surface connected, subsurface connected and disconnected channels (Tockner et al., 2002). Surface connected channels included the main channel and channels that were permanently connected to the main channel at their upstream end. Subsurface connected channels were fed by alluvial ground water. Disconnected channels were fed by hillslope ground water (both channel types lack an upstream surface connection with the main channel). SiO_2 and NO_3-N exhibited trends of increasing concentrations from connected (surface water-fed) toward disconnected channels (hillslope groundwater-fed), whereas concentrations of PP decreased along this connectivity gradient (Table 3). SiO_2 concentrations were up to 3.8 times higher in disconnected channels compared to surface-connected channels. Concentrations of TOC, TP and TN were significantly different among channel types. There were also significant interactions between channel type and hydrological period, indicating that spatial patterns varied as a function of season. Posthoc comparisons showed that TN and TOC were significantly higher in disconnected channels, whereas TP was highest in connected channels. Results demonstrated that on average subsurface and disconnected channels rather than the main channel were a major source of nutrients, particularly for dissolved organic carbon and nitrate (see Tockner et al., 2002).

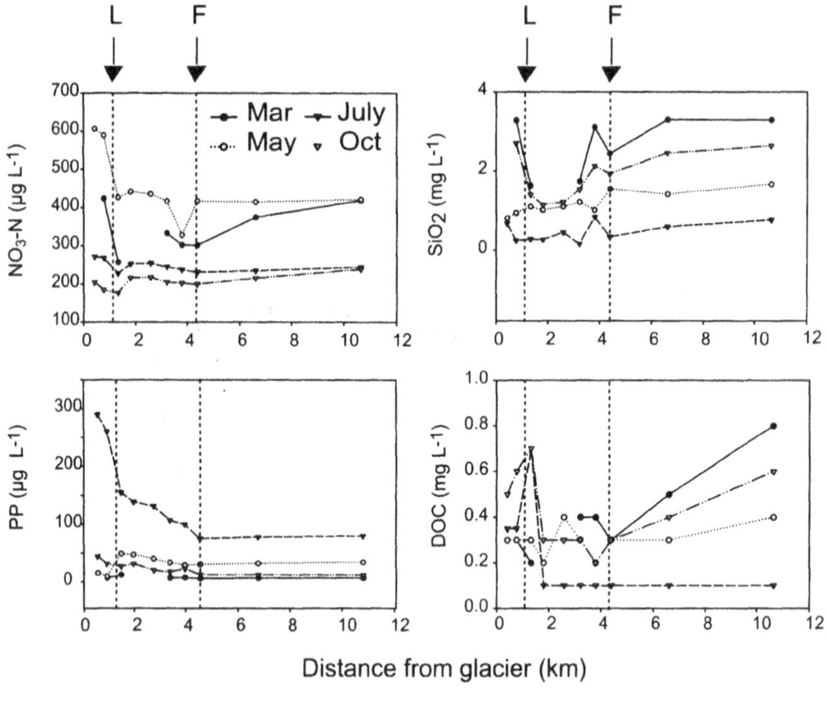

Figure 2. Longitudinal patterns of selected nutrients in main channel sites from the Tschierva Glacier to the end of the catchment in March, May, July and October 1997, representing four major hydrological periods (see chapter 4). L = Lake inflow; F = Knickpoint at the lower end of the flood plain. In March, several sites in the upper part of the flood plain lacked surface water. Modified from Tockner et al. (2002), Fig. 4, p. 272.

3.3. *Differences between surface and hyporheic water*

We compared surface water with interstitial water at a sediment depth of 80 cm in selected channel types (Fig. 3). In a channel fed by hillslope ground water (site 83), hyporheic water was much richer in nitrate compared to surface water (average seasonal difference: 90 µg L^{-1}). Silicate exhibited a similar trend. That suggests that hyporheic flowpaths transferred nutrient-rich ground water to nutrient-poor surface channel habitats. In the main channel (site 80), concentrations of both nitrate and silicate were very similar in surface water and in interstitial water, indicating substantial vertical mixing.

Table 3. Average concentrations of selected nutrients in surface-connected, subsurface-connected (seepage-fed) and disconnected (hillslope-groundwater fed) floodplain channels during the four hydrological seasons (n = 9 per channel type and season).

Variable	*Season*	*Surface-Connected*	*Subsurface-Connected*	*Disconnected*
SiO_2 ($mg\ L^{-1}$)	Low water	3.7	2.6	5.7
	Expansion	2.6	2.7	5.9
	Glacial ablation	1.6	2.7	6.1
	Contraction	2.2	2.2	6.3
DOC ($mg\ L^{-1}$)	Low water	0.4	0.4	0.5
	Expansion	0.4	0.3	0.5
	Glacial ablation	0.2	0.2	0.6
	Contraction	0.4	0.5	0.5
POC ($mg\ L^{-1}$)	Low water	0.17	0.17	0.11
	Expansion	0.13	0.11	0.10
	Glacial ablation	0.26	0.09	0.12
	Contraction	0.17	0.16	0.15
NO_3-N ($\mu g\ L^{-1}$)	Low water	265	301	263
	Expansion	353	393	432
	Glacial ablation	205	173	244
	Contraction	190	191	228
PP ($\mu g\ L^{-1}$)	Low water	1.8	1.1	1.6
	Expansion	7.2	3.4	2.8
	Glacial ablation	54.0	7.4	3.8
	Contraction	11.5	4.1	5.3

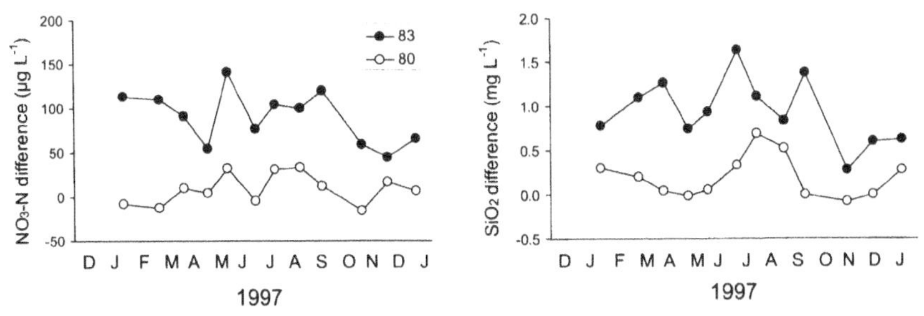

Figure 3. Seasonal differences in nitrate and silicate concentrations between interstitial water (80 cm sediment depth) and surface water at a main channel site (80) and a groundwater channel (83). Positive values indicate higher concentrations in 80 cm sediment depth.

4. TEMPORAL DYNAMICS

4.1. Seasonal patterns

The seasonal shift of water from the different sources (ground water, snowmelt water, and glacial meltwater) created a distinct sequence of nutrient peaks (Fig. 4A-D; Tockner et al., 2002). SiO_2 and DOC peaked in winter; NO_3-N in early spring, followed by peaks in DON, PP, ISS (inorganic suspended solids) and NH_4-N. As a consequence of the shifting sequence of individual nutrient and organic matter fractions, resource availability changes during a hydrological cycle. Two ecological windows, one in early spring and one in autumn, were identified when resource availability coincides with benign environmental conditions (elevated temperature, low turbidity and no bedload transport). Both periods are of short duration (a few weeks), but are critical for the accrual of benthic algae and the development and diversity of invertebrates (Chapters 10 and 12).

Plotting nutrient concentrations as a function of discharge revealed marked differences between expansion and contraction phases (Fig. 4E-H). Concentrations of NO_3-N and DON were much higher during the expansion than during the contraction period. Concentrations of NH_4-N and PP exhibited a reverse pattern. Concentrations of SiO_2 and DOC decreased with increasing discharge, with only minor differences between expansion and contraction phases. Differences detected between the expansion and the contraction periods result from differences in the relative contribution of water sources to total discharge in spring (snowmelt) and autumn (primarily subglacial and alluvial ground water, see Chapter 3).

4.2. Temporal coherency

The calculation of synchrony[1], or temporal coherence, allowed us to identify spatiotemporal differences in the behavior of individual nutrient fractions. High synchrony indicates similar behavior of different locations through time, whereas low synchrony indicates individual site responses to variations in regional factors. In the Val Roseg, the average synchrony for all variables (list of variables: see Tockner et al., 2002) was 0.66 ± 0.15 for the longitudinal transect and 0.37 ± 0.16 for the lateral transect. These averages were taken across all pairs of sites along both transects. A value of 0.66 is much higher than values reported for similar variables from other sites (Kling et al., 2000, and references therein). Longitudinally, synchrony of variables ranged from 0.28 for PN to 0.89 for silicate. Six out of the fourteen variables exhibited a significant synchrony across most sites. Laterally, synchrony ranged from 0.10 for NH_4-N to 0.87 for DON. DON and NO_3-N were the only variables that were in average significantly correlated across all sampling sites. A high synchrony indicated catchment control for DON and NO_3-N that resulted from the melting of the snowpack. Hence, snowmelting was a primary catchment process that superceded local processes within individual channel types. Glacial

[1] Synchrony or temporal coherence is a measure of the similarity between a pair of sampling sites.

melting, however, controlled nutrient resources primarily in the main channel and in channels that were connected at their upstream end to the main channel (high synchrony along the main stem).

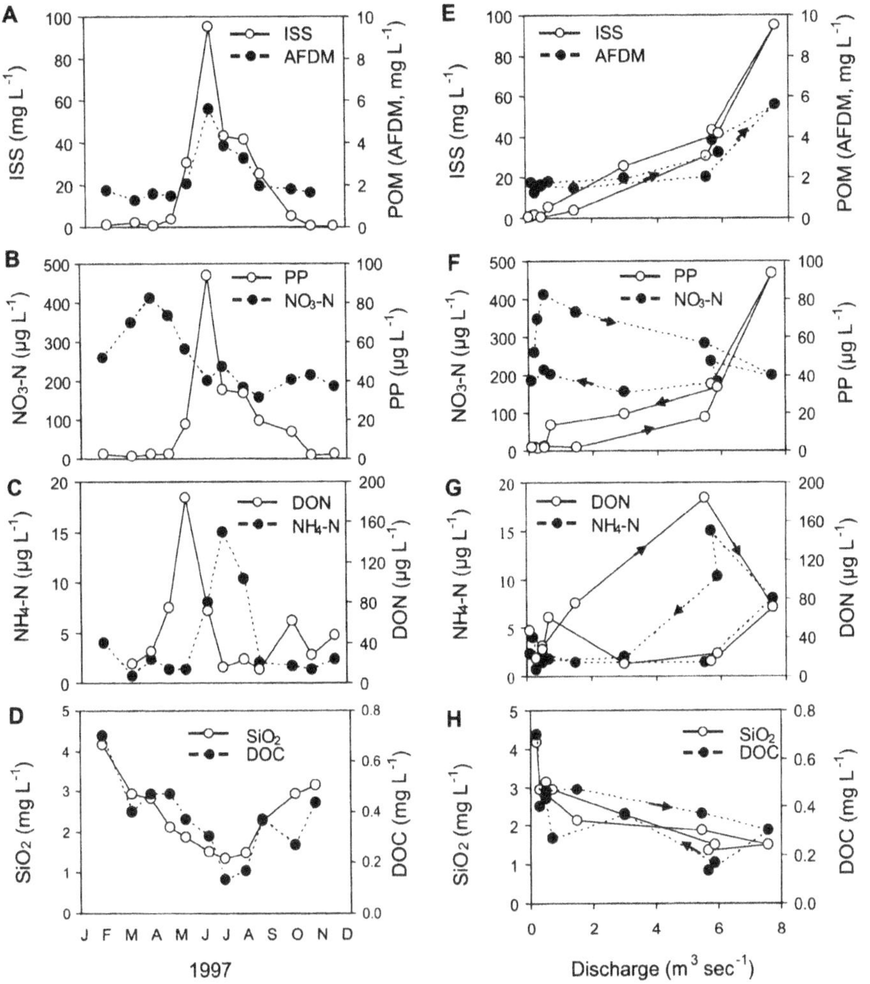

Figure 4. Seasonal dynamics of A) inorganic suspended solids (ISS) and particulate organic matter (POM, expressed as AFDM), B) Nitrate-N (NO_3-N) and particulate phosphorous (PP), C) ammonium (NH_4-N) and dissolved organic nitrogen (DON), D) silicate (SiO_2) and dissolved organic carbon (DOC). E-H: Relationship between discharge and concentrations of the above variables. Arrows pointing to the right indicate the expansion phase, whereas arrows pointing to the left indicate the contraction phase. All samples were taken between 0600 and 0900 near the lower end of the flood plain (site 90). Modified from Tockner et al. (2002), Fig. 3, p. 271.

Disconnected channels exhibited their own dynamics decoupled from processes controlled by glacial melting. Synchrony decreased with distance. Sites close to one another had higher synchrony values than sites more separated in space (longitudinal transect) or sites arranged along the lateral ground water-surface water gradient (Fig. 5). Within the flood plain, synchrony was much lower among groundwater-fed sites (0.35 ± 0.12) than among subsurface-connected (0.52 ± 0.02) or surface-connected waters (0.79 ± 0.05). Lower synchrony values for groundwater channels reflect the high spatial heterogeneity that characterizes this channel type (Chapter 4).

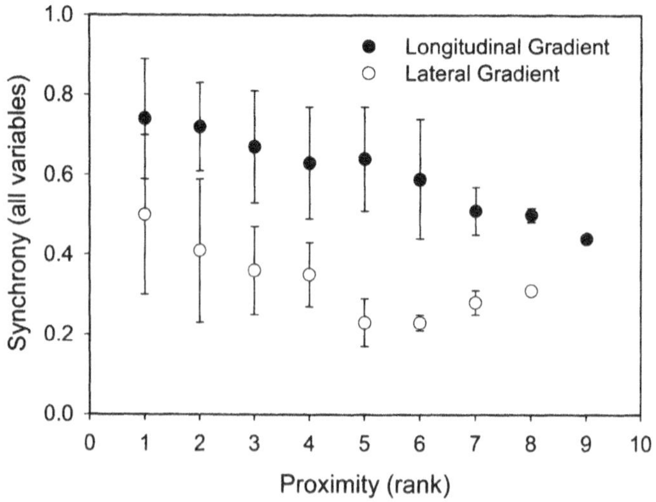

Figure 5. Average synchrony (r, average and SD) for 14 nutrient and organic matter variables after Tockner et al. (2002) combined versus distance (proximity) between pairs of sampling sites along the longitudinal and lateral gradients (see text).

5. NUTRIENT FLUX

The Val Roseg is a sink for some nutrients, a source for others and a transition area for yet others. At the catchment scale, the Val Roseg was a net sink for nitrate since inputs exceeded outputs by a factor of 3 (about 12 vs. 4 kg ha^{-1} yr^{-1}, Tockner et al., 2002). The catchment was a major source for phosphorous. Precipitation input rates of TP were not available for the Val Roseg but commonly range between 0.05-0.4 kg ha^{-1} yr^{-1} (Likens et al., 1977) which was much lower compared to output fluxes from the Val Roseg catchment (4.3 kg ha^{-1} yr^{-1}). For most catchments phosphorous seemed to be in balance, with exports matching inputs (Gibson et al., 1995, and references therein). The high output of phosphorous from the Val Roseg catchment is primary in the form of inorganic PP (98% of TP) derived from weathering of apatite bedrock (see Malard et al., 1999).

5.1. Floodplain retention

At the floodplain scale, the Val Roseg was primarily a transfer area for nutrients and organic matter originating from different catchment sources. The flood plain was never a sink or a source for dissolved nutrient fractions or for particulate nitrogen and carbon. The flood plain was, however, a sink for PP (and inorganic suspended sediments) as a result of sedimentation processes (Tockner et al., 2002). The limited role of the flood plain as a sink or source for nutrients and organic matter is explained by the lack of lentic water bodies, rapid transit times of surface waters (in summer about 40 min, U. Uehlinger, unpublished data) and reduced biological activities. In contrast to most temperate and tropical flood plains, such as the Amazon, Orinoco or the Danube (Forsberg et al., 1988; Lewis et al., 1990; Tockner et al., 1999), the Val Roseg flood plain lacks significant biological uptake of DIN and SRP. Differences in nutrient concentrations upstream and downstream of the flood plain were explained solely by mixing processes from different water sources.

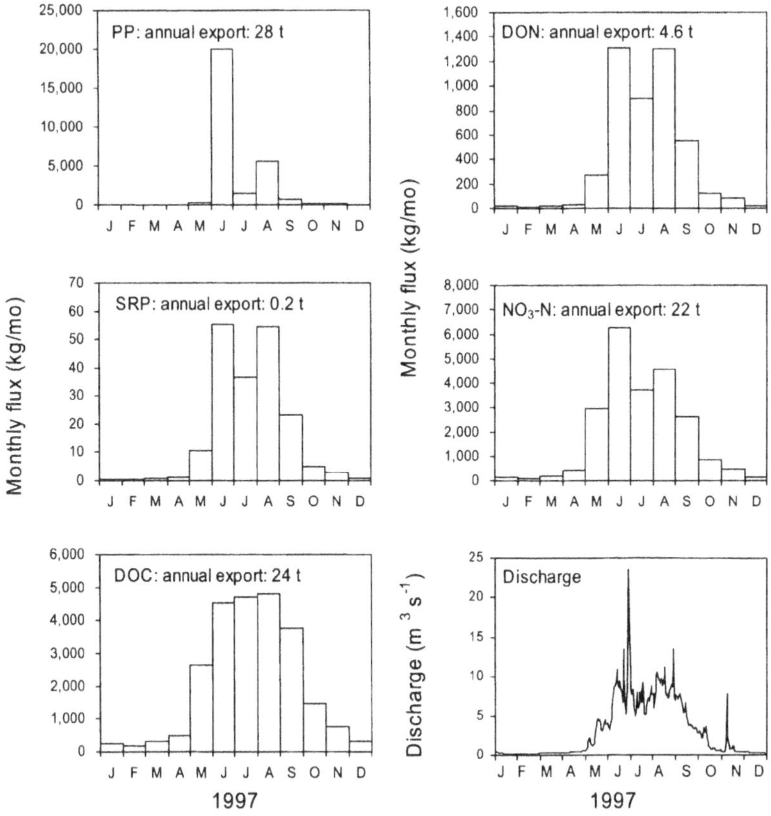

Figure 6. Discharge and monthly fluxes of selected nutrient and organic matter fractions for the Val Roseg flood plain.

5.2. Catchment export rates

Total and area-specific export rates for individual nutrient fractions were calculated from a set of 77 samples (covering diel and seasonal discharge fluctuations) collected near the downstream end of the Val Roseg plain (site 90) (Tockner et al., 2002; Table 4). On an annual basis, the majority of N was transported as NO_3-N (78%), P as PP (98%) and TOC as POC (53%). Dissolved organic nitrogen (DON) accounted for a much greater percentage of N flow than NH_4-N (16% versus 1.4%). Significant seasonal variation was apparent in both the magnitude and composition of nitrogen and phosphorous. For all nutrient fractions, monthly fluxes were highest during summer. For example, 58% of DOC, 65% of NO_3-N, about 75% of DON and SRP and >95% of PP were transported during the main ablation period between June and August (Fig. 6). Seasonally, there was a major shift in the relative proportion of nutrient fractions, from a dominance of dissolved fractions in winter to a dominance of particulate fractions in summer.

Table 4: Specific export rates (kg ha^{-1} yr^{-1}) for individual nutrient fractions in the Val Roseg river compared to selected rivers throughout the world. Nd = Not determined.

Catchment	*SRP*	*PP*	*TP*	*NO_3-N*	*NH_4-N*	*DON*	*Reference*
Val Roseg glacial catchment	0.03	4.2	4.3	3.3	0.07	0.7	Tockner et al., 2002
Six undisturbed headwaters in temperate USA	Nd	Nd	Nd	0.2	0.02	Nd	Lewis et al., 1999
Small rivers in Northern Ireland	0.09	0.08	0.22	0.64	0.63	Nd	Gibson et al., 1995
River Swale, Yorkshire, UK	0.54	n.d.	2.63	2.34	Nd	Nd	House and Warwick, 1998
Severen, a mid-sized river, UK	0.82	0.70	1.62	15.04	0.21	2.85	Russel et al., 1998
Eleven rivers, York, UK	0.3-3.3	0.3-2.2	0.8-6.4	12-34	Nd	Nd	House et al., 1997
Small forested catchment, Central Amazon, Brazil	0.02	0.01	0.08	2.67	0.18	0.8	Williams & Melack, 1997
Twenty-five small to very large tropical catchments	Nd	Nd	Nd	2.43	0.39	2.53	Lewis et al., 1999
Richmond River in subtropical Australia*	Nd	2.1 (22)	2.9 (28)	3.4 (27)	2.5 (20)	Nd	McKee et al., 2000

* Values in parentheses indicate maximum potential rates (25 year record).

In the Val Roseg catchment, NO_3-N yield was an-order-of-magnitude higher than average values reported from undisturbed temperate headwater catchments in North America (Lewis et al., 1999, and references therein). Despite its remote location, Val Roseg receives high aerial deposition of anthropogenic nitrogen (in average 12 kg ha^{-1} yr^{-1}, Rihm, 1996), that are five times above baseline deposition in undisturbed catchments (Lewis et al., 1999). As expected, DON, NH_4-N and SRP yields were low compared to most tropical and temperate streams. However, particulate phosphorous (PP) rates and the proportion of total phosphorous that was transported as PP (98%) were very high, similar to fluxes from agricultural catchments (Table 4). TP flux was almost as high as TN flux. In most lowland rivers TN fluxes are several times higher than for TP (e.g., 11-19 times for UK rivers, Russel et al., 1998).

DOC export rates (3.6 kg ha^{-1} yr^{-1}) were low compared to most stream studies. Buffam et al. (2001) summarized data from a wide range of rivers where annual DOC fluxes ranged from 5.1 kg ha^{-1} yr^{-1} in small forested rivers to more than 60 kg ha^{-1} yr^{-1} in humic rivers. In small arctic tundra streams total carbon fluxes ranged from 13 to 31 kg ha^{-1} yr^{-1} (Oswood et al., 1996) with > 95% being DOC (Val Roseg: > 50% DOC). Much lower values were reported from a tiny brook in Antarctica, where DOC and POC fluxes of 0.04 and 0.14 kg ha^{-1} yr^{-1}, respectively, were calculated (McKnight & Tate, 1997); low rates resulted from the low average flow (20 L sec^{-1}) and a flow duration of only 62 days during an annual cycle.

6. NUTRIENT LIMITATION AND ORGANIC MATTER QUALITY

6.1. Molar N:P ratio

For the individual water sources, the molar N:P ratio (TN, TP) was high for snow meltwater and hillslope ground water but low for glacial meltwater. The low ratio for glacial meltwater reflected very high PP concentrations. N:P ratios indicated phosphorous limitation for all water sources except glacial meltwater (moderate nitrogen limitation).

Within the flood plain, the N:P ratio exhibited a pronounced seasonality with an average value of 129 (range: 6-510). Lowest values corresponded to the main ablation period when glacial meltwater was the primary water source. Generally, a N:P ratio <10 indicates N-limitation and a ratio >17 P-limitation. Therefore, N:P ratios indicate strong P-limitation for all floodplain channels. Experiments with nutrient diffusing substrata performed in different floodplain channels also suggested that phosphorous rather than nitrogen was a limiting resource (Robinson et al., 2002; Chapter 14).

Only during the main ablation period in summer, with an N:P ratio of about 6 in the main channel, was there any evidence of nitrogen limitation. For the Val Roseg, we expect that there has been a major change in the N:P ratio during the past decades as a result of increasing aerial deposition of DIN (dissolved inorganic nitrogen) from anthropogenic inputs, resulting in higher N:P ratios and a stronger P-limitation than occurred in the past. Care has been taken, however, in the interpretation of N:P ratios since the ratio greatly depends upon the potential biological availability of the individual P and N fractions and the transformation between organic and inorganic forms, which is virtually unknown (e.g. Edwards et al., 2000). In the Val Roseg, PP was almost entirely in the inorganic form, which contrasts to most aquatic sites with up to 95% of PP being bound to living organisms (e.g. bacteria, Wetzel, 2001).

6.2. Molar C:N ratio

The molar C:N ratio can be used as a food quality indicator (low values representing high quality) and as an indicator for differentiating between allochthonous (high C:N) and autochthonous (low C:N) origin of organic matter. In the Val Roseg, the average C:N ratio was <6 for all water sources, well below 15, the critical value separating labile from refractory components (Hyne, 1978). Similar to the N:P ratio, the C:N ratio indicated no nitrogen deficiency for individual water sources (critical value: 8.3; Wetzel, 2001). Within the Val Roseg flood plain, however, the C;N ratio of dissolved organic matter (DOM) ranged from 8.6 to 14.5 (range of average annual values for individual channel types). These values were elevated compared to the initial values for individual water sources, which means that the quality of DOM changes as it moves through the flood plain.

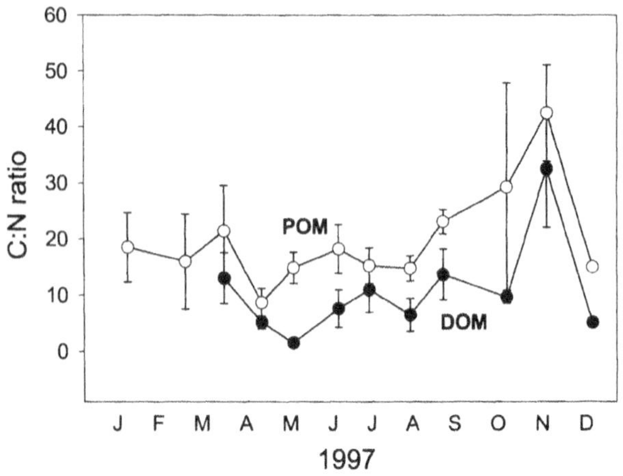

Figure 7. Seasonal dynamics of the C:N ratio (average ± SD) for dissolved (DOM) and particulate (POM) organic matter in surface-connected water bodies (n=3).

The C:N ratio exhibited a distinct seasonality in all channel types (Fig. 7). Lowest values were measured during the spring season (C:N <5) and highest values in autumn (>30). C:N ratios of particulate matter (PN, POC) ranged between 16 and 26 (average values for individual channel types) and showed a similar seasonal pattern for DOM (Fig. 7). For comparison, the average C:N ratio in the bed sediments (size fraction: <1.0 mm) of Val Roseg channels was about 13 (Tockner et al., 1997). In small forested catchments C:N ratios were higher, with values between 20 and 50 (Buffam et al., 2001). In the Val Roseg flood plain high C:N ratios for

POM probably resulted from the predominance of terrestrial sources throughout the year (release of aeolian organic matter by glacier melting and leaf litter input from lateral tributaries, cf. Teeri & Barrett, 1977; Zah & Uehlinger, 2001). Additional inputs of allochthonous terrestrial organic matter (e.g. needles) in autumn led to a further increase in the C:N ratio (see Chapter 12). Low C:N ratios for DOM during the spring season resulted from peaks in organic nitrogen rather than from higher carbon concentrations and reflected potential autochthonous origin of organic matter (benthic algae and biofilm).

7. SUMMARY

The study of the Val Roseg flood plain produced one of the few year-round data sets on nutrients fractions available for alpine stream ecosystems (cf. Table 2). Results highlighted that hydrology (sources and pathways of water) was the major determinant of the availability of nutrients within the floodplain-catchment complex. Overall, the interaction between catchment processes (snow melting), regional processes (glacial melting) and local processes (local upwelling/downwelling of ground water within the flood plain) created very heterogeneous nutrients patterns, both in space and time. The Val Roseg flood plain is not nutrient poor, as we may have expected for a remote alpine ecosystem. There are two major reasons for the comparable high nutrient concentrations in the Val Roseg: glacial scouring released high concentrations of phosphorous and atmospheric deposition resulted in nitrogen levels about 5-times higher than under pristine conditions. However, the availability of nutrient resources to stream biota remains somewhat contentious. Phosphorous, the limiting nutrient in the Val Roseg flood plain, is primarily released during the short glacial ablation period in summer when transfer of water is rapid and biotic uptake is low. In contrast to most lowland flood plains, the Val Roseg flood plain served primarily as a mixing zone, rather than as a transformation area for nutrients and organic matter. Further, we have limited information on the reactivity of particulate phosphorous (by far the most important P-fraction) and dissolved organic nutrients.

We may expect that the rapid retreat of the glaciers will lead to major future changes in resource dynamics and availability. During glacial recession freshly exposed sediments result in increased sediment and particulate phosphorous yields. With ongoing retreat, the hydrology will shift from a glacier- to a snow-dominated pattern with a less predictable seasonal flow regime, larger single floods (rain-on-snow events) and, therefore, to a major seasonal shift in the amount and temporal availability of nutrients and organic matter.

8. REFERENCES

Bretschko, G. (1966). Untersuchungen zur Phosphatführung zentralalpiner Gletscherabflüsse. *Archiv für Hydrobiologie, 62*, 327-334.

Buffam, I., Galloway, J. N., Blum, L. K. & McGlathery, K. J. (2001). A stormflow/baseflow comparison of dissolved organic matter concentrations and bioavailability in an Appalachian stream. *Biogeochemistry, 53*, 269-306.

Duff, J. H., & Triska, F. J. (2000). Nitrogen biogeochemistry and surface-subsurface exchange in streams. In J. B. Jones, and P. J. Mulholland, (Eds.), *Streams and Groundwater* (pp. 197-220). London: Academic Press.

Edwards, A. C., Cook, Y., Smart, R. & Wade A. J. (2000). Concentration of nitrogen and phosphorous in streams draining the mixed land-use Dee catchment, north-east Scotland. *Journal of Applied Ecology, 33*, 159-170.

Forsberg, B. R., Devol, A. H., Richey, J. E., Martinelli, L. A. & dos Santos, H. (1988). Factors controlling nutrient concentrations in Amazon floodplain lakes. *Limnology and Oceanography, 33*, 41-56.

Friberg, N., Milner, A. M., Svendsen, L. M., Lindegaard, C. & Larsen, S. E. (2001). Macroinvertebrate stream communities along regional and physico-chemical gradients in Western Greenland. *Freshwater Biology, 46*, 1753-1764.

Füreder, L. (1999). High alpine streams: cold habitats for insect larvae. In R. Margesin & F. Schinner (Eds.), *Cold adapted organisms: Ecology, physiology, enzymology and molecular biology* (pp. 181-196). Berlin: Springer.

Füreder, L., Schütz, C., Wallinger, M., & R. Burger (2001). Physico-chemistry and aquatic insects of a glacier-fed and a spring-fed alpine stream. *Freshwater Biology, 46*, 1673-1690.

Gibson, C. E., Wu, Y. & Pinkerton, D. (1995). Substance budgets of an upland catchment: the significance of atmospheric phosphorous inputs. *Freshwater Biology, 33*, 385-392.

Hieber, M., Robinson, C. T., Uehlinger, U. & Ward (2002). Are alpine lake outlets less harsh than other alpine streams. *Archiv für Hydrobiologie, 154*, 199-223.

House, W. A., Leach, D., Warwick, M. S., Whitton, B. A., Pattinson, S. N., Ryland, G., Pinder, A, Ingram, J., Lishman, J. P., Smith, S. M., Rigg, E., & Denison, F. H. (1997). Nutrient transport in Humber rivers. *The Science of the Total Environment, 194/195*, 303-320.

House, W. A. & Warwick, M. S. (1998). A mass-balance approach to quantifying the importance of in-stream processes during nutrient transport in a large river catchment. *The Science of the Total Environment, 210/211*, 139-152.

Hyne, N. J. (1978). The distribution and sources of organic matter in reservoir sediments. *Environmental Geology* 2, 279-285.

Kling, G. W., Kipphut, G. W., Miller, M. M. & O'Brian, W. J. (2000). Integration of lakes and streams in a landscape perspective: the importance of material processing on spatial patterns and temporal coherence. *Freshwater Biology, 46*, 477-497.

Lewis, W. M., Weibezahn, F. H., Saunders III, J. F. & Hamilton, S. K. (1990). The Orinoco River as an ecological system. *Interciencia, 15*, 346-357.

Lewis, W. M., Melack, J. H., McDowell, W. H., McClain, M. & Richey, J.E.. (1999). Nitrogen yields from undisturbed watersheds in the Americas. *Biogeochemistry, 46*, 149-162.

Likens, G. E., Bormann, F. H., Pierce, R. S., Eaton, J. J. & Johnson, N. M. (1977). *Biogeochemistry of a forested ecosystem*. New York: Springer.

Lock, M. A., Ford, T. E., Hullar, A. J., Kaufman, M., Vestal, J. R., Volk, G. S. & Ventullo, R. M. (1990). Phosphorous limitation in an Arctic river biofilm – a whole ecosystem experiment. *Water Research, 24*, 1545-1549

Malard, F., Tockner, K., Ward, J. V. (1999). Shifting dominance of subcatchment water sources and flow paths in a glacial flood plain (Val Roseg, Switzerland). *Arctic, Antarctic and Alpine Research, 31*, 135-150.

McKee, L., Eyre, B., & Hossain, S. (2000). Intra- and interannual export of nitrogen and phosphorous in the subtropical Richmond River catchment, Australia. *Hydrological Processes, 14*, 1787-1809.

McKnight, D. M. & Tate, C. M. (1997). Canada Stream: a glacial meltwater stream in Taylor Valley, South Victoria Land, Antarctica. *Journal of the North American Benthological Society, 16*, 14-17.

Meybeck, M. (1982). Carbon, nitrogen and phosphorous transport by world rivers. *American Journal of Science, 282*, 401-450.

Meybeck, M., Chapman, D. V. & Hlemer, R. (1989). *Global Freshwater Quality*. Oxford: Basil Blackwell

Milner, A. M. & Petts, G. E. (1994). Glacial rivers: physical habitat and ecology. *Freshwater Biology, 32*, 295-307.

Nikus, U., Kuhn, M., Baltensberger, U., Delmas, R., Gäggeler, H., Kasper, A., Kromb-Kolb, H., Maupetit, F. Novo, A., Pichlmayer, F., Preunkert, S., Puxbaum, H., Rossi, G., Schöner, W., Schwikowski, M., Seibert, P., Staudinger, M., Trockner, V., Wagenbach, D., & Winiwarter, D. (1997). SNOSP: Ion deposition and concentration in high alpine snow packs. *Tellus, 49B*, 56-71.

Oswood, M. W., Irons III, J. G. & Schell, D. M. (1996). Dynamics of dissolved and particulate carbon in an arctic stream. In J. F. Reynolds, and J. D. Tenhunen, (Eds.). *Landscape function and disturbance in arctic tundra* (pp. 275-289). New York: Springer.

Pandey, K. D., Kashyap, A. K. & Gupta, R. K. (1995). Nutrient status, algal and cyanobacterial flora of six fresh water streams of Schirmacher Oasis, Antarctica. *Hydrobiologia, 299*, 83-91.

Peterson, B. J., Deegan, L., Helfrich, J., Hobbie, J. E., Hullar, M., Moller, B., Ford, T. E., Hershey, A., Hiltner, A., Kipphut, G., Lock, A. M., Fiebig, D. M., McKinley, V., Miller, M. C., Vestal, J. R., Ventullo, R., & Volk, G. (1993). Biological responses of a tundra river to fertilization. *Ecology, 74*, 653-672.

Rihm, B. (1996). *Critical loads of nitrogen and their exceedances*. Environmental Series No. 275. BUWAL, Bern, Switzerland.

Robertson, A.I., Bunn, S. E., Boon, P. I. & Walker, P. F. (1999). Sources, sinks and transformations of organic carbon in Australian floodplain rivers. *Marine and Freshwater Research, 50*, 813-29.

Robinson, C. T. & Gessner, M. O. (2000). Nutrient addition accelerates leaf breakdown in an alpine springbrook. *Oecologia, 122*, 258-263.

Robinson, C. T., Uehlinger U., Schenkel, P. & Skvarc, R. (2002). Limitation and retention of nutrients in alpine streams of Switzerland. *Verhandlungen der Internationalen Vereineinigung für Theoretische und Angewandte Limnologie*, *28*, 263-272.

Robinson, C. T., Uehlinger, U. & Hieber, M. (2001). Spatio-temporal variation in macroinvertebrate assemblages of glacial streams in the Swiss Alps. *Freshwater Biology*, *46*, 1663-1672.

Russell, M. A., Walling, D. E, Webb, B. W. & Bearne, R. (1998). The composition of nutrient fluxes from contrasting UK river basins. *Hydrological Processes, 12*, 1461-1482.

Sickman, J. O., Leydecker, A. & Melack, J. M. (2001). Nitrogen mass balances and abiotic controls on N retention and yield in high-elevation catchments of the Sierra Nevada, California, United States. *Water Resource Research, 37*, 1445-1461.

Stottlemyer, R. (1992). Nitrogen mineralisation and streamwater chemistry, Rock Creek Watershed, Denali National Park, Alaska, U.S.A. *Arctic and Alpine Research, 24*, 291-303.

Teeri, J. A. & Barrett, P. E. (1977). Detrital transport by wind in a high arctic terrestrial ecosystem. *Arctic and Alpine Research* 7, 387-391.

Tockner, K., Malard, F., Burgherr, P., Robinson, C. T., Uehlinger, U., Zah, R. & Ward, J. V. (1997). Physico-chemical characterization of channel types in a glacial floodplain ecosystem (Val Roseg, Switzerland). *Archiv für Hydrobiologie, 140*, 433-463.

Tockner, K., Pennetzdorfer, D., Reiner, N., Schiemer, F. & Ward, J. V. (1999). Hydrologic connectivity and the exchange of organic matter and nutrients in a dynamic river-floodplain system (Danube, Austria). *Freshwater Biology, 41*, 521-535.

Tockner, K., Malard, F., Uehlinger, U. % Ward, J. V. (2002). Nutrients and organic matter in a glacial river-floodplain. *Limnology and Oceanography, 47*, 266-277.

Ward, J. V. (1994). Ecology of alpine streams. *Freshwater Biology, 32*, 277-294.

Wetzel, R. G. Limnology: *Lakes and river ecosystems*. 3rd Edition. New York: Academic Press.

Williams, M. R. & Melack, J. M. (1997). Solute export from forested and partially deforested catchments in the central Amazon. *Biogeochemistry, 38*, 67-102.

Williams, M. W. & Melack, J. M. (1989). Effects of spatial and temporal variation in snow melt on nitrate ion and sulfate ion pulses in melt waters within an alpine basin. *Annals of Glaciology, 13*, 285-288.

Wissmar, R.C., Swanston, D. N., Bryant, M. & McGee, K. (1997). Factors influencing stream chemistry in catchments on Prince of Wales Island, Alaska. *Freshwater Biology, 38*, 301-314.

Zah, R. & Uehlinger, U. (2001). Particulate organic matter inputs to a glacial stream ecosystem in the Swiss Alps. *Freshwater Biology, 46*, 1597-1608.

CHAPTER 7

C. WELLSTEIN, U. UEHLINGER & R. ZAH

TERRESTRIAL FLOODPLAIN VEGETATION

1. INTRODUCTION

Ecological studies of glacier forelands[1] in the Alps already began at the end of the 19th century (Coaz, 1887). Primary succession has been studied using the method of space-for-time substitution (Matthews, 1992). Investigations of the foreland vegetation in the Bernina massif included several studies (Burga, 1999; Flütsch, 1930; Lüdi, 1958; Richter, 1994; Rübel, 1912).

The glacier foreland and the flood plain directly interact with the Roseg River ecosystem. These environments have undergone dramatic changes during the past 150 years. The newly-formed landscape in front of the rapidly receding valley glaciers exemplifies this dynamic process. The relatively short time span between the end of the Little Ice Age and the present usually precludes major morphological alterations of the deglacierized zone. Nevertheless, forelands can be subject to continuous modifications resulting from the formation of persistent but dynamic drainage networks (Price, 1980). This may locally result in distinct changes such as the lateral expansion of the uppermost part of the Roseg flood plain during the past 100 years (Figs. 9 and 10 in Chapter 2).

Flood plains constitute a particularly diverse form of riverine landscapes providing a mosaic of terrestrial and aquatic habitats (Ward et al., 2002; Wiens, 2002). Flood plains encompass an area of dynamic transition from wet to dry that is characterized by alternating periods of low flow and floods with erosion and alluvial deposition. The fluvial dynamics change the configuration of riverine landscapes at different spatial and temporal scales, and initiate and reset succession of terrestrial and aquatic communities (Hughes, 1997). Under the severe environmental constraints near the alpine tree line (see Chapter 1) the terrestrial floodplain vegetation is expected to recover slowly, which affects its role as a source of energy for the heterotrophic communities of the aquatic floodplain habitats.

The main focus of this chapter is on the spatial patterns of plant communities in a landscape context. A provisional list of the terrestrial plant taxa is provided in Appendix II. The chapter is mainly based on recent studies of the plant communities in Roseg flood plain and the foreland of the Tschierva Glacier (Wellstein, 2002). Additional information on the vegetation of the upper Roseg valley was found in Richter (1994).

---

[1] Glacier foreland is the area of newly-formed landscape in front of a glacier, which was ice covered but has since been exposed by glacier retreat (Matthews, 1992).

J.V. Ward & U. Uehlinger (eds.), Ecology of a Glacial Flood Plain, 109-121.
© 2003 *Kluwer Academic Publishers. Printed in the Netherlands.*

Domains of geomorphic processes in the flood plain and the foreland of the Tschierva/Roseg glacier were mapped at the scale of 1:2500 using a GIS based map and GPS. Historic information on the location of the glacier termini since 1850 (Schollenberger, 1976) and changes in the channel network of the flood plain since 1947 (Zah et al., 2001) were also considered. Two strata, chronosequence[2] and topology (Fig. 1), were employed in a stratified design; sampling sites within each stratum were selected to assess as much as possible all typical vegetation types. The size of the sampling plots was adjusted to meet the criterion of quasi-homogeneity (Dierschke, 1994). Sampling took place from July to September 2001 and comprised 129 vegetation relèves in the flood plain and 139 relèves in the foreland.

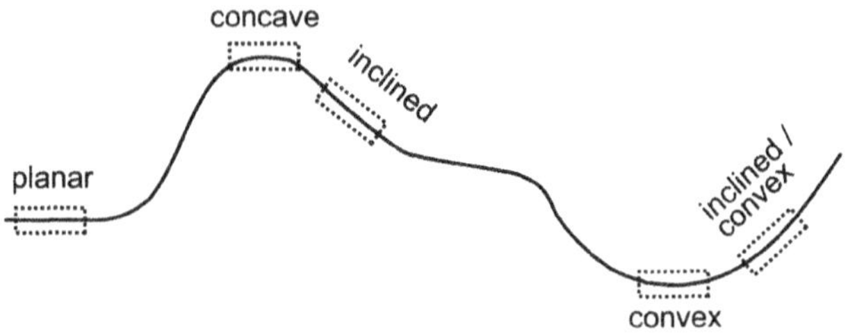

Figure 1. Schematic diagrams of topological structures in the flood plain.

2. PHYSICO-CHEMICAL HABITAT TEMPLATE

Fluvial, glacial, periglacial, and gravitational processes, in concert with climatic factors, create and influence the particular terrestrial habitats of the Tschierva and Roseg glacier forelands and the Roseg flood plain. These processes operate across a range of spatio-temporal scales. Grain size distribution, for example, is an important micro-scale (<1 m) factor. Sediments are poorly sorted immediately after deglaciation but runoff can arrange particles by size, form or specific weight. Determination of grain size distribution in the foreland and the flood plain indicated minor differences in the sorting degree except for finer sediments, which were more frequent in the flood plain (Fig. 2). At the mesoscale (1 – 10 m) fluvial processes create topologic diversity by erosion, transport and sedimentation. Typical topological structures include concave banks, flat areas, channels, and slopes, and at the macroscale (100 – 1000 m) geomorphic and hydrological processes create and differentiate the landscape (Chapter 2).

---

[2] Chronosequences are spatial sequences in which environmental factors other than time are unimportant either because they are invariant or because they are relatively ineffective (Matthews, 1992). In the foreland of a receding glacier or in a flood plain with laterally shifting channels time is a major factor controlling spatial vegetation patterns.

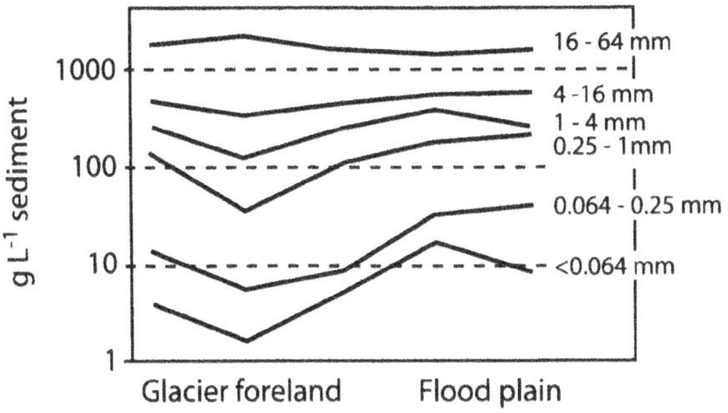

Figure 2. Grain size distribution along the main channel of the Roseg River between the Tschierva Glacier and the end of the flood plain. Sediments were sampled by freeze coring (Rainer Zah, unpublished data).

2.1. Foreland

The foreland extending from the uppermost part of the flood plain to the termini of the Roseg and Tschierva glaciers is embedded in an alpine environment. It is covered by rock debris and alpine vegetation (Fig. 6 in Chapter 1). Close to the glacier this debris is unconsolidated, which results in highly unstable melt water channels. Moraines and aerial photographs provide information to estimate the age of ice free areas (Fig. 7 in Chapter 2; Schollenberger, 1976). In close proximity to the Tschierva glacier, local climatic gradients are superimposed on the regional climate. Cold glacier surfaces result in steep vertical temperature gradients and cool air slides down the glacier surface and affects temperatures in the glacier foreland, in particular close to the ground (Matthews, 1992). In the upper Roseg valley these downslope katabatic glacier winds may offset the up-valley wind during high pressure weather (Richter, 1994).

2.2. Flood plain

The terrestrial environment of Val Roseg flood plain is influenced by hydrological factors such as floods, water table elevation, erosion, transport, and deposition of sediments. Lateral channel shifting (Chapter 2) eliminates plots of formerly stable vegetation and simultaneously creates areas of bare gravel and sand. Most morphological structures in the active flood plain are <55 years old (Fig. 3). Sediments deposed during floods can affect soil development in the active flood plain. These sediments, which are low in organic matter, provide only minor amounts of nutrients compared to low elevation floodplain systems (Müller & Scharms, 1996).

The upper part of the flood plain is a major downwelling zone that falls dry from November to March (Chapter 3; Malard et al., 1999). Even during summer high flow, the alluvium below many channels is unsaturated (Chapter 3) indicating a relatively low water table elevation. In the lower part of the flood plain (below transect 10), the groundwater table intersects with the soil surface at many locations.

Shallow ground water also affects the duration of snow cover in spring; in the lower part of the flood plain snow disappears about one week earlier than in the upper part.

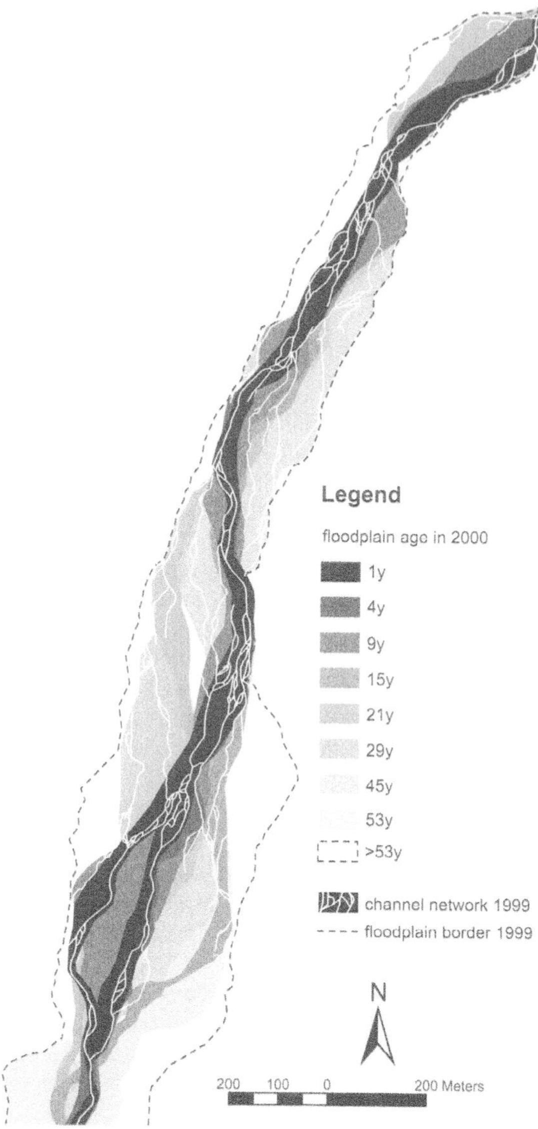

Figure 3. Age of the flood plain habitats estimated from the location of the main channel area. The information is based on aerial photographs (Zah et al., 2001).

The uppermost part of the present flood plain was glacierized at the end of the Little Ice Age (Chapter 2), and thus also belongs to the foreland. Today substantial parts of this area are apparently no longer influenced by fluvial processes because the main channel is deeply incised in glacial till.

2.3 Influence of side slopes

The terrestrial vegetation on the valley side slopes is a source of diaspores, thereby providing potential colonizers and influencing succession on the flood plain and foreland. Side slopes are covered mainly by alpine grassland (*Seslerio-Caricetum sempervirentis* and *Caricetum curvulae* associations). Snowbed communities (*Salicetum herbaceae* and *Luzuletum alpinopilosae*), as well as rock and rock debris communities (*Androsacetum alpinae* and *Oxyrietum digynae*), occur toward the sub-nival vegetation zone. The relatively dense hill slope vegetation along the floodplain border includes subalpine grassland (the *Aveno-Nardetum* association), part of which are dominated by *Deschampsia caespitosa* (tufted hair-grass), *Caricetum frigidae*, occurring in small tributaries, subalpine shrubs (*Alnetum virdis*) and larch-Arolla pine forests (the *Larici-Pinetum cembrae* association).

The age of the still expanding forelands ranges from 1 to 150 years and that of floodplain habitats from <1 to about 70 years. But in contrast to the foreland, the flood plain has existed for more than 2,000 years (Chapter 2), except for the uppermost parts that have been subject to cycles of glacier advance and retreat. Climate variation during this time may have changed the flow regime and sediment supply and as a consequence lateral channel shift (Chapter 2) and turnover of terrestrial floodplain habitats.

3. TERRESTRIAL VEGETATION PATTERNS

3.1. Family and life form spectrum

In the foreland and flood plain 30% of all the vascular plants species belong to the families Poaceae and Asteracea. These and Salicaceae, Rosaceae, Caryophyllaceae, Saxifragaceae, Scrophulariaceae, Cyperaceae, and Fabacea represent two thirds of the species (Fig. 4). The species list of vascular plants occurring in the flood plain and foreland is in Appendix II. The life form[3] spectrum of vascular plants is dominated by hemi-cryptophytes and chamaephytes, which collectively include 80% of the species (Fig. 4); the remaining 20% fell in the categories of phanerophytes, geophytes and therophytes.

[3] Life forms after Raunkiaer (1934) characterize plants based on strategies evolved to persist during dormant periods such as cold or droughts. For example, hemicryptophytes have buds near the soil surface and die back to these buds in autumn/winter.

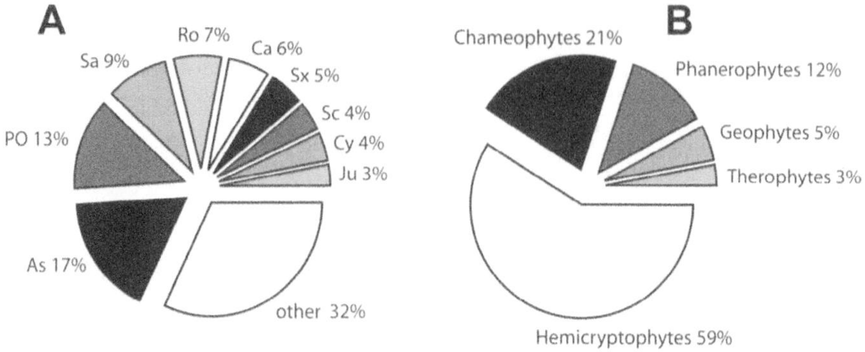

Figure 4. A) Family spectrum: Ju = Juncaceae, Cy = Cyperaceae, Sc = Scophulariaceae, Sx = Saxifragaceae, Ca = Caryophyllaceae, Ro = Rosaceae, Sa = Salicacea, Po = Pocaceae, As = Asteracea. B) Life form spectrum of the vascular plants. Combined data from the flood plain and glacier foreland.

In the deglacierized and frequently disturbed terrain, rock debris communities ("Schuttfluren Gesellschaften") prevail during the first successional decades. These formations are typically dominated by hemicryptophytes and a few chamaephytes (Wilmanns, 1998). Plants growing on rock debris have to cope with coarse unstable substrata and low content of fine organic and inorganic components. The root system is typically enlarged; the long tap root acts as an anchor while side roots grow laterally through the fine soil (root dimorphism, Wilmanns, 1998) as for example *Rumex scutatus*. Moreover, the regeneration potential after injury is remarkably high (Reisigl & Keller, 1994).

3.2. Terrestrial vegetation patterns

The mosaic of habitats observed in a landscape at any point in time is the result of disturbances across different spatial and temporal scales (Walker et al., 1989). Fig. 5 in chapter 1 delimits domains of the prevailing processes controlling the disturbance regimes in the upper Val Roseg. Lateral channel migration is the major agent of disturbance in the flood plain; active channel margins tend to be more frequently reworked than other parts of the flood plain (Hughes, 1997). Fluvial processes affect only minor parts of the forelands. Gravitational processes are more important in the most recently deglacierized steep areas below the Tschierva glacier.

3.2.1. Flood plain

The vegetation of the active flood plain is a mosaic of different vegetation units (Fig. 5), reflecting successional stages and local environmental conditions. The

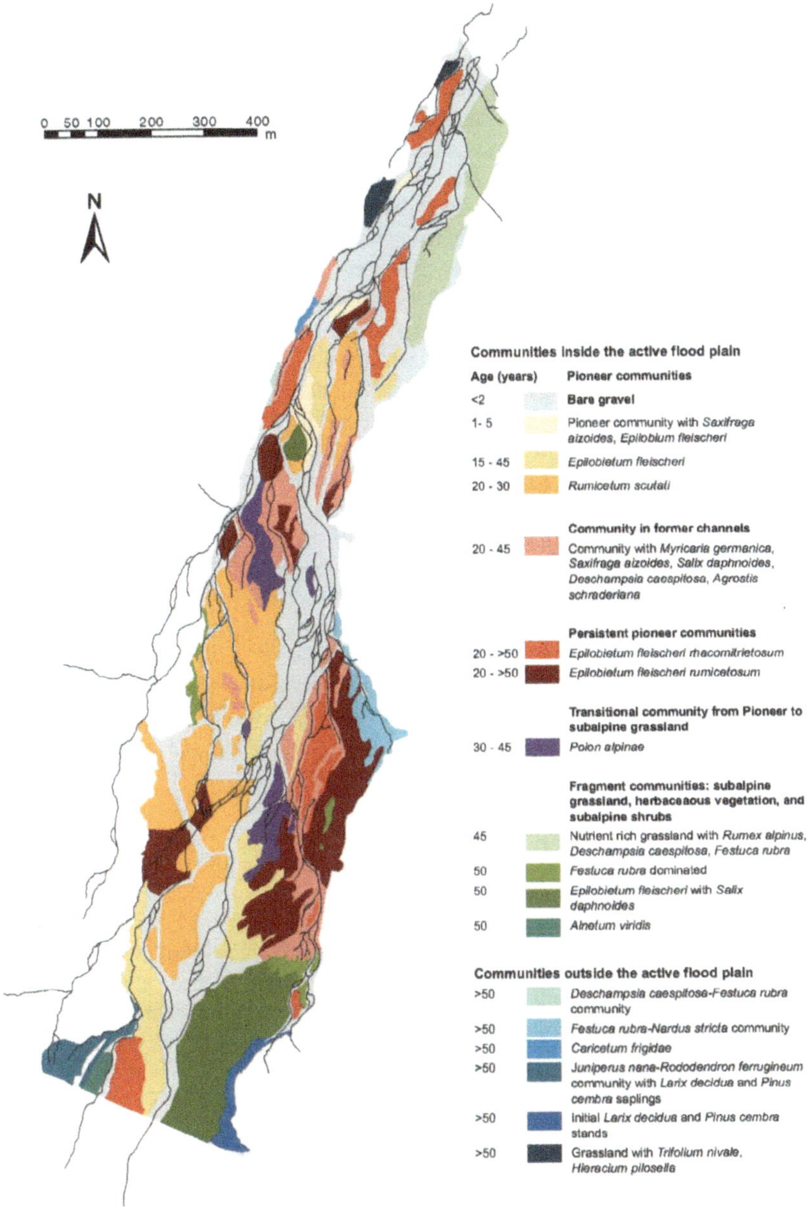

Figure 5. Plant communities in the Roseg flood plain in summer 2001.

initial stage of succession[4] in habitats freshly created by laterally shifting channels lasts <2 years and is characterized by a few pioneer plants such as *Saxifraga aizoides*, *Oxyria digyna*, *Linaria alpina* and *Epilobium fleischeri* (Fig. 6a). During the second year, additional species become established and form pioneer and fragment communities[5]. After the second year diverging succession trajectories become evident at sites differing in topology. For example, *Epilobietum fleischeri* and *Rumicetum scutati* communities develop at dry concave sites (Fig. 6b). These communities persist during the entire second successional stage, which lasts for about 50 years, before they partly shift to subalpine pasture (*Poion alpinae*, Fig. 6c). Species assemblages comprising *Rumex scutatus* and *Epilobium fleischeri* as well as Asteraceae such as *Leontodon hispidus*, the grass-like *Poa nemoralis*, and *Phleum alpinum* form a link between the pioneer communities and subalpine pasture (Fig. 6d). *Agrostietum agrostiflorae* with Cyperaceae such as *Carex bicolor* coincides with convex topology as for example sites adjacent to recent channels subject to sand deposition during high flow. The typical vegetation of these relatively humid areas also includes *Myricaria germanica* (German false tamarisk, Fig. 6e), *Saxifraga aizoides*, *Deschampsia caespitosa* (tufted hair grass), and *Salix daphnoides*. The age of most floodplain habitats is less than 53 years. Communities of the third successional stage occur in areas older than 53 years near the floodplain boundary. They include dwarf-shrub heath with *Juniperus communis* ssp. *nana* (common juniper, Fig. 6f), *Leontodon hispidus* (rough hawkbit), as well communities dominated by *Nardus stricta* (mat grass) and *Festuca rubra* (red fescue). However, these areas are affected by cattle grazing. The grazing impact is reflected by the formation of grasslands with frequent occurrence of *Nardus stricta.* Vegetation coverage in the left part of the upper flood plain reflects minor fluvial influence. *Epilobietum fleischeri* with few *Salix daphnoides* dominate the sparse vegetation of the uppermost flood plain, which was covered by ice in 1850 but is not part of the active river corridor.

3.2.2. Foreland

In the foreland, the initial stage of succession comprises a period of about 15 years but seres of the second successional stage are not homogenous. On the left part, for example, *Anthyllis alpestris* meadows dominate, whereas in other areas the lichen *Stereocaulon alpinum* prevails. This divergence of succession could be attributed neither to topology (meso scale) nor to grain size (micro scale). In the nearby foreland of the Morterastch Glacier, Burga (1999) found that differences in the preferences of species with respect to grain size resulted in divergent succession. In areas deglacierized for <53 years, shrubs such as *Salix helvetica*, *Rhododendron ferrugineum*, and *Juniperus communis* were dominant. Subalpine forest (*Larici-*

---

[4] Stages of succession were identified by the classification of species into species groups (Wellstein, 2002). Reducing the numbers of descriptors (species) is commonly used in studies of vegetation dynamics (e.g., Achermann et al., 2000).

[5] Fragment communities: Fragments of taxonomic vegetation types with a repetitional species composition which can be classified but is attributed only to a higher taxonomic level (Wilmanns, 1998).

Cembretum and *Junipero-Laricetum*) occurred at sites which were ice free for >80 years and not affected by fluvial and gravitational processes.

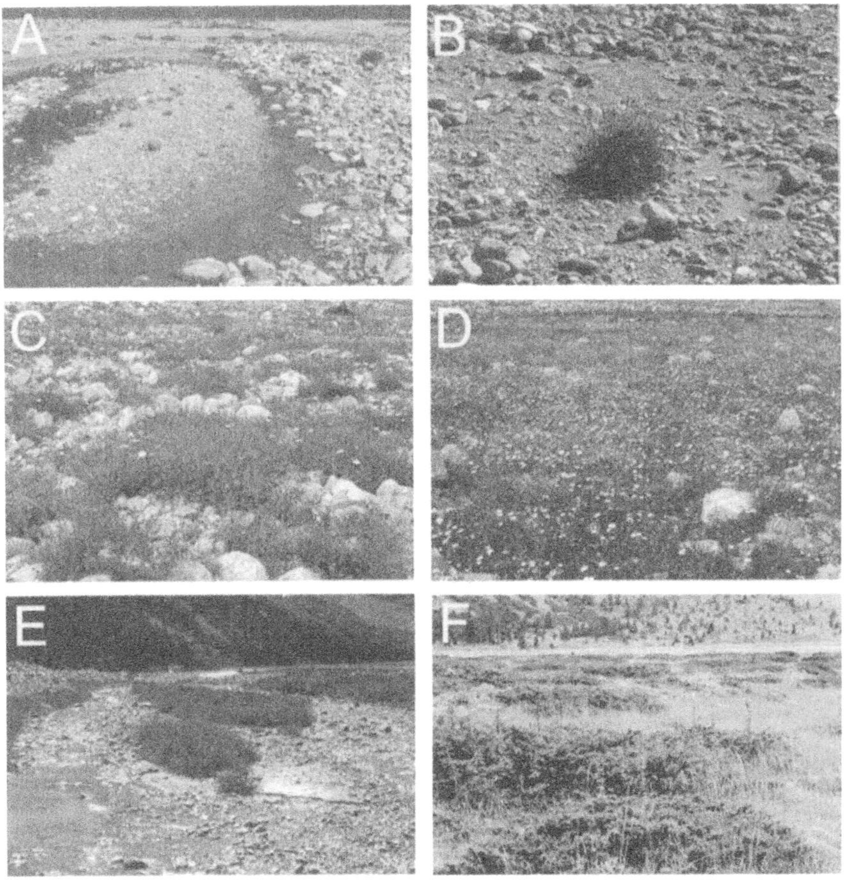

Figure 6. A) Initial stage of succession, B) Rumex scutatus *individual on a dry concave site, C) transitional stage between the pioneer community (*Rumicetum scutati*) and the subalpine grassland, D) Subalpine grassland community (*Poion alpinae*), E) stands of Myricaria germanica (German false tamarisk), and F) dwarf-shrub heath with common juniper.*

3.3. Plant cover

Plant cover (Fig. 7) reflects temporal differences in the succession between flood plain and foreland. At sites <15 years old, vegetation densities are up to three times higher in the flood plain than in the foreland, presumably reflecting differences in climate and distance to vegetated patches. Only at the oldest floodplain sites (>53 years old) is the vegetation cover more than 90%.

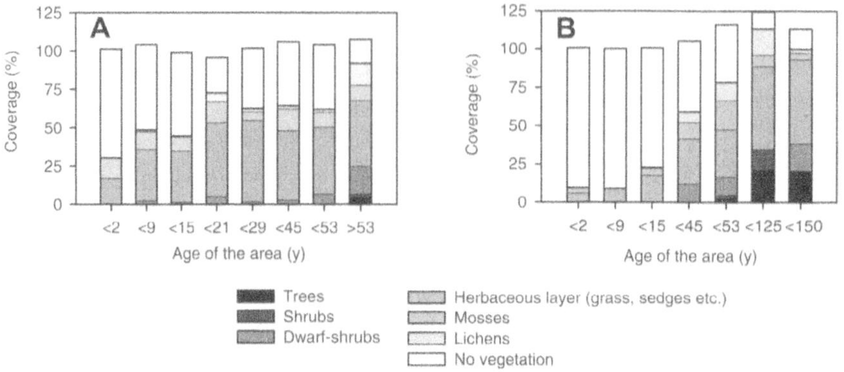

Figure 7. Plant cover along chronosequences in A) the flood plain and B) the foreland. Adding the coverage of different vegetation layers may result in values exceeding 100% due to overlap of layers.

4. PLANT DIVERSITY

The species inventory of flood plain habitats <53 years old includes 38 species occurring only at alluvial sites and 75 species growing in the flood plain as well in the glacier foreland. Considering older successional stages (>53 years) in the recent flood plain, which include *Juniperus* and *Festuca rubra - Nardus stricta* communities, and in the foreland, which comprise subalpine grasslands and forests (*Larici-Cembretum*), floristic differences between flood plain and foreland decreases (about 100 common species). Richter (1994) found 220 species in the foreland of the Tschierva Glacier. The cumulative number of species in the foreland of the nearby Morteratsch Glacier is about 202 (Bäumler, 1988; Burga, 1999; Maisch et al., 1999; Ungricht, 1995).

Frequency and intensity of disturbance affect diversity. Comparing five major terrain units in the Arctic foothills of northern Alaska, Walker et al (1989) reported higher diversity (Shannon index) in flood plains, which are more frequently disturbed than glacial till and basin colluvium. In the Val Roseg, alpha diversity of 4 m plots was higher at alluvial sites than at sites in the foreland of the Tschierva Glacier. The fluvial disturbance results in a relatively stochastic habitat mosaic, which is driven by the laterally shifting channels and where patches of different successional stages often share common boundaries (Fig. 5). The spatial configuration of vegetation patches influencing the dispersal and recruiting of plants is different in the foreland, where large-scale gradients are more important (e.g. distance from glacier, influence of glacier wind, distance from diaspore reservoirs). This is reflected by differences in the average number of species between flood plain and foreland along the chronosequence (Fig. 8).

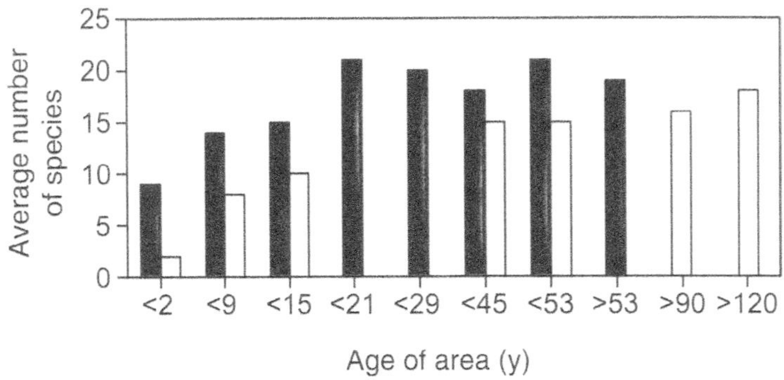

Figure 8. Average number of plant species in 4 m^2-plots of the flood plain (filled bars) and foreland (open bars) along the chronosequence of both landscape elements. In the foreland, the age classes <21 and <29 years are missing because of the advance of the Tschierva Glacier between 1960 and 1985 (see Chapter 2).

The diversity of life forms parallels to some extent species diversity in both flood plain and foreland (Fig. 9). With increasing number of species there is a corresponding increase in the proportion of hemicryptophytes. In the foreland, herbaceous chamaephytes are maximum at intermediate successional stages (<45 years). Phanerophytes appear at sites >2 years old and reach maximum numbers during middle successional stages.

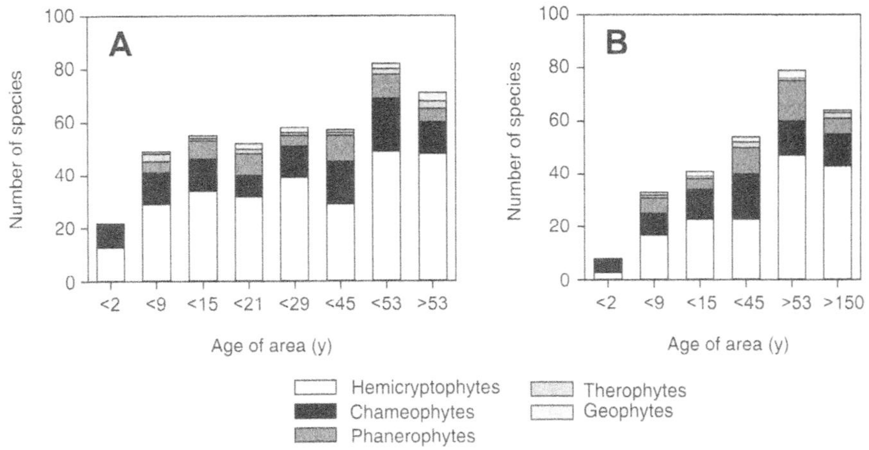

Figure 9. Life forms along the chronosequences of A) the flood plain and B) the foreland.

5. SUMMARY

The vegetation of the Roseg flood plain forms a highly diverse, shifting mosaic of different vegetation patches. This relatively stochastic mosaic results from the local erosion and deposition of sediments during summer high flow, and the subsequent succession in the disturbed patches, which is influenced by grain size distribution, local topography, and distance to vegetated patches. Diversity of species and life forms increase relatively fast with time since disturbance and levels out after about 50 years. The present vegetation pattern reflects the physico-chemical habitat template of the flood plain, which, we assume, was relatively constant during the past 60 years. In the forelands, the effect of fluvial and gravitational processes on vegetation patterns is less important with respect to the area affected than influences resulting from the recession of the valley glaciers. As a consequence, species and life form diversity change directionally from the present ice margin to the terminus of 1850.

Besides fluvial disturbance, climate is presumably the major constraint for the development of the terrestrial floodplain vegetation. Successional stages at the oldest floodplain sites do not go beyond dwarf-shrub heath. This strongly contrasts with the vegetation of the highly dynamic Tagliamento River flood plain (floodplain turnover presumably <30 years), which is located at an elevation <200 m a.s.l. In the flood plain of this large Alpine river, islands with lush woody vegetation may develop within only 22 years (Edwards et al., 1999). In the upper Val Roseg, climate change mainly resulted in a rapid glacier recession. Today, the oldest parts of the forelands (>120 years) are covered by a larch-Arolla pine forest, except in the uppermost flood plain. This successional trend is expected to continue along a trajectory controlled by local climate, soil development, geomorphic, glacial and periglacial processes. The present vegetation patterns in the flood plain are expected to persist until the climate change is sufficient to appreciably affect hydrology and plant (tree) growth, and as a consequence channel-riparian vegetation interactions and floodplain turnover rates (Gurnell et al., 1999).

6. ACKNOWLEDGEMENTS

We thank the "Stiftung zur Förderung der Pflanzenkenntnis" (Basel, Switzerland), "Badischer Landesverein für Naturkunde und Naturschutz" (Freiburg, Germany) and "Verband der Freunde der Albert Ludwigs Universität Freiburg" (Freiburg, Germany) for financial support of field work of Camilla Wellstein in the Val Roseg.

7. REFERENCES

Achermann, G., Schütz, M., Krüsi, B. O., & Wildi, O. (2000). Tall-herb communities in the Swiss National Park: Long-term development of the vegetation. In M. Schütz, B. O. Krüsi & P. J. Edwards (Eds.), Succession research in the Swiss National Park. *Nationalparkforschung in der Schweiz, 89,* 67-88).

Bäumler, E. (1988). *Untersuchung zur Besiedelungsdynamik und Populationsbiologie einiger Pionierpflanzen im Morteratschgletschervorfeld.* PhD thesis, University of Basel, Basel.

Burga, C. A. (1999). Vegetation development on the glacier forefield Morteratsch (Switzerland). *Applied Vegetation Science, 2*, 17-24.

Coaz, J. (1887). Erste Ansiedlung phanerogamischer Pflanzen auf von Gletschern verlassenem Boden. *Mitteilung der Naturforschenden Gesellschaft Bern*, 3-12.

Dierschke, H. (1994). *Pflanzensoziologie. Grundlagen und Methoden*. Stuttgart: Ulmer.

Edwards, P. J., Kollmann, J., Gurnell, A. M., Petts, G. E., Tockner, K., & Ward, J. V. (1999). A conceptual model of vegetation dynamics on gravel bars of a large Alpine river. *Wetlands Ecology and Management, 7*, 141-153.

Flütsch, P. (1930). Über die Pflanzengesellschaften der alpinen Stufe des Berninagebietes. *Jahresberichte der Naturforschenden Gesellschaft Graubündens, 68*, 37-88.

Gurnell, A. M., Edwards, P. E., Petts, G. E., & Ward, J. V. (1999). A conceptual model for alpine proglacial river channel evolution under changing climatic conditions. *Catena, 38*, 223-242.

Hughes, F. M. R. (1997). Floodplain biogeomorphology. *Progress in physical geography, 21*, 501-529.

Lüdi, W. (1958). Beobachtungen über die Besiedlung von Gletschervorfeldern in den Schweizeralpen. *Flora, 146*, 386-407.

Maisch, M., Burga, C. A., & Fitze, P. (1999). *Lebendiges Gletschervorfeld. Von schwindenden Eisströmen, schuttreichen Moränenwällen und wagemutigen Pionierpflanzen im Vorfeld des Morteratschgletschers. Führer und Begleitbuch zum Gletscherlehrpfad Morteratsch* (2. Auflage). Pontresina: Gemeinde Pontresina.

Malard, F., Tockner, K., & Ward, J. V. (1999). Shifting dominance of subcatchment water sources and flow paths in a glacial floodplain, Val Roseg, Switzerland. *Arctic, Antarctic, and Alpine Research, 31*, 135-150.

Matthews, J. A. (1992). *The ecology of recently-deglaciated forelands and primary succession.* Cambridge: Cambridge University Press.

Müller, N., & Scharms, S. (1996). Zur Ökologie der Vegetation von Flussauen. In D. Brandes (Ed.), *Ufervegetation von Flüssen. Tagungsbericht der Braunschweiger Kolloquiums über Ufervegetation und - flora von Flüssen* (Vol. 4, pp. 269-296).

Price, R. J. (1980). Rates of geomorphological changes in proglacial areas. In R. A. Cullingford, D. A. Davidson & J. Lewin (Eds.), *Timescales in geomorophology* (pp. 79-93). Chichester: Wiley.

Raunkiaer, C. (1934). *The life forms of plants*. Oxford: Oxford University Press.

Reisigl, H., & Keller, L. (1994). *Alpenpflanzen im Lebensraum* (2. Auflage). Stuttgart: Fischer.

Richter, M. (1994). Die Pflanzensukzession im Vorfeld des Tschierva-Gletschers/Oberengadin. *Geoökodynamik, 15*, 55-88.

Rübel, E. (1912). *Pflanzengeographische Monographie des Berninagebietes*. Leipzig: Wilhelm Engelmann Verlag.

Schollenberger, B. (1976). *Moränenwälle im hinteren Rosegtal.* Diploma thesis, University of Zürich, Zürich.

Ungricht, S. (1995). *Die Flora der Gletscherinsel Isla Pers (Kanton Graubünden).* Diploma thesis, Swiss Federal Institute of Technology Zürich (ETH), Zürich.

Walker, D. A., Binnian, E., Evans, B. M., Lederer, N. D., Norstrand, E., & Webber, P. J. (1989). Terrain, vegetation and landscape evolution of the R4D research site, Brooke Range Foothills, Alaska. *Holarctic Ecology, 12*, 238-261.

Ward, J. V., Tockner, K., Arscott, D. B., & Claret, C. (2002). Riverine landscape diversity. *Freshwater Biology, 47*, 517-539.

Wellstein, C. (2002). *Vegetationsdynamik auf Moränenflächen eines Gletschervorfeldes (Val Roseg, Schweiz) im Vergleich zu anschliessenden Schwemmebene.* Diploma Thesis, Albert-Ludwigs-Universität Freiburg im Breisgau, Freiburg im Breisgau, Germany.

Wiens, J. A. (2002). Riverine landscapes: taking landscape ecology into the water. *Freshwater Biology, 47*, 501-515.

Wilmanns, O. (1998). *Ökologische Pflanzensoziologie* (6. Auflage). Wiesbaden: Quelle and Meyer Verlag.

Zah, R., Niedröst, M., Rinderspacher, H., Uehlinger, U., & Ward, J. V. (2001). Long-term dynamics of the channel network in a glacial floodplain, Val Roseg, Switzerland. *Arctic, Antarctic, and Alpine Research, 33*, 440-446.

CHAPTER 8

M. O. GESSNER & C. T. ROBINSON

AQUATIC HYPHOMYCETES IN ALPINE STREAMS

1. INTRODUCTION

In the mid-twentieth century Ingold (1942) discovered a diverse community of stream fungi that he called aquatic hyphomycetes. Although not well defined either taxonomically or ecologically, aquatic hyphomycetes can be broadly delimited by their thin-walled and hyaline (nonpigmented) conidia with characteristic shapes (Fig. 1) and their association with submerged decaying leaves and wood in streams. An important trait of aquatic hyphomycetes is their prolific sporulation under water. The capacity of at least the non-tropical species to germinate, to grow and to reproduce at relatively low temperatures is another noteworthy feature (Ingold, 1975; Webster & Descals, 1981; Bärlocher, 1992a).

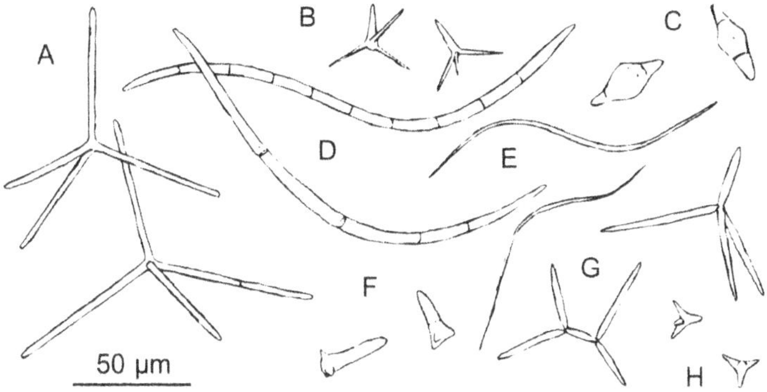

Figure 1. Conidia of some aquatic hyphomycetes identified in streams of the Val Roseg flood plain: Lemonniera aquatica *(A),* Clavatospora longibrachiata *(B),* Tumularia aquatica *(C),* Anguillospora longissima *(D),* Flagellospora curvula *(E),* Heliscus lugdunensis *(F),* Articulospora tetracladia *(G),* Heliscella stellata *(H). Modified from Chauvet (1990).*

Identification and cultivation of many species of aquatic hyphomycetes is straightforward, making these fungi attractive model organisms for ecological studies at the individual, population, community and ecosystem level (Bärlocher, 1992b; Suberkropp, 1992a, 1998). As a result, the critical roles of aquatic

J.V. Ward & U. Uehlinger (eds.), Ecology of a Glacial Floodplain, 123-137
© 2003 *Kluwer Academic Publishers. Printed in the Netherlands.*

hyphomycetes in leaf decomposition (Gessner et al., 1997; Suberkropp, 1998; Gessner & van Ryckegem, 2003) and in the nutrition of stream detritivores (Bärlocher, 1985; Suberkropp, 1992b; Graça, 2001; Gessner & van Ryckegem, 2003) have become firmly established.

Alpine, arctic and antarctic ecosystems are commonly perceived as harsh environments imposing serious constraints on organisms (Chapin & Körner, 1995; Körner, 1999). Alpine streams, glacial ones in particular, are no exception (Ward, 1994; Milner et al., 2001, see Chapter 16). Harshness relates primarily to the physical habitat template and encompasses features such as temperatures near freezing, high shear stress and low stream-bed stability (Ward, 1994; Milner et al., 2001). For heterotrophic organisms, the shortage of organic matter is an additional limiting factor. Estimated inputs of particulate organic matter to streams of the Val Roseg range from only about 1 to 34 g m^{-2} yr^{-1} (Zah & Uehlinger, 2001, see Chapter 12), compared to >500 g of plant litter m^{-2} typically received per year by canopied forest streams at lower elevations (Benfield, 1997; Abelho, 2001). Dead leaves, needles and wood, which are the primary resources of aquatic hyphomycetes, are particularly scarce.

In view of the paucity of particulate organic matter in alpine streams, do aquatic hyphomycetes occur in these environments? The small size of their propagules and their occurrence in streams worldwide suggest effective dispersal of propagules. Aquatic hyphomycetes may also be able to survive periods of unfavorable environmental conditions in the teleomorph (sexual) state or as sexual resistant spores, in spite of their delicate conidia (asexual spores) and mycelia. Therefore, the main determinants governing the abundance of these fungi in alpine streams are likely the availability of suitable resources and the general ability of aquatic hyphomycetes to cope with the harsh alpine environment.

This chapter summarizes results of studies on aquatic hyphomycetes conducted between 1996 and 1998 in five contrasting streams of the Val Roseg flood plain, Switzerland (Gessner et al., 1998; Robinson & Gessner, 2000a; Robinson et al., 2000). Leaf and needle litter of alder (*Alnus viridis* (Chaix) Dc.) and larch (*Larix decidua* Mill.), the two most common woody plants in the upper Roseg catchment, were placed in streams in litter bags during autumn, retrieved periodically for up to 7.5 months, and analyzed for sporulating fungi (Gessner et al., 1998; Robinson et al., 2000). Fungal biomass was estimated with the ergosterol technique (Gessner & Newell, 2002). The objectives of theses studies were to assess whether aquatic hyphomycetes are present in a range of alpine stream types and whether they can colonize leaf litter to a similar extent as in woodland streams despite the generally harsh environmental conditions, including the paucity of organic substrates. The chapter begins with an account of fungal community structure on litter introduced to different stream types, presents information on colonization rates and biomass accumulation, describes successional patterns of fungi on the decomposing litter, and examines the effects of leaf-shredding invertebrates and litter type on fungal communities.

2. DIVERSITY AND ABUNDANCE

Fungi are a consistent component of the biota in all streams examined in the Val Roseg. More than 21 species were encountered, 16 of which could be identified (Table 1, Fig. 1). A minimum of 10 species was recorded in any one stream. More diverse communities have been described from temperate forest streams (sometimes >50 species; e.g., Shearer and Webster, 1985; Descals & Chauvet, 1992; Dubey et al., 1994; Garnett et al., 2000), but several subarctic streams in Sweden and on

Table 1. Aquatic hyphomycetes found on alder leaves (A) and larch needles (L) decomposing in coarse-mesh litter bags in five streams of the Val Roseg flood plain. MP =Main channel, proglacial reach; ML = Main channel, lake outlet; MF = Main channel, flood plain; X = floodplain channel fed by water of two or more sources (mixed channel); T = Tributary fed by hillslope ground water. Site refers to the number of streams in which a species was found. See Appendix III for authorities.

Taxon	MP		ML		MF		X		T		Sites	
	A	L	A	L	A	L	A	L	A	L	A	L
Alatospora acuminata	•	•	•	•	•			•	•	•	4	4
Anguillospora longissima	•	•		•							1	2
Articulospora tetracladia			•	•	•		•	•			3	2
Clavariopsis aquatica		•		•					•	•	1	3
Clavatospora longibrachiata	•			•	•		•	•	•	•	4	3
Flagellospora curvula	•		•	•	•	•	•	•	•	•	5	4
Heliscella stellata	•	•			•		•		•		4	1
Heliscus lugdunensis	•					•	•		•	•	3	2
Lemonniera aquatica	•	•	•	•	•	•	•	•	•	•	5	5
Lemonniera terrestris						•		•	•		1	2
Taeniospora gracilis		•		•			•	•	•	•	2	4
Tetrachaetum elegans									•		1	0
Tetracladium marchalianum										•	0	1
Tricladium curvisporum	•	•			•	•	•	•	•	•	4	4
Tricladium patulum									•		1	0
Tumularia aquatica									•		1	0
ND1	•				•						2	0
ND2	•										1	0
ND3			•				•				2	0
ND4	•										1	0
Unidentified sigmoid		•		•	•	•	•			•	2	4
Unidentified others			•		•		•				3	0
Number of species	11	8	6	9	10	6	11	8	13	10	20	14

Greenland, where organic matter was much more abundant than in the Val Roseg streams, had either a similar or lower fungal species richness. Furthermore, the average species richness on both types of litter examined in the Val Roseg streams, alder leaves and larch needles, was generally greater than 5 at least on one sampling date (Fig. 2). Given that in forested mountain streams species numbers on a single leaf have generally been reported to range from about 4 to 16 (e.g., Gönczöl, 1989; Bärlocher, 1992b; Suberkropp, 1992b; Dubey et al., 1994; Chauvet et al., 1997), fungal diversity in the alpine streams of the Val Roseg was not exceptionally low. Detailed taxonomic surveys involving isolation of species are likely to result in even higher numbers, since we encountered a number of rare forms whose identity could not be determined (Gessner et al., 1998) and since we limited our survey to the examination of leaf and needle litter, although analyses of foam and water samples often yield higher species numbers.

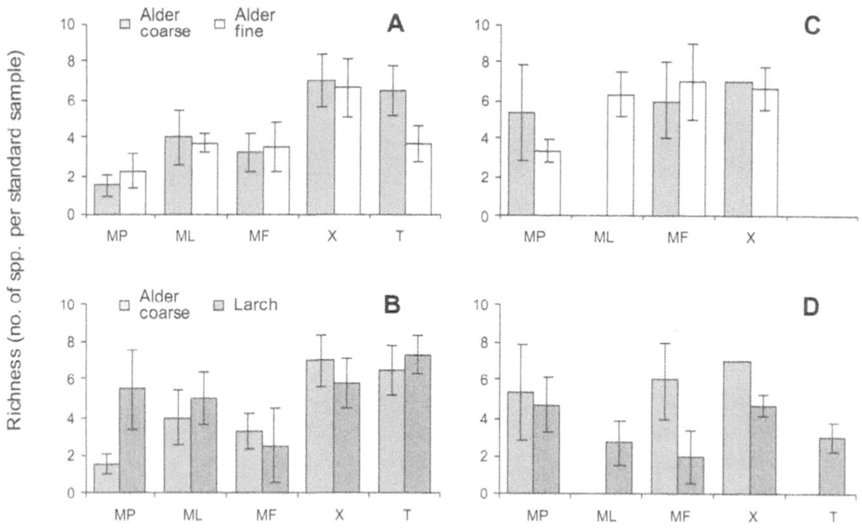

*Figure 2. Species richness of sporulating aquatic hyphomycetes on decaying alder (*Alnus viridis*) leaves and larch (*Larix decidua*) needles in five contrasting streams of the Val Roseg flood plain. Litter was placed in the streams in coarse-mesh and fine-mesh (alder only) bags, and retrieved in the fall after 6 (alder) or 11 (larch) weeks (A, B) and in the following spring after about 7 months (C, D). Error bars denote ±1 SD. MP = Main channel, proglacial reach (site 7); ML = Main channel, lake outlet (site 10); MF = Main channel, flood plain (site 60); X = floodplain channel fed by water of two or more sources (mixed channel, site 59); T = Tributary fed by hillslope ground water (site 22). (Based on data from Gessner et al., 1998; Robinson et al., 2000; and Robinson & Gessner, 2000a).*

Colonization of leaves by aquatic hyphomycetes was generally slow, as indicated by the fact that fungi did not show notable sporulation during the first 3 weeks following submergence of leaves and needles in the streams (Gessner et al., 1998; Robinson et al., 2000; Robinson & Gessner, 2000a). For comparison, in forest streams at lower elevation, aquatic hyphomycetes may start to sporulate within less than a week and maximum sporulation rates have been observed after only 2-4 weeks (e.g., Gessner et al., 1993; Hieber & Gessner, 2002). The delay in sporulation on leaves in the Val Roseg streams was only partly reflected in the dynamics of fungal biomass. Ergosterol concentrations increased significantly during the first three weeks in all streams, although levels still remained moderate during that initial period.

At later stages, however, ergosterol concentrations exceeded 350 μg/g leaf dry mass and in most streams even 500 μg/g (Fig. 3), corresponding to a fungal biomass of about 6-9% of the total litter dry mass (see conversion factors in Gessner et al., 1993; and Gessner & Newell, 2002). This biomass level is in the upper range of values observed in woodland streams (see compilations in Gessner, 1997; and Abelho, 2001). Similarly, in the majority (but not all) of the Roseg streams,

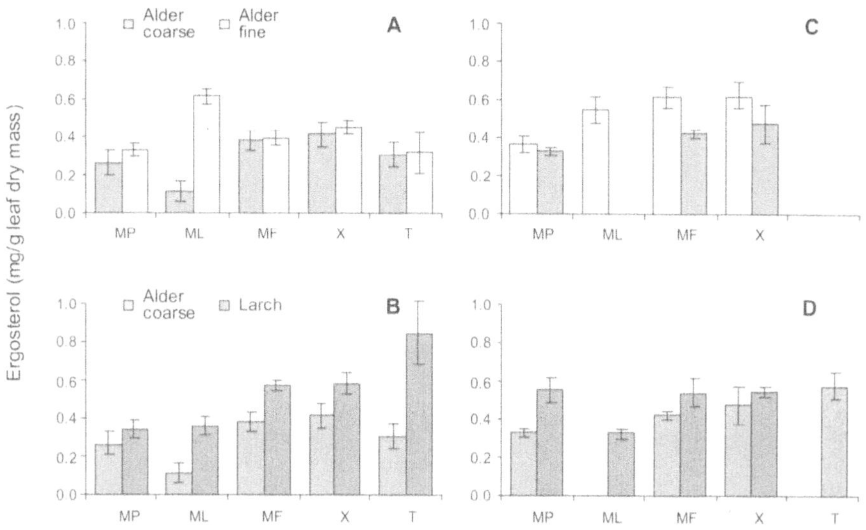

*Figure 3. Fungal biomass associated with decaying alder (*Alnus viridis*) leaves and larch (*Larix decidua*) needles in five contrasting streams of the Val Roseg flood plain (for code of stream types see legend of Fig. 2). Litter was placed in the streams in coarse-mesh and fine-mesh (alder only) bags, and retrieved in the fall after 6 (alder) and 11 (larch) weeks (A, B) and in the following spring after about 7 months (C, D). One μg ergosterol is equivalent to about 180 mg fungal dry mass (Gessner & Chauvet, 1993; Gessner & Newell, 2002). Error bars denote ±1 SD. (Based on data from Gessner et al., 1998; Robinson et al., 2000; and Robinson & Gessner, 2000a).*

sporulation rates eventually reached or exceeded levels of 100 conidia mg^{-1} leaf dry mass d^{-1} (Fig. 4), which are not too greatly depressed compared to rates observed in forest streams (Gessner, 1997). The sporulation rates observed on alder leaves in the tributary (up to 2000 conidia mg^{-1} d^{-1}) are even rather high compared to literature values.

Leaf packs in streams constitute discrete resource patches whose colonization by aquatic hyphomycetes is initiated by the attachment and germination of conidia (Read et al., 1992). That neither fungal biomass nor sporulation activity was detected in alder leaves before submergence (Gessner et al., 1998) indicates, therefore, that leaves had not been colonized by fungi before they were placed in streams. Similarly, no sporulation was initially observed on larch needles during laboratory incubations for up to 5 days, although ergosterol levels were about 100 µg/g needle dry mass (Robinson et al., 2000), probably as a result of colonization by terrestrial fungi before the fully brown needles were collected from the ground. These findings suggest that aquatic hyphomycetes are capable of completing their full asexual life cycle – comprising attachment of conidia, germination, mycelial growth, and release of conidia – in the streams of the Val Roseg flood plain.

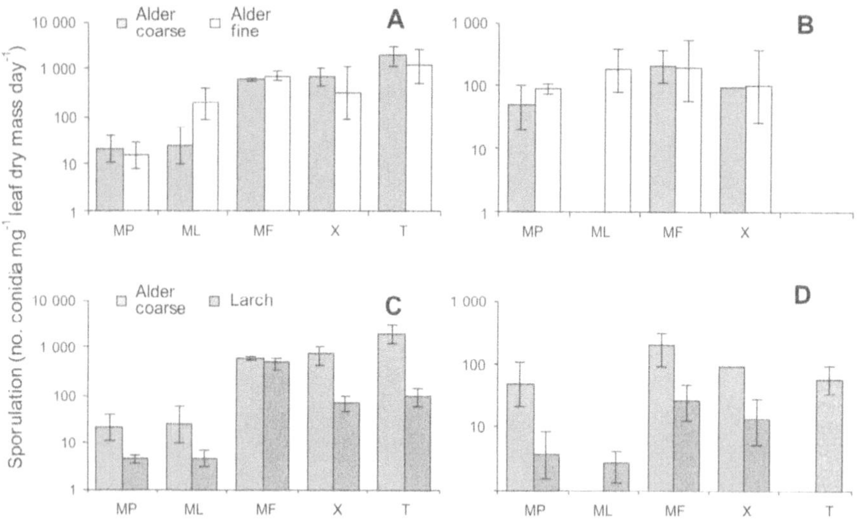

*Figure 4. Sporulation of aquatic hyphomycetes on alder (*Alnus viridis*) and larch (*Larix decidua*) leaves decaying in five contrasting streams of the Val Roseg flood plain (for code of stream types see legend of Fig. 2). Litter was placed in the streams in coarse-mesh and fine-mesh (alder only) bags, retrieved in the fall after 6 (alder) or 11 (larch) weeks (A, B) and in the following spring after about 7 months (C, D), and incubated for 5 days at 4 °C in filtered stream water to induce sporulation. Error bars denote ±1 SD. (Based on data from Gessner et al., 1998; Robinson et al., 2000; and Robinson & Gessner, 2000a).*

Alpine streams thus offer suitable habitat for aquatic hyphomycetes provided an appropriate source of organic matter is accessible. The inoculum potential is sufficient and, with some exceptions discussed below, growth and reproduction on both leaf and needle litter appear to be only moderately constrained. Discharge-related physical constraints such as periodically high flow, associated scour, and channel instability are most prominent during summer as a result of diel freeze-thaw cycles. Conditions tend to be less severe in the fall, early winter and spring, when our studies were carried out (Malard et al., 2000, see Chapter 16). Although temperature is very low in most alpine streams year round (Ward, 1994; Milner et al., 2001, see Chapter 5), aquatic hyphomycetes are well-adapted to cold water

(Koske & Duncan, 1974; Suberkropp, 1984; Graça & Ferreira, 1995) and may even survive temperatures well below zero in the short term (Sridhar & Bärlocher, 1993, 1994). Adaptation to cold water is in fact critical to aquatic hyphomycetes, because even in temperate forest streams temperatures often drop to near freezing or below in the fall and early winter, when aquatic hyphomycetes show their best development in response to autumn leaf fall (Chauvet, 1991; Bärlocher, 1992b; Suberkropp, 1992b; Gessner & van Ryckegem, 2003).

3. SUCCESSIONAL PATTERNS ON DECOMPOSING LITTER

Aquatic hyphomycete communities have been shown to follow broadly predictable successional patterns on decomposing leaves in streams (see overview in Bärlocher, 1992b). Gessner et al. (1993) and Gönczöl (1989), for example, described patterns on alder (*A. glutinosa* (L.) Gaertn.) leaves in mountain streams below tree line. These successional trajectories are characterized by shifts in the relative abundance of conidia released by different species rather than by species replacements. In the study by Gessner et al. (1993), *Flagellospora curvula* was the dominant early colonizer, and four additional species, including *L. aquatica*, also occurred at higher frequencies at this stage. *H. stellata, C. longibrachiata* and, at lower frequencies, *A. acuminata* characterized later colonization stages.

Similar successional trends were observed on alder leaves decomposing in the alpine streams of the Val Roseg (Gessner et al., 1998). Species richness was reduced and changes in species composition were delayed, but *F. curvula* was the dominant sporulating fungus. It colonized the leaves in an early stage and tended to decline in abundance at later sampling dates (Gessner et al., 1998; Gessner & Robinson, 2000). *L. aquatica* also occurred at early stages, but showed no consistent decline as leaf decomposition proceeded. In contrast, both *C. longibrachiata* and, even more so *H. stellata*, were consistently restricted to late sampling dates. Finally, the abundance of *A. acuminata* also tended to increase at advanced decomposition stages, although this trend was less pronounced than for *C. longibrachiata* and *H. stellata*, just as has been found by Gessner et al. (1993). Successional patterns on larch needles were less pronounced, partly because *F. curvula* dominated the fungal communities even more strongly than on alder leaves, accounting for >75% of all released conidia even 7 months after needles were submerged in the streams (Robinson et al., 2000). As a consequence, other species appeared more sporadically. However, neither *H. stellata* nor *C. longibrachiata* or *A. acuminata* ever occurred on larch needles at early sampling dates.

The species complements determining these successional patterns and the overall species richness in the Roseg streams broadly match those of conidial assemblages found in foam and water samples and on glass slides in streams draining boreal forests near the arctic circle in northern Sweden (Nilsson, 1964; Müller-Haeckel, 1977; Müller-Haeckel & Marvanová, 1977, 1979; Marvanová & Müller-Haeckel, 1980) and in tundra streams along the coast of southern Greenland (Engblom et al., 1986). *F. curvula* has been found as a dominant species in many, although not all, of these streams. Conditions in the subarctic streams are more benign than in most of the alpine streams of the Val Roseg. Those streams lack glacial meltwater influence (Engblom et al. 1986), experience water temperatures near (Engblom et al., 1986) or even above 10 °C (Müller-Haeckel & Marvanová, 1977) in the summer, albeit close to zero for six months in the winter (Müller-Haeckel & Marvanová, 1979), and the catchments are mostly covered by vegetation (Müller-Haeckel & Marvanová, 1979; Engblom et al., 1986) with some streams lined by deciduous trees or shrubs (Müller-Haeckel & Marvanová, 1977, 1979; Engblom et al., 1986). Similar species of aquatic hyphomycetes were also observed in a mountain stream at lower elevation (850 m a.s.l.) further west in the Alps (del Frate, 1983; del Frate & Caretta, 1983).

4. EFFECTS OF HABITAT CHARACTERISTICS

Five stream types were selected in the Val Roseg flood plain to study aquatic hyphomycete colonization of leaf and needle litter along a putative harshness gradient (Robinson & Gessner, 2000b; Robinson et al., 2000). Sites included the main channel in the proglacial reach (site 7), the outlet of a glacial lake (site 10), the main channel of the Roseg River in the floodplain reach 1500 m downstream from the confluence of the proglacial reach with the lake outlet (site 60), a channel carrying water of mixed origin (mixed channel, site 59), and a tributary (site 22) (see Chapter 1 and Appendix I). The main channel in the proglacial reach in many respects represents the harsh end of the gradient, whereas the tributary was expected to be the most benign environment. Characteristics of these stream types and the Roseg catchment are described in detail in Chapters 1 and 4.

Four variables were used to describe the relation of aquatic hyphomycete communities to the environmental gradient across the five selected streams in the Val Roseg flood plain. These included species richness, species composition, sporulation rate, and biomass. In the fall, after 6 weeks of colonization, species richness on alder leaves was lowest in the proglacial reach, higher in the lake outlet and the main channel, and highest in the mixed channel and tributary, and thus followed the anticipated harshness gradient (Fig. 2a). This pattern was not seen on larch needles, however, where species richness was lowest in the main channel and tended to increased towards both the benign and harsh end of the gradient (Fig. 2b). It is not clear why especially low species numbers were recorded on larch needles in the main channel, although, in the fall, this observation may have been partly due to a delay in colonization and decomposition of needle compared to the other sites. Needles had lost only 16% of their initial dry mass after 6 weeks (compared to 32-61% in the other streams; Gessner et al., 1998; Robinson et al., 1998, see Chapter 13) and the abundant sporulation of *F. curvula* (Fig. 4) indicated, moreover, that fungal colonization was in an early stage (Gessner et al., 1993, 1998). The similar species richness in the four remaining streams may reflect the longer residence in streams of the recalcitrant needles compared to alder leaves (11 versus 6 weeks at the time of sampling), which could have allowed a greater number of species to colonize the litter also under less favorable habitat conditions. In line with this idea, the pattern of increasing species richness in more benign habitats had disappeared in spring, seven months after placement of the litter in streams (Fig. 1c, d).

The proglacial stream was the only site at which the species composition (Table 1) and successional pattern of the aquatic hyphomycete community differed noticeably from the general pattern described above. *Lemonniera aquatica* and, on needles only, an unidentified sigmoid species were the early colonizers, whereas *Anguillospora longissima* and *Taeniospora gracilis* dominated late colonization stages on larch needles, although not on alder leaves. *F. curvula* was not recorded from needles in the proglacial stream (Robinson et al., 2000). On alder leaves in coarse-mesh bags, this species ranked sixth only, accounting for 1% of all conidia. Only on alder leaves in fine-mesh bags did *F. curvula* co-dominate the community, together with *L. aquatica* and *H. stellata* (Gessner et al., 1998).

Sporulation rates on alder leaves in the fall showed a similar pattern as fungal species richness, with an increase in more benign habitat conditions (Fig. 4a). Sporulation rates on larch needles also followed this trend broadly, with higher rates in the mixed channel and tributary, and very low rates (<5 conidia mg^{-1} needle dry mass day^{-1}) in both the proglacial reach and the lake outlet (Fig. 4b). Although lake outlets are normally considered comparatively benign stream habitats (Hieber et al., 2002), they may in fact be rather unfavorable environments for aquatic hyphomycetes. Colonization of leaves by these fungi relies on propagules from upstream, which supposedly are as scarce in glacial lake outlets as in proglacial stream reaches. This idea is consistent with observations by Müller-Haeckel & Marvanová (1979) that among a total of eight sites in a subarctic stream system, the lowest numbers of aquatic hyphomycete conidia were recorded in the two lake outlets, although these outlets were lined by dense riparian tree vegetation. Highest sporulation rates on larch needles in the Val Roseg streams were found in the main channel. This is in contrast to species richness and breaks the trend of increasing sporulation rates along the harshness gradient from the proglacial reach to the tributary. As discussed above, this finding may be related to the unusually early colonization and decomposition stage of the needles collected in the main channel after six weeks. However, it must also be borne in mind that sporulation of aquatic hyphomycetes on decomposing litter is highly dynamic, begetting the risk of confounding temporal dynamics with spatial patterns.

In the spring, about seven months after submergence of litter in the streams, sporulation rates of aquatic hyphomycetes had generally declined (Fig. 4c, d), which is typical of decomposing leaves in streams (Gessner, 1997; Gessner et al., 1997; Suberkropp, 1998). The exception was alder leaves in the proglacial reach, where rates had actually increased. As a result, differences among streams had leveled out, as was seen with species richness. The low sporulation on larch needles in the proglacial stream and lake outlet nevertheless was still clearly apparent.

Ergosterol concentrations, a measure of fungal biomass, reflected the stream harshness gradient only in the fall on larch needles (Fig. 3a), whereas differences on alder leaves were generally small among streams, litter types and seasons (Fig. 3b-d). An exception was the alder leaves in coarse-mesh litter bags in the lake outlet where rapid consumption by leaf-shredding caddisfly larvae prevented any substantial fungal colonization (Robinson et al., 1998, see below and Chapter 13). Concentrations of ergosterol on litter decaying in the proglacial reach tended to be lower than at the other sites, but even here, <300 m downstream of the glacier snout and at temperatures continually near freezing, fungal biomass attained >4% of litter dry mass. The seemingly weak response of fungal biomass to differences in environmental harshness may be related to the fact that under more favorable conditions aquatic hyphomycetes channelled a much greater part of the produced biomass into the production of conidia, which are subsequently released into the flowing water, thus roughly maintaining the same biomass concentration in leaves. This conclusion is supported by evidence from forest streams, indicating that aquatic hyphomycetes respond to differences in environmental conditions (e.g., nutrient availability and heavy metal contamination) most strongly by adjusting sporulation rates while maintaining mycelial biomass at a much more constant level (e.g., Suberkropp, 1998b; Sridhar & Bärlocher, 2000; Grattan & Suberkropp, 2001; Sridhar et al., 2001).

5. EFFECTS OF DETRITIVORE ACCESS TO LEAVES

Alder leaves were exposed in the Val Roseg streams in both coarse-mesh and fine-mesh bags to assess the effects of leaf-shredding detritivores accessing the litter. The largest differences in the colonization of coarse- and fine-mesh bags by invertebrates were observed in the lake outlet and tributary. Leaves in coarse-mesh bags in the outlet attracted a limnephilid caddisfly, *Acrophylax zerberus*, in high densities, resulting in a maximum biomass of >100 mg g^{-1} litter dry mass (Robinson et al., 1998, see Chapter 13), whereas nemourid stoneflies, *Nemoura* spp. and *Protonemura* spp. , were the dominant leaf-shredding detritivores in the tributary (Robinson & Gessner, 2000a, see Chapter 13).

Access of invertebrates to leaves had strong effects on the fungal communities associated with alder leaves in these two streams. In the outlet, *A. zerberus* fed so rapidly on leaves that fungi failed to build up a significant biomass (Gessner et al., 1998). Ergosterol concentrations remained below 100 µg/g litter dry mass, 5 times lower than in fine-mesh bags in the lake outlet and considerably lower than in other streams of the Val Roseg (Fig. 3a). Sporulation rates also were greatly depressed (on average 5 times lower in coarse- than in fine-mesh bags; Fig. 4a), but no reduction in species richness was noted during the period when leaves in both coarse- and fine-mesh bags remained in the stream (Fig. 2a; Gessner et al., 1998). However, because leaves in fine-mesh bags persisted in the stream for a much longer period, the total number of species observed in fine-mesh bags was eventually 2.2 fold greater than in coarse-mesh bags (13 versus 6 species).

The increase in total richness when invertebrates were absent contrasts with our observations in the tributary, where fungal species diversity was greatly reduced when invertebrates were prevented from colonizing the leaves (Fig. 2a, b; Robinson & Gessner, 2000a). Sporulation rates were slightly higher on leaves in coarse-mesh bags, although not significantly so. Most strikingly, however, only 4 species of aquatic hyphomycetes were recorded in fine-mesh bags, compared to 11 species in the coarse-mesh bags – even though decomposition was less advanced in fine than in coarse-mesh bags and ergosterol levels were very similar (Robinson & Gessner,

2000a). These findings hint that the fungal community was diversified and activity possibly stimulated by the moderate feeding pressure (disturbance) exerted by the detritivores in the tributary, whereas intense feeding, as in the outlet, results in a near-total suppression of fungal activity. Consequently, the net outcome of interactions between aquatic hyphomycetes and leaf-shredding invertebrates may be either synergistic or antagonistic, depending on the invertebrate's feeding pressure relative to the rates of fungal growth and sporulation.

6. EFFECTS OF LITTER TYPE

Alder leaves provide an excellent resource for aquatic hyphomycetes (e.g., Ingold, 1942; Gessner & Chauvet, 1994) and have been widely used to examine community dynamics of these fungi on decomposing leaves in streams (e.g., Chamier, 1987; Gönczöl, 1989; Gessner et al., 1993; Fabre & Chauvet, 1998). Larch needles have been studied much less, but although more recalcitrant than alder leaves, they clearly can also serve as a resource for aquatic hyphomycetes in woodland streams (Bärlocher, 1982). As discussed above, this conclusion holds true even under the harsh environmental conditions experienced in the streams of the Val Roseg, including a reach just downstream of a glacier with temperatures permanently near freezing (Robinson et al., 2000).

Larch needles are colonized more slowly than alder leaves in the Val Roseg streams, but neither species diversity (Fig. 2b, d) nor composition of the fungal communities differed greatly in our studies (Table 1; Gessner et al., 1998; Robinson & Gessner, 2000b; Robinson et al., 2000). Exceptions include the presence of a fungus with sigmoid conidia that was dominant during early colonization of larch needles, but was not found on alder leaves, and different dominants in later colonization stages at the proglacial site compared to the other four study streams. Additional differences, which were manifest across all streams, were the significantly lower sporulation rates (Fig. 4b, d) and consistently higher fungal biomass on needle litter (Fig. 3b, d). The seemingly contradictory pattern of ergosterol and sporulation rates on the two types of litter probably reflects differences in the allocation of assimilated resources to mycelial expansion and reproduction on the more recalcitrant larch needles compared to the more easily degradable alder leaves (see above). Neither this divergence nor the other observed difference should be overrated, however, because our experiments with alder and larch litter – except for the one in the tributary – were conducted in different years, so that effects of litter type and interannual variations cannot be neatly separated. For example, it is not clear whether the large effect of the caddisfly, *A. zerberus*, on fungi associated with alder leaves in the lake outlet was lacking on larch needles (where fungi developed much better than on alder leaves) because needles were less palatable or because *A. zerberus* was less abundant when the experiment with needles was conducted. Quantitative samples of benthic macroinvertebrates in both years suggest, however, that densities of *A. zerberus* (range of 16-110 ind m^{-2}) were not markedly different between the two study years (Burgherr, 2000).

7. CONCLUSION

In conclusion, aquatic hyphomycetes are an integral component of the biota of alpine streams in the Val Roseg, including a reach immediately downstream of the glacier margin. Even glacial streams thus provide suitable habitat for aquatic hyphomycetes, with the abundance of these fungi probably being controlled more strongly by limited access to suitable organic resources than by the physically harsh setting of streams in alpine environments. Completion of the life cycle of aquatic hyphomycetes depends critically on the availability of allochthonous litter. This type of resource is exceedingly scarce in alpine environments, but litter is rapidly colonized and exploited by aquatic hyphomycetes once it does become available. This is true even for relatively recalcitrant conifer needles. The patterns of fungal colonization and exploitation do not differ fundamentally from those known from forest streams at lower elevations. This observation suggests that as glaciers retreat and tree lines are shifted upward as a result of the milder climate anticipated by global change scenarios (e.g., Beniston et al., 1994), aquatic hyphomycetes will rapidly assume their roles as litter decomposers and food for detritivores as soon as notable amounts of leaves and needles are supplied to streams.

8. REFERENCES

Abelho, M. (2001) From litterfall to breakdown in streams: a review. *The ScientificWorld* 1, 656-680.

Bärlocher, F. (1982). Conidium production from leaves and needles in four streams. *Canadian Journal of Botany*, *60*, 1487-1494.

Bärlocher, F. (1985). The role of fungi in the nutrition of stream invertebrates. *Botanical Journal of the Linnean Society*, *91*, 83-94.

Bärlocher, F. (Ed). (1992a). *The ecology of aquatic hyphomycetes*, *Ecological Studies 94*. Berlin: Springer.

Bärlocher, F. (1992b). Community organization. In F. Bärlocher (Ed.), *The ecology of aquatic hyphomycetes, Ecological Studies 94* (pp. 38-76). Berlin: Springer.

Benfield, E. F. (1997). Comparison of litterfall inputs to streams. *Journal of the North American Benthological Society*, *16*, 104-108.

Beniston, M. Rebetez, M., Giorgi, F., & Marinucci, M. R. (1994). An analysis of regional climate change in Switzerland. *Theoretical and Applied Climatology*, *49*, 135-159.

Burgherr, P. (2000). Spatio-temporal community patterns of lotic zoobenthos across habitat gradients in an alpine glacial stream ecosystem. PhD Thesis, Swiss Federal Institute of Science and Technology (ETH), Zurich, Switzerland.

Chamier, A.-C. (1987). Effect of pH on microbial degradation of leaf litter in seven streams of the English Lake District. Oecologia, *71*, 491-500.

Chapin, F. S. III, & Körner, C. (1995). *Arctic and alpine biodiversity: patterns, causes, and ecosystem consequences*. New York: Springer.

Chauvet, E. (1990). Hyphomycètes aquatiques du sud-ouest de la France. *Gaussenia*, *6*, 3-31.

Chauvet, E. (1991). Aquatic hyphomycete distribution in south-western France. *Journal of Biogeography*, *18*, 699-706.

Chauvet, E., Fabre, E., Elósegui, A., & Pozo, J. (1997). The impact of eucalypt on the leaf-associated aquatic hyphomycetes in Spanish streams. *Canadian Journal of Botany*, *75*, 880-887.

Chauvet, E., & Suberkropp, K. (1998). Temperature and sporulation of aquatic hyphomycetes. *Applied and Environmental Microbiology*, *64*, 1522-1525.

Del Frate, G. (1983). Ifomiceti acquatici di corsi d'acqua di pianura (Pavia) e di montagna (Alpi Pennine e Lepontine). *Archivio botanico e biogeografico*, *59*, 169-174.

Del Frate, G., & Caretta, G. 1983. Aquatic Hyphomycetes of a mountain stream in Valsesia (Piemonte). *Hydrobiologia*, *102*, 69-71.

Descals, E., & Chauvet, E. (1992). Diversité des champignons Ingoldiens de quelques rivières du sud-ouest de la France. *Nova Hedwigia, 54*, 83-96.
Dubey, T, Stephenson, S. L., & Edwards, P. J. (1994). Effect of pH on the distribution and occurrence of aquatic fungi in six West Virginia mountain streams. *Journal of Environmental Quality, 23*, 1271-1279.
Engblom, E., Lingdell, P.-E., Marvanová, L., & Müller-Haeckel, A. (1986). Foam spora in running waters of southern Greenland. *Polar Research, 4*, 47-51.
Fabre, E., & Chauvet E. (1998). Leaf breakdown along an altitudinal stream gradient. *Archiv für Hydrobiologie, 141*,167-179.
Garnett, H., Bärlocher, F., & Giberson, D. (2000). Aquatic hyphomycetes in Catamaran Brook: colonization dynamics, seasonal patterns, and logging effects. *Mycologia, 92*, 29-41.
Gessner, M. O. (1997). Fungal biomass, production and sporulation associated with particulate organic matter in streams. *Limnetica, 13*, 33-44.
Gessner, M. O., & Chauvet, E. (1993). Ergosterol-to-biomass conversion factors for aquatic hyphomycetes. *Applied and Environmental Microbiology, 59*, 502-507.
Gessner, M. O., Thomas, M., Jean-Louis, A.-M., & Chauvet, E. (1993). Stable successional patterns of aquatic hyphomycetes on leaves decaying in a summer cool stream. *Mycological Research, 97*, 163-172.
Gessner, M. O. and Chauvet, E. (1994). Importance of stream microfungi in controlling breakdown rates of leaves. *Ecology, 75*, 1807-1817.
Gessner, M. O., Suberkropp K., & Chauvet, E. (1997). Decomposition of plant litter by fungi in marine and freshwater ecosystems. In D. T. Wicklow & Söderström, B. (Eds.), *The Mycota, Vol. 4, Environmental and microbial relationships* (pp. 303-322). Berlin: Springer.
Gessner, M. O., Dobson, M., & Chauvet, E. (1999). A perspective on leaf litter breakdown in streams. *Oikos, 85*, 377-384.
Gessner, M. O., Robinson, C. T., & Ward J. V. (1998). Leaf breakdown in streams of an alpine glacial floodplain: Dynamics of fungi and nutrients. *Journal of the North American Benthological Society, 17*, 403-419.
Gessner, M. O., & Newell, S. Y. (2002). Biomass, growth rate, and production of filamentous fungi in plant litter. In: Hurst C. J., Crawford, G., Knudsen, R. L., McInerney, M., & Stetzenbach, L. D. (Eds.), *Manual of Environmental Microbiology*, 2nd edition (pp. 390-408). Washington D. C.: American Society for Microbiology Press.
Gessner, M. O., & van Ryckegem, G. (2003). Water fungi as decomposers in freshwater ecosystems. In G. Bitton (Ed.) *Encyclopedia of Environmental Microbiology*. New York: John Wiley & Sons. (corrected online version)
Gessner, M. O., Bärlocher, F., & Chauvet, E. (2003). Qualitative and quantitative analyses of aquatic hyphomycetes in streams. In Tsui, C. K. M., Hyde, K. D., & Ho, W. H. (Eds.), *Freshwater fungi: a practical approach* (in press). Hong Kong: Fungal Diversity Press.
Gönczöl, J. (1989). Longitudinal distribution patterns of aquatic hyphomycetes in a mountain stream in Hungary – experiments with leaf packs. *Nova Hedwigia*, 48, 391-404.
Graça, M. A. S. (2001). The role of invertebrates on leaf litter decomposition in streams – A review. *International Review of Hydrobiology, 86*, 383-393.
Graça, M. A. S., & Ferreira, R. C. F. (1995). The ability of selected aquatic hyphomycetes and terrestrial fungi to decompose leaves in freshwater. *Sydowia, 47*, 167-179.
Grattan, R. M. II, & Suberkropp, K. (2001). Effects of nutrient enrichment on yellow poplar leaf decomposition and fungal activity in streams. *Journal of the North American Benthological Society, 20*, 33-43.
Hieber, M., & Gessner, M. O. (2002). Contribution of stream detrivores, fungi, and bacteria to leaf breakdown based on biomass estimates. *Ecology, 83*, 1026-1038.
Hieber, M, Robinson, C. T., Uehlinger, U., & Ward, J. V. 2002. Are alpine lake outlets less harsh than other alpine streams? *Archiv für Hydrobiologie, 154*, 199-223.
Ingold, C. T. (1942). Aquatic Hyphomycetes of decaying alder leaves. *Transactions of the British Mycological Society, 25*, 339-417.
Ingold, C. T. (1975). An illustrated guide to aquatic and waterborne hyphomycetes (Fungi Imperfecti) with notes on their biology. *Freshwater Biological Association, Scientific Publication, 30*, 1-96.
Körner, C. (1999). Alpine plant life. Berlin: Springer.

Koske, R. E., & Duncan, I. W. (1974). Temperature effects on growth sporulation and germination of some aquatic hyphomycetes. *Canadian Journal of Botany*, *52*,1387-1391.
Malard, F., Tockner, K., Ward, J. V. (2000). Physico-chemical heterogeneity in a glacial riverscape. Landscape Ecology, 15, 679-695.
Marvanová, L., & Müller-Haeckel, A. (1980). Water-borne spores in foam in a subarctic stream system in Sweden. *Sydowia*, *33*, 210-220.
Milner, A. M., Brittain, J. E., Castella, E., & Petts, G. E. (2001). Trends of macroinvertebrate community structure in glacier-fed rivers in relation to environmental conditions: a synthesis. *Freshwater Biology*, *47*, 1833-1847.
Müller-Haeckel, A. (1977). Annual production of Hyphomycetes conidia in a subarctic stream. *Oikos*, 29, 396-397.
Müller-Haeckel, A., &; Marvanova, L. (1977). Konidienproduktion und -kolonisation von Süsswasser-Hyphomyceten im Kaltisjokk (Lappland). *Botaniska Notiser*, *129*, 405-409.
Müller-Haeckel, A., & Marvanová, L. (1979). Periodicity of aquatic hyphomycetes in the Subarctic. *Transactions of the British Mycological Society*, *73*, 109-116.
Nilsson, S. (1964). Freshwater hyphomycetes: taxonomy, morphology and ecology. *Symbolae Botanicae Upsaliensis*, *18*, 5-130.
Read, S. J., S. T. Moss, & E .B. G. Jones (1992). Attachment and germination of conidia. In F. Bärlocher (Ed.), *The ecology of aquatic hyphomycetes, Ecological Studies 94* (pp. 135-151). Berlin: Springer.
Robinson, C. T., Gessner, M. O., & Ward J. V. (1998). Leaf breakdown and associated macroinvertebrates in alpine glacial streams. *Freshwater Biology*, *40*, 215-228.
Robinson, C. T., & Gessner, M. O. (2000a). Nutrient addition accelerates leaf breakdown in an alpine springbrook. *Oecologia*, 122, 258-263.
Robinson, C. T., & Gessner, M. O. (2000b). Leaf breakdown in an alpine spring brook. *Verhandlungen der Internationalen Vereinigung für theoretische und angewandte Limnologie*, *27*, 744-747.
Robinson, C. T., Gessner, M. O., Callies, K. A., Jolidon, C., & Ward, J. V. (2000). Larch needle breakdown in contrasting streams of an alpine glacial floodplain. *Journal of the North American Benthological Society*, *19*, 250-262.
Shearer, C. A., & Webster, J. (1985). Aquatic hyphomycete communities in the River Teign. I. Longitudinal distribution patterns. *Transactions of the British Mycological Society*, *84*, 489-501.
Sridhar, K. R., & Bärlocher, F. (1993). Seasonal changes in microbial colonization of fresh and dried leaves. *Archiv für Hydrobiologie*, *128*, 1-12.
Sridhar, K. R., & Bärlocher, F. (1994). Viability of aquatic hyphomycete conidia in foam. *Canadian Journal of Botany*, *72*, 106-110.
Sridhar, K. R., & Bärlocher, F. (2000). Initial colonization, nutrient supply, and fungal activity on leaves decaying in streams. *Applied and Environmental Microbiology*, *66*, 1114-1119.
Sridhar, K. R., Krauss, G., Bärlocher, F., Raviraja, N. S., Wennrich R, Baumbach R, Krauss G. J. (2001). Decomposition of alder leaves in two heavy metal-polluted streams in central Germany. *Aquatic Microbial Ecology*, *26*, 73-80.
Suberkropp, K. 1984. Effect of temperature on seasonal occurrence of aquatic hyphomycetes. *Transactions of the British Mycological Society*, *82*, 53-62.
Suberkropp, K. (1992a). Interactions with invertebrates. In F. Bärlocher (Ed.), *The ecology of aquatic hyphomycetes, Ecological Studies 94* (pp. 118-133). Berlin: Springer.
Suberkropp, K. (1992b). Aquatic hyphomycete communities. In G. C. Carroll & Wicklow, D. T. (Eds.), *The fungal community. Its organization and role in the ecosystem*, 2nd edition (pp. 729-747). New York: Marcel Dekker.
Suberkropp, K. (1998a). Microorganisms and organic matter decomposition. In R. J. Naiman & Bilby, R. E. (Eds.), *River ecology and management: Lessons from the Pacific Coastal Ecoregion* (pp. 120-143). New York: Springer.
Suberkropp, K. (1998b). Effect of dissolved nutrients on two aquatic hyphomycetes growing on leaf litter. *Mycological Research*, *102*, 998-1002.
Suberkropp, K. (2001). Fungal growth, production, and sporulation during leaf decomposition in two streams. *Applied and Environmental Microbiology*, *67*, 5063-5068.
Suberkropp, K., Gessner, M.O., & Chauvet, E. (1993). Comparison of ATP and ergosterol as indicators of fungal biomass associated with leaf litter in streams. *Applied and Environmental Microbiology*, *59*, 3367-3372.
Ward, J. V. (1994). Ecology of alpine streams. *Freshwater Biology*, *32*, 277-294.

Webster, J., & Descals, E. (1981). Morphology, distribution, and ecology of conidial fungi in freshwater habitats. In G. T. Cole & Kendrick, B. (Eds.), *Biology of Conidial Fungi* (pp. 295-355). New York: Academic Press.

Zah, R., & Uehlinger, U. (2001). Particulate organic matter inputs to a glacial stream ecosystem in the Swiss Alps. *Freshwater Biology*, *46*, 1597-1608.

CHAPTER 9

H.R. BÜRGI, P. BURGHERR & U. UEHLINGER

AQUATIC FLORA

1. INTRODUCTION

The ecology of alpine streams, in particular glacier-fed systems, has attracted major attention in the past few years (e.g., Brittain & Milner, 2001; Castella et al., 2001; Milner et al., 2001; Milner & Petts, 1994; Robinson et al., 2001; Ward, 1994). There exists, however, surprisingly little information on algal communities in alpine and glacial streams (e.g., Elster & Svoboda, 1996; Hieber et al., 2001; Kawecka, 1980; Kownacki et al., 1997; Vavilova & Lewis, 1999). Alpine streams, especially those fed by glaciers, are harsh environments for aquatic autotrophs. Light availability is reduced by an extended period of snow cover and attenuation by glacial flour; essential nutrients typically occur in low concentrations; low temperatures eliminate all but cold-adapted taxa; suspended sediment and ice action scour surfaces; and dramatic flow fluctuations result in desiccation and unstable substratum. Rooted macrophytes are absent or poorly represented. Allochthonous organic matter input, so important in forested headwaters, is limited in glacial-fed streams (see Chapter 12; Milner & Petts, 1994; Ward, 1994). The major energy base of alpine streams is normally restricted to benthic algae and mosses. Although mosses may be abundant in some alpine streams (Ward, 1994), they were not encountered in the glacial flood plain of the Val Roseg, even in groundwater sources. This chapter therefore, deals exclusively with benthic algae.

Following a brief description of sampling and analytical procedures, the taxonomic composition of benthic algae in the Roseg River is presented. The spatial and temporal distribution patterns of common taxa and chlorophyll *a* are described, first for sites along the main channel then for five other channel types within the floodplain reach.

Algae were collected during four seasons from 5 reaches along the main channel (16 sites) and from the following 5 channel types (total 11 sites) in the floodplain reach: (1) tributaries entering the flood plain from the hillslope, (2) groundwater channels originating as upwelling groundwater sources and connected downstream to the main channel, (3) mixed channels fed by water from 2 or more sources, (4) intermittently-connected channels occasionally upstream-connected to the main channel, and (5) side channels with both upstream and downstream connections to the main channel. Habitat characteristics and spatial distribution of these channel types are described in Chapter 4. At each site and date 5 to 10 rocks (B-diameter or breadth 3 –12 cm) were randomly collected and stored at –22°C. In addition,

J.V. Ward & U. Uehlinger (eds). Ecology of a Glacial Flood Plain, 139-151.
© 2003 *Kluwer Academic Publishers. Printed in the Netherlands.*

qualitative algal samples were taken from larger rocks (mean diameter >30 cm). Epilithon was removed using a wire brush and a wash bottle containing water. Aliquots of the resulting suspension were used for the determination of chlorophyll *a*. A subsample was fixed with Formalin: five ml of the suspension were transferred to Utermöhl settling chambers and the abundance of taxa was estimated under an inverted microscope. This estimated abundance (EA) was ranked in 6 classes (Table 1) and the relative abundance (RA) was calculated as $RA = 2^{EA} - 1$.

Table 1. Ranking of the estimated abundance (EA) and relative abundance (RA)

EA		RA (%)
0	not observed in the 5ml chamber	0
1	single specimens in the chamber	1
2	rare	3
3	medium	7
4	abundant	15
5	very abundant, dominant	31

Chlorophyll *a* was determined according to Meyns et al. (1994) and areal-based concentrations (mg chlorophyll *a* m^{-2}) were calculated as described by Uehlinger (1991).

2. TAXONOMIC COMPOSITION

More than 60 algal taxa (Appendix IV) were recorded in the main channel of the Roseg River and the diverse floodplain channels. The major components were diatoms and cyanobacteria (Figs. 1 and 2) and the chrysophyte *Hydrurus foetidus* (Fig. 1C). *Hydrurus foetidus* is a characteristic macroalga of high mountain streams

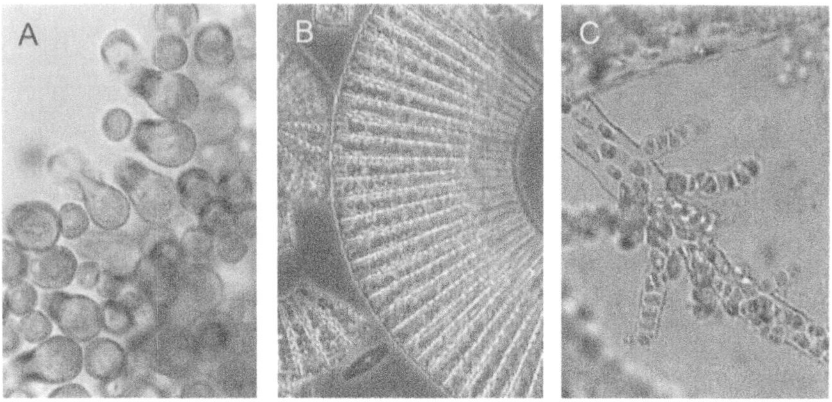

Figure 1. Microphotographs of algae occurring in the Val Roseg. A) The crust-forming cyanobacterium Chamaesiphon polonicus, *B) the diatom* Meridion circulare, *and C) the chrysophyte* Hydrurus foetidus.

throughout the Holarctic (Ward, 1994). In the Val Roseg, *Hydrurus foetidus* exhibited high spatial and seasonal variability; during certain times at some locations this alga was extremely abundant, its filaments completely covering the stream bed (Fig. 2 in Chapter 12). Chlorophytes appeared rarely and did not provide much biomass. Red algae (Rhodophyta) were not encountered.

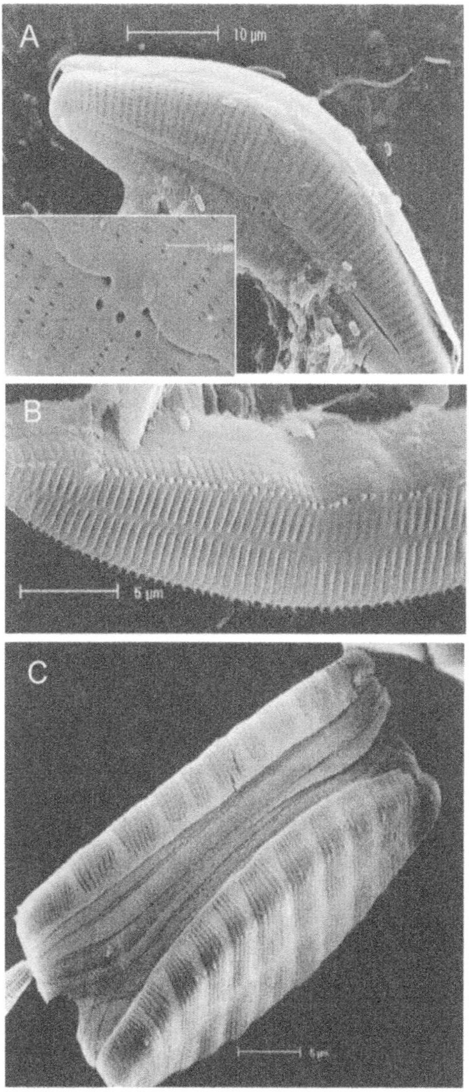

Figure 2. Scanning electron micrographs of some common diatoms of Val Roseg. A) Cymbella cistula, *B)* Fragilaria arcus, *and C)* Diatoma mesodon.

3. SPATIAL AND TEMPORAL DISTRIBUTION

3.1. Main channel

Epilithon was sampled seasonally at sites distributed along the entire length of the Roseg River, from near the glaciers downstream to the confluence with the Bernina River. Relative abundance values of algae for the 5 reaches along the longitudinal gradient are shown in Fig. 3; chlorophyll *a* values are shown in Fig. 4.

The biomass (measured as chlorophyll *a*) of epilithic algae in the proglacial reach (MP) below the Tschierva glacier was low throughout most of the year, except in fall. This is attributable to the highly unstable bed, cold water (daily mean water-temperatures <1.0°C in July and August), high turbidity and extreme flow fluctuations in summer. A few dense patches of algae occurred only near the air-water interface on large stable rocks protruding out of the water; high shear stress, the abrasive impact of suspended solids and bedload during high flow usually constrain the formation of epilithic algae on submerged stable substrata. Species of *Chamaesiphon* and some diatoms can survive in the boundary layer at high flow and are able to tolerate freezing and drying. *Chamaesiphon polonicus*, for example, persists in the amphibiotic zone of streams because thick sheaths protected cells from drying (Kann, 1978). In autumn, when discharge markedly declined, biomass increased from <1 to >50 g ash free dry mass m^{-2} (September 1996, see Chapter 12). The large standard deviations of chlorophyll *a* (Fig. 4) reflect the high spatial heterogeneity of periphyton at the reach scale. The chrysophyte *Hydrurus foetidus* is the most abundant alga in the proglacial reach during this time of the year. *Hydrurus foetidus* is considered as an ubiquitous alpine species occurring in streams with water temperatures of 0.4 to 12.5 °C (Kawecka, 1971). The diatom *Achnanthes minutissima* is an indicator of nutrient poor waters with moderate alkalinity (Cholnoky, 1968).

In the lake outlet stream (ML) average algal biomass (chlorophyll *a*) was the highest of all main channel reaches (Fig. 4). This may be attributed to the high channel stability, the lack of bedload transport, damped diel discharge fluctuations, and daily mean summer temperatures that are about 3°C higher than in the adjacent proglacial reach. Major taxa in the lake outlet are diatoms (*Achnanthes* spp., *Fragilaria* spp. , *Navicula* spp.) and cyanobacteria (*Homoeothrix janthina, Lyngbya* spp.) and during winter also *Hydrurus foetidus*.

In the incised reach (MI), seasonal patterns of chlorophyll *a* were almost the same as in the proglacial reach and presumably reflect the similar environment with respect to flow, turbidity, and bedload. Dominant taxa are *Chamaesiphon* spp., *Achnanthes* spp., *Fragilaria* spp. (including *Ceratoneis* spp.) and Ulotrichales. *Hydrurus foetidus* occurred year round, but never dominated.

In the main channel of the flood plain (MF), environmental conditions during summer are similar to those in the proglacial and incised reaches but temperatures at the end of the flood plain were up to 4.5°C higher than in the upper proglacial reach. From December to March, the main channel in the upper flood plain falls dry, whereas the lower main channel is fed by clear ground water (Chapter 4); during this period, chlorophyll *a* in the lower flood plain peaked at 34 ± 43 mg m^{-2}. During summer high flow, the algal community was dominated by *Fragilaria* spp. followed by *Hydrurus floetidus* and the crustforming *Chamaesiphon* spp. . *Hydrurus foetidus* produced only short filaments under high flow conditions but formed large streamers during winter (Fig. 2 in Chapter 12). *Diatoma* spp. and the monoraphide diatoms *Achnanthes* spp. *and Cocconeis* spp. were less important than at other sites along the main channel.

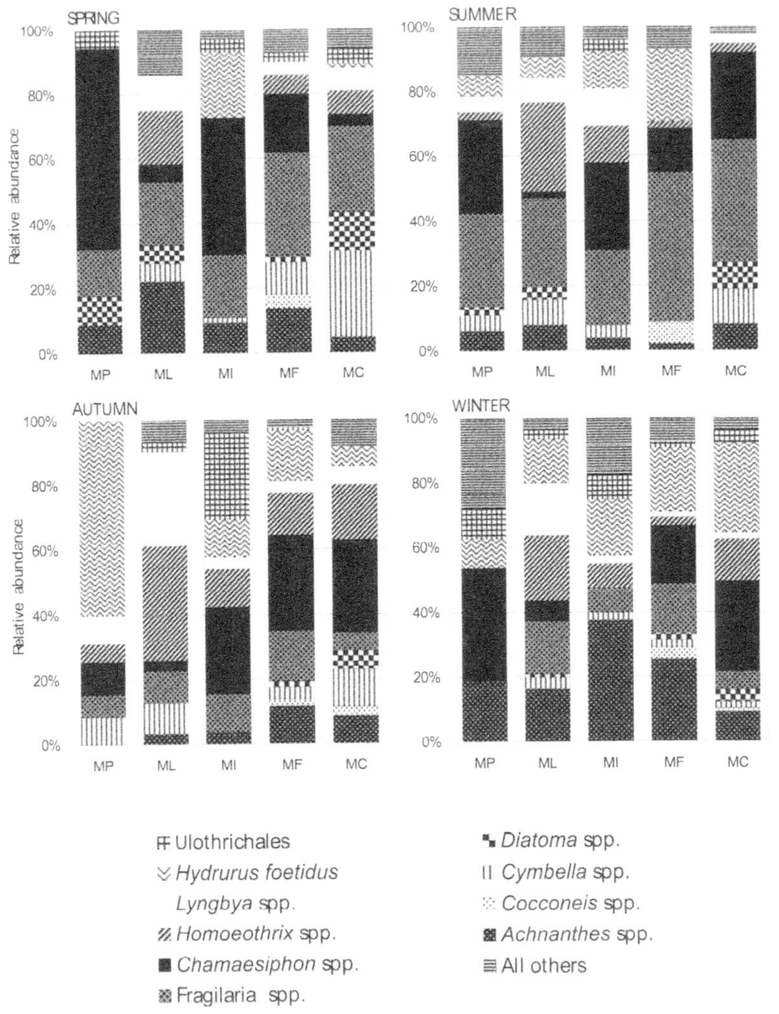

Figure 3. Relative abundance of algal communities in the five reaches of the main channel system of Val Roseg. MP = Proglacial reach; ML = Lake outlet; MI = Incised reach; MF = Flood plain; MC = Constrained reach.

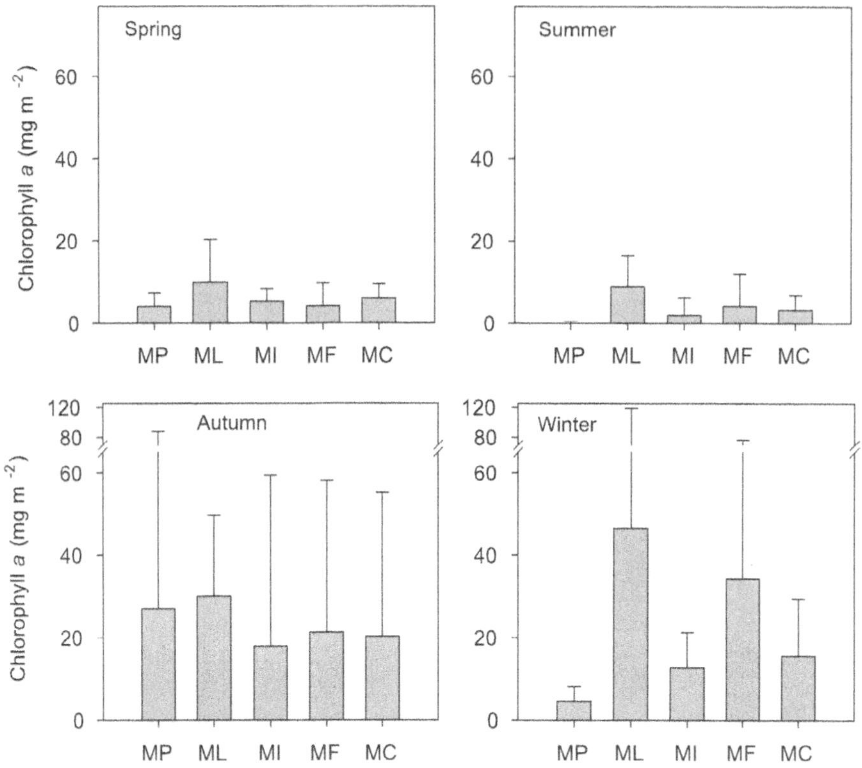

Figure 4. Concentrations of chlorophyll a *(mean + 1 SD) in the main channel of the Roseg River during four seasons. MP = Main channel, proglacial reach; ML = Main channel, lake outlet; MI = Main channel, incised reach; MF = Main channel, flood plain; MC = Constrained reach.*

The cyanobacterium *Chamaesiphon* spp. and the diatoms *Cymbella* spp. and *Fragilaria* prevailed in the constrained reach (MC), whereas *Hydrurus foetidus* was never abundant and was virtually absent during summer. At MC and all other main channel sites, chlorophytes were restricted to periods with low and moderate discharge (September – April). The highest chlorophyll *a* levels in this reach were measured during fall and winter.

3.2. *Floodplain channels*

During the course of the annual expansion and contraction cycle of the floodplain channel network, side channels, tributaries and mixed channels start to fall dry at the end of September. The extent of the channel network is minimum in January/February (Chapter 4).

The tributaries (T) are dominated by the diatoms *Achnanthes* spp., *Fragilaria* spp. and . spp.; however, the latter two genera were largely missing from winter samples at which time *Cocconeis* spp. was the most abundant taxon (Fig. 5). After the dry period in winter, the revitalization of algal flora starts with the spores remaining from the autumnal period. In contrast to the dried channels in the flood plain, tributaries lack the perennial upstream reaches which support recolonization after flow has resumed in spring. The diversity of algal flora of the tributaries is, therefore, limited to pioneer species, which can survive under harsh conditions. *Fragilaria arcus* and *Cymbella* spp., as well as early succession species from other diatom genera (*Achnanthes* spp., *Meridion* spp.), dominate the algal community in spring. Cyanobacteria are present from spring to fall but they are never predominant. *Hydrurus foetidus* and chlorophytes are not important in flood plain tributaries of the Val Roseg. Within several weeks following resumption of flow the algal biomass of the tributaries increased to about 20 mg chlorophyll *a* m^{-2} and remained more or less stable for the whole vegetation period until fall (Fig. 6).

Groundwater channels (G) are the most productive habitats in the Roseg River system with 55 and 65 mg chlorophyll *a* m^{-2} in autumn and winter. Although a given groundwater channel exhibits high temporal stability of environmental conditions, groundwater channels as a group are variable because the sources of water feeding these channels can be quite different (Chapter 3). As the water table declines, some smaller groundwater channels may stop flowing entirely. The algal community is characterized by a high diversity of diatoms (dominated by *Achnanthes* spp.) and cyanobacteria (*Homeothrix* spp. and *Chamaesiphon* spp.). *Hydrurus foetidus* was present during all seasons but was never abundant. Stagnant waters near groundwater channels contain many conjugates such as *Zygnema* spp., *Mougeotia* spp., *Spirogyra* spp., *Pleurotaenium* spp., *Cosmarium* spp., *Staurastrum* spp. and colonies of the diatom *Meridion circulare.* Conjugates may survive the winter period by forming zygotes but vegetative cells may also survive the long frozen period in situ (Hawes, 1989).

Mixed channels (X) have relatively high channel stability but are highly variable, having upstream connection with the main channel (mainly in summer) and fed by ground water (during winter). During winter more than 50% of these channels are dry. During summer, when chlorophyll *a* content peaks in some of these channels, cyanobacteria (*Lyngbya* spp. and *Homeothrix* spp.) contributed the most essential fraction to the biomass. Similar to the groundwater channels, the mixed channels exhibit markedly more biomass than the main channel in the flood plain except during winter.

From January to May all intermittently-connected channels (I) are dry. Despite high diel and seasonal fluctuations of flow, algal biomass was relatively constant on a mean level throughout the wet period. In summer, when disturbance by glacial melt water is high, chlorophyll a in intermittently-connected channels was higher than in all main channels sites, but in fall and early winter chlorophyll *a* concentrations were clearly below the levels occurring in the main channel within the flood plain reach. Diatoms (*Fragilaria* spp., *Diatoma* spp., *Cymbella* spp. and *Achnanthes* spp.) provided about 50% of the algal biomass, the rest belonging mainly to different cyanobacteria. *Hydrurus foetidus* and chlorophytes were never abundant in this channel type.

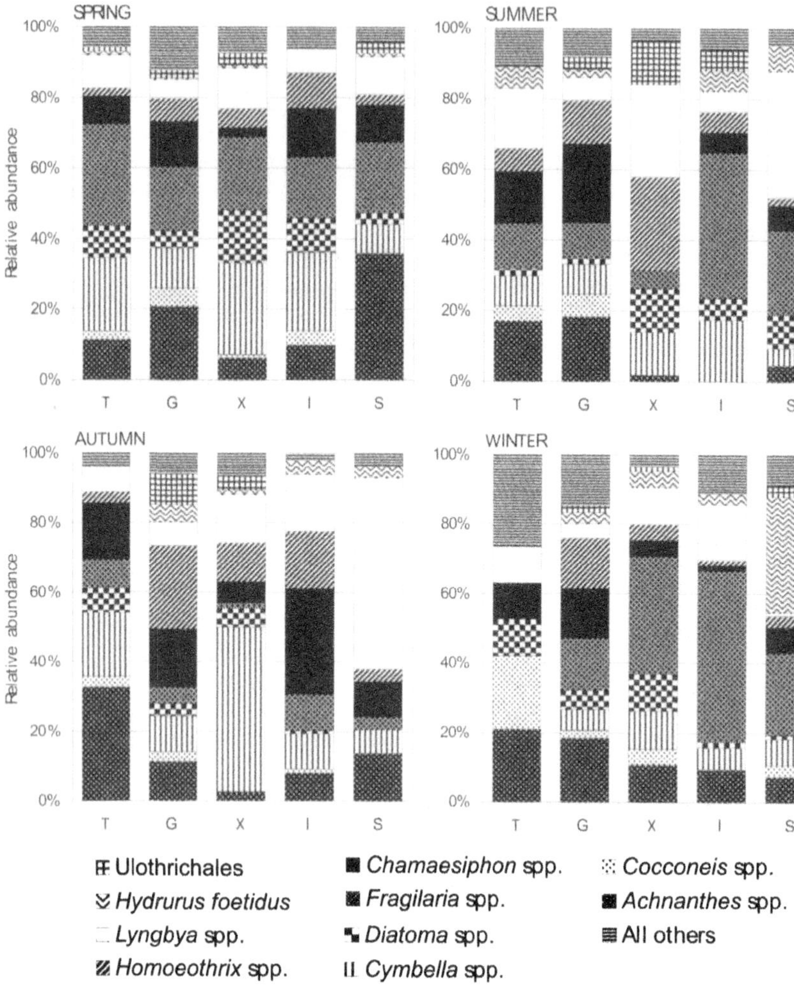

Figure 5. Relative abundance of algal communities in various channel types (main channel not included) in the flood plain of Val Roseg. T =Tributaries; G = Groundwater channels; X = Mixed channels; I = Intermittently-connected channels; S = Side channels.

In the side channels (S) environmental conditions are similar but less harsh than in the adjacent main channel as, for example, with respect to shear stress and bed stability. This is reflected by chlorophyll *a* values, which are lower than in T, G, I, and X channels, but significantly higher than in the main channel except at the lake outlet. The succession of dominant periphyton species started in spring with diatoms

(*Achnanthes* spp. and *Fragilaria arcus*), continued during summer and autumn with cyanobacteria (*Lyngbya* spp. and *Chamaesiphon* spp.) and ended in winter with *Hydrurus foetidus*, *Cymbella* spp. and other diatoms that were present throughout the annual cycle.

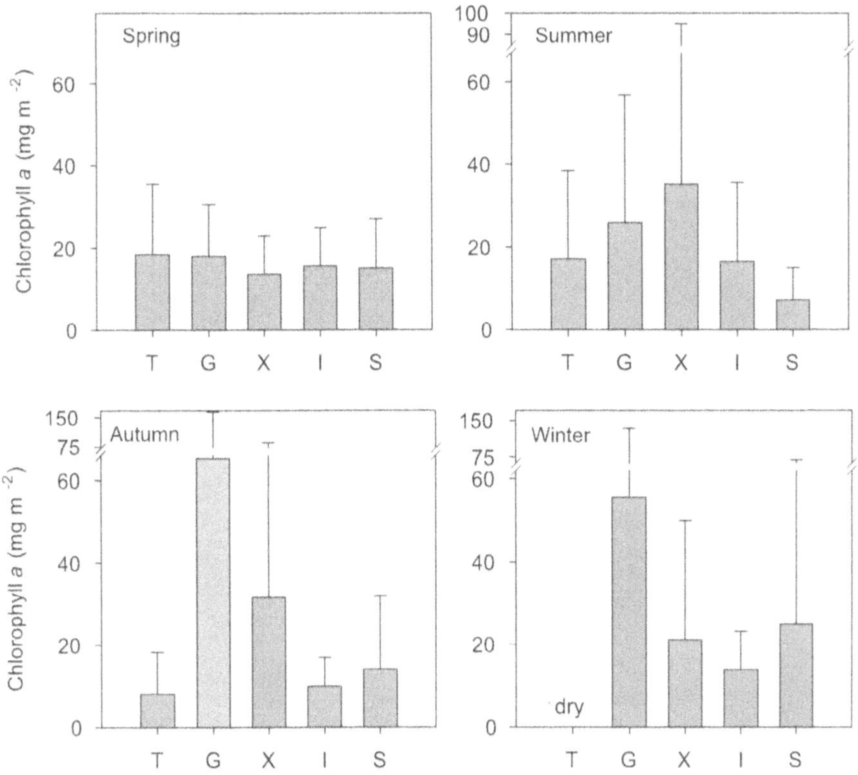

Fig. 6. Concentrations of chlorophyll a *(mean + 1 SD) in various channel types (except main channel) in the flood plain of Val Roseg. T = Tributaries; G = Groundwater channels; X = Mixed channels; I = Intermittently-connected channels; S = Side channels.*

3.3. Differences between major habitats

The variability of algal communities along the corridor of the Roseg River is very high in space and time. To examine the spatial and seasonal patterns of algal taxa (rare taxa excluded) we used centered (covariance) principal components analysis (Fig. 7). The first two axes of the PCA explained 48.1% and 15.9% of the total variance. The PCA clearly distinguished three groups of sites along axis F1, formed by (i) MP_U, MP_L and MF_U, (ii) MI, MF_L, MC, ML, S, I, T, X, and (iii) G. Further separation within groups was attributable to axis F2 (Fig. 7A). Algal assemblages in the main channel exhibited a high spatio-temporal variation because of distinct seasonal changes in glacial influence. Among floodplain habitats, groundwater channels displayed the least difference in epilithic assemblage structure over time as suggested by the low dispersion of sampling dates (short line lengths) around its average score (centroid).

Axis F1 was best explained by *Achnanthes* spp., *Chamaesiphon polonicus*, *Homoeothrix* spp., *Fragilaria arcus* and *Fragilaria* spp., whereas axis F2 was primarily related to *Hydrurus foetidus*, *Cymbella* spp., *Lyngbya* spp., and *Diatoma mesodon* (Fig. 7B). The same direction of "species-vectors" is indicative of a strong "size effect", i.e., a simultaneous increase of species abundance in some sampling units.

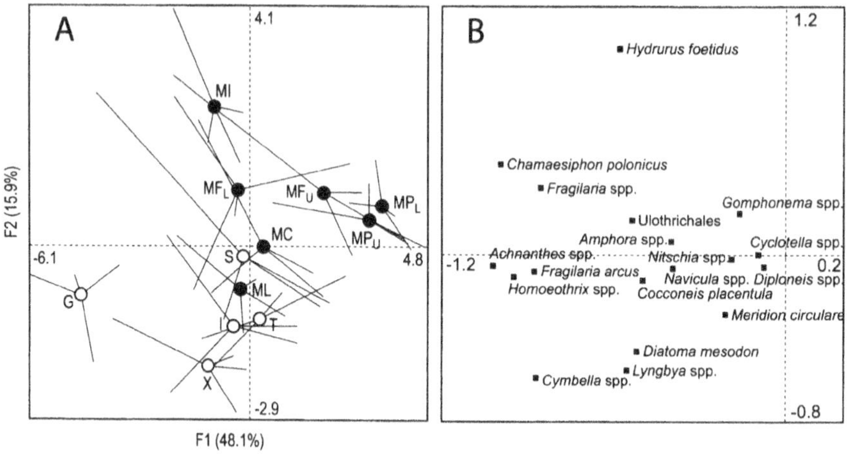

Figure 7. Ordination plot of the principal components analysis (PCA). Axes F1 and F2 explained 48.1% and 15.9% of the total variance, respectively. A) F1xF2 factorial map of the sites based on their periphyton assemblages. Small circles represent average scores for sampling sites (black = main channel and lake outlet, open = floodplain channel types), and are connected by lines to actual scores for each season (seasons are not labelled for visual reasons). Abbreviations: MP_U = Main channel, upper proglacial reach; MP_L = Main channel, lower proglacial reach; MI = Main channel, constrained reach; MF_U = Main channel, upper flood plain; MF_L = Main channel, lower flood plain; MC = Main channel, constrained reach; ML = Main channel, lake outlet; T = Tributaries; S = Side channels; X = Mixed channels; I = Intermittently-connected channels; G = Groundwater channels. B) F1xF2 plot of 18 periphyton taxa used in the analysis.

3.4. Taxa richness in the main habitats

A relatively small number of taxa (≤9) characterized the algal community of the proglacial (kryal) reach (MP in Table 2); low taxa richness is a common feature of harsh kryal streams (e.g., Hieber et al., 2001; Kawecka, 1980). The number of taxa averaged 17 in the lake outlet stream (ML), where environmental conditions with respect to temperature, channel stability and transport of coarse sediments were less harsh than in the adjacent proglacial reach. Taxa richness in floodplain channels, excluding the main channel, was constant throughout the annual cycle (FC in Table 2). Hieber (2001) found little seasonality of taxa richness in 10 alpine streams, which included several glacier fed systems. In contrast, the number of taxa in the main channel of the flood plain (MF) peaked in autumn (31 taxa) and was lowest in summer (9 taxa). This decline in taxa coincides with the summer flow pulse, when harsh conditions, in particular bed load transport, prevail. During this period, taxa richness within the other habitats within the flood plain persists at a relatively high level due to the fact that many floodplain channels are relatively benign habitats for benthic algae.

Table 2. Number of algal taxa in the proglacial reach (MP), in the lake outlet (ML), in the main channel in the flood plain (MF), and in floodplain channels excluding the main channel (FC).

Season	Channel types				Sum
	MP	ML	MF	FC	MF + FC
Spring	7	21	21	18	26
Summer	9	18	9	18	18
Autumn	9	16	31	18	36
Winter	7	14	22	18	25

4. SUMMARY AND CONCLUSIONS

The spatial and temporal patterns of periphyton biomass measured as chlorophyll *a* apparently reflect different habitat conditions. We suspect that high flow, high turbidity, transport of coarse sediments and light exclusion by snow cover during winter are major constraints to biomass accrual (the physico-chemical habitat template and its implication for benthic algae are discussed in more detail in Chapter 12). Algal biomass in floodplain channels less affected by glacial melt water was usually higher in spring and particularly in summer; in autumn, when glacial melt declines and groundwater influence starts to dominate, differences between main channel reaches and other channels become less distinct. In the Roseg River corridor, diatoms and cyanobacteria generally dominated algal communities. The relative abundance of diatoms increased by >50% along the continuum of the main channel except during summer high flow. Corresponding longitudinal patterns of cyanobacteria or other taxa were not evident. The chrysophyte *Hydrurus foetidus*, a species typically occurring in kryal streams of the Alps (Hieber et al., 2001), proliferated in the proglacial reach during autumn but was also found in most floodplain channels. In the High Tatra Mountains Kawecka (1971) distinguished three zones based on of the most abundant algal species: Zone I, (1550-1780 m a.s.l.) with *Chamaesiphon polonicus* prevailing, zone II, (890-1550 m a.s.l.) with large biomass of *Hydrurus foetidus, Homoeothrix janthina* and *Diatoma mesodon*, and finally in zone III (540-890 m a.s.l.) mass development of diatoms: *Diatoma vulgare*, *Cymbella affinis* and *Fragilaria ulna*. These species usually occur simultaneously in different channel types within the Roseg flood plain. Local influences such as spatial variation in channel stability, relative contribution of ground water and glacial water, and the relatively small altitudinal gradient (about 400 m) presumably confound the expression of more distinct longitudinal and lateral patterns. The environmental harshness of the proglacial reach, characterized by extended periods of bed load transport and snow cover, is reflected by low taxon richness. The highest number of algal taxa was found in the main channel of the flood plain, where environmental conditions vary seasonally from a turbid glacial stream to a clear groundwater stream. One may infer that the proglacial lake and the flood plain markedly increase the diversity of the algal community compared to a confined glacial stream lacking lake influence.

5. REFERENCES

Brittain, J. E., & Milner, A. M. (2001). Ecology of glacier-fed rivers: current status and concepts. *Freshwater Biology, 46*, 1571-1578.

Castella, E., Adalsteinsson, H., Brittain, J. E., Gislason, G. M., Lehmann, A., Lencioni, V., Lods-Crozet, B., Maiolini, B., Milner, A. M., Olafsson, J. S., Saltveit, S. J., & Snook, D. L. (2001). Macrobenthic invertebrate richness and composition along a latitudinal gradient of European glacier-fed streams. *Freshwater Biology, 46*, 1811-1831.

Cholnoky, B. J. (1968). *Die Ökologie der Diatomeen in Binnengewässern*. Weinheim: J. Cramer Verlag.

Elster, J., & Svoboda, J. (1996). Algal diversity, seasonality and abundance in, and along glacial stream in Sverdrup Pass, 79°N, Central Ellesmere Island, Canada. *Memoirs of National Institute of Polar Research, Special Issue, 51*, 99-118.

Hawes, I. (1989). Filamentous green algae in freshwater streams on Signy Island, Antarctica. *Hydrobiologia, 172*, 1-18.

Hieber, M., Robinson, C. T., Rushfort, S. R., & Uehlinger, U. (2001). Algal communities associated with different alpine stream types. *Arctic, Antarctic, and Alpine Research, 33*, 447-456.

Kann, E. (1978). Systematik und Ökologie der Algen österreichischer Bergbäche. *Archiv für Hydrobiologie/Supplement, 53*, 405-643.

Kawecka, B. (1971). Zonal distribution of alga communities in streams of the Polish High Tatra Mountains. *Acta Hydrobiologica, 13*, 393-414.

Kawecka, B. (1980). Sessile algae in European mountain streams 1. The ecological characteristics of communities. *Acta Hydrobiologica, 22*, 361-420.

Kownacki, A., Dummincka, E., Galas, J., Kawecka, B., & Wojtan, K. (1997). Ecological characteristics of a high mountain lake-outlet stream (Tatra Mts., Poland). *Archiv für Hydrobiologie, 139*, 113-128.

Meyns, S., Illi, R., & Ribi, B. (1994). Comparison of chlorophyll-*a* analysis by HPLC and spectrophotometry: Where do the differences come from? *Archiv für Hydrobiologie, 132*, 129-139.

Milner, A. M., Brittain, J. E., Castella, E., & Petts, G. E. (2001). Trends of macroinvertebrate community structure in glacier-fed rivers in relation to environmental conditions: a synthesis. *Freshwater Biology, 46*, 1833-1847.

Milner, A. M., & Petts, G. E. (1994). Glacial rivers: physical habitat and ecology. *Freshwater Biology, 32*, 295-307.

Robinson, C. T., Uehlinger, U., & Hieber, M. (2001). Spatio-temporal variation in macroinvertebrate assemblages of glacial streams in the Swiss Alps. *Freshwater Biology, 46*, 1663-1672.

Uehlinger, U. (1991). Spatial and temporal variability of the periphyton biomass in a prealpine river (Necker, Switzerland). *Archiv für Hydrobiologie, 123*, 219-237.

Vavilova, V. V., & Lewis, W. M. (1999). Temporal and altitudinal variations in the attached algae of a mountain stream in Colorado. *Hydrobiologia, 390*, 99-106.
Ward, J. V. (1994). Ecology of alpine streams. *Freshwater Biology, 32*, 277-294.

CHAPTER 10

P. BURGHERR, B. KLEIN, C.T. ROBINSON & K. TOCKNER

SURFACE ZOOBENTHOS

1. INTRODUCTION

Based on their local water sources and temperature regimes, three principal types of streams, each with distinctive zoobenthic assemblages, occur in alpine environments (Ward, 1994):

- kryal stream segments fed by glacial meltwater
- krenal segments fed by ground water
- rhithral segments dominated by seasonal snowmelt/rainfall

Although environmental harshness (low temperatures and scarce food resources) is a general feature of alpine streams, habitat conditions vary considerably between the three types. High elevation flood plains with complex channel networks often form a complex mosaic of kryal, krenal and rhithral segments (Tockner et al., 1997, Chapter 4). Kryal stream segments are characterized by the following attributes:

- year-round low temperatures ($T_{max} \leq 4^{o}C$)
- large diel flow fluctuations in summer with peaks in late afternoon from glacial melting
- highly turbid water from glacial flour
- low channel stability

Furthermore, instream environmental conditions depend on distance from the glacier terminus, season, and the contribution of non-glacial water sources. Downstream of the kryal segment, a glacio-rhithral zone, fed by a mixture of water sources, can extend for a considerable distance, with temporal changes in discharge and temperature reflecting the relative proportion of glacial influence (Füreder, 1999).

In contrast, groundwater-fed streams or krenal biotopes typically exhibit the following characteristics:

- constant flow regimes
- winter-warm temperatures
- physico-chemical constancy
- high water clarity

The environmental conditions in alpine glacial streams exert a severe constraint on successful colonization and persistence for aquatic macroinvertebrate species. Therefore, the occurrence of a particular species is a direct expression of this

J.V. Ward & U. Uehlinger (eds.), Ecology of a Glacial Flood Plain, 153-173.
© 2003 *Kluwer Academic Publishers. Printed in the Netherlands.*

species' ability to tolerate or adapt to the existing conditions (Füreder, 1999; Ward, 1992). Although the stream fauna of glacial-fed streams is greatly reduced, downstream faunal changes are distinct and predictable (Milner & Petts, 1994; Ward, 1994).

Survival under the harsh environmental conditions in high alpine streams requires physiological and/or life cycle adaptations. Some *Diamesa* species are able to complete a one-year life cycle at very low temperatures (Koshima, 1984; Milner & Petts, 1994), whereas other species have a two-year life cycle close to the glacier terminus but one generation per year at lower altitudes (Sæther, 1968). Growth and development in the same species may also vary depending on location. For example, the mayfly *Rhithrogena loyolaea* usually has a univoltine life cycle in the Swiss Alps (Sartori & Landolt, 1999), but requires three years to complete larval development in the French Pyrenees (Lavandier, 1981). Benthic macroinvertebrates also have to deal with overwintering, either through avoidance or tolerance of freezing (Irons III et al., 1993; Olsson, 1981; Oswood et al., 1991). The often sparse food resources in glacial streams impose another constraint on successful dispersal and persistence. Results from studies in the Val Roseg imply that feeding plasticity may play an important role in the feeding behavior of benthic macroinvertebtrates (Burgherr et al., 2001; Zah et al., 2001).

In the Val Roseg, two strategies to persist in glacial streams were observed: Chironomid midges of the genus *Diamesa* have adapted to the harsh environmental conditions during glacial melting in summer, whereas most Ephemeroptera and Plecoptera avoid these conditions and are most abundant during periods of favorable environmental conditions in spring and late autumn/early winter (Burgherr & Ward, 2001).

This chapter summarizes research on the distribution patterns of benthic macroinvertebrates in different glacial streams of the Val Roseg. Specifically, spatio-temporal dynamics of zoobenthic assemblages and their relationships to environmental gradients were investigated along the main glacial channel and in a variety of different floodplain habitats.

2. COMPOSITION OF THE ZOOBENTHOS

The inventory of benthic macroinvertebrate taxa for the Val Roseg is based on a total of 364 samples collected in 1997. A total of 177 samples were taken in the main glacial channel and the lake outlet channel of proglacial Lake Roseg, and 187 from different floodplain channel types. Note that analyses of zoobenthic distribution patterns along the main glacial channel and across a habitat gradient used specific subsets of the total number of samples. Zoobenthic samples from randomly selected locations in riffle/run habitats were taken using a Hess sampler (area 0.043 m^2, 100 μm mesh size), and preserved in 4% formalin. In the laboratory, macroinvertebrates were sorted, counted and identified to the lowest practical taxonomic level for most groups using a dissecting microscope.

In total, 146 benthic macroinvertebrate taxa, most identified to genus or species level, have been found in the Val Roseg (Appendix V). Figure 1 gives an overview of the most abundant and common groups occurring in surface zoobenthos. Non-Insecta comprised 35% of the total taxa, with Oligochaeta, Hydrachnellae and Ostracoda having the highest proportions. Insecta are dominated by Chironomidae, whereas the species richness of other groups is distinctly lower.

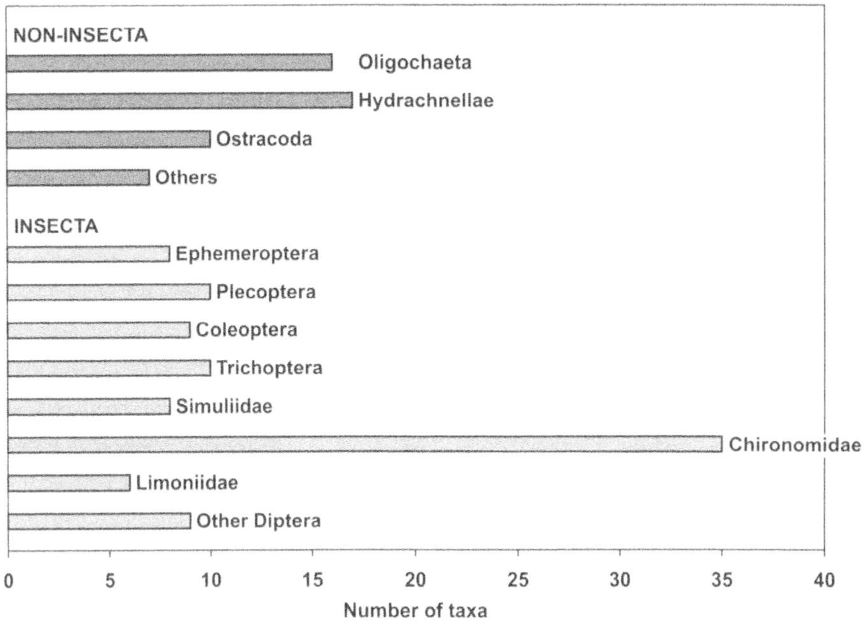

Figure 1. Diversity of various taxonomic groups in the Val Roseg flood plain.

Research on the zoobenthos of glacier-fed streams has often focused on insect groups with special emphasis on Chironomidae (Füreder et al., 2001; Kownacki, 1991; Kownacki & Kownacka, 1973; Rossaro & Bettinetti, 2000; Steffan, 1971). However, our studies in Val Roseg have shown that non-insect taxa may significantly contribute to overall biodiversity of glacial streams (Klein & Tockner, 2001; Malard et al., 2001). Non-insect taxa generally were most abundant and diverse in groundwater channels and springbrooks of the glacial flood plain of Val Roseg. Oligochaeta, however, could also become abundant in the main channel during certain periods. Malard discusses in chapter 11 how the hyporheic zone may act as a source from which suitable benthic habitats could be colonized.

Most insect groups exhibited similar numbers of species except for Chironomidae. Ephemeroptera and Plecoptera were more common in the main channel, whereas Coleoptera and Trichoptera were more diverse in other channel types. Diptera richness was dominated by Chironomidae (35 taxa), followed by Simuliidae (8) and Limoniidae (6). Nevertheless, a diversity of 8 Simuliidae species was higher than expected because glacier-fed streams are considered poor habitats for these filter feeders; this number represents over 20% of the species currently known for Switzerland. In the main channel the genus *Diamesa* was the most frequent chironomid taxon, the only one close to the glacier terminus. Further downstream *Diamesa* spp. decreased at the expense of Orthocladiinae. However, maximum diversity of Orthocladiinae occurred in other habitats. Chironomidae diversity in Val Roseg is comparable to findings from studies conducted in the French Pyrénées (Snook & Milner, 2001), Italian Alps (Maiolini & Lencioni, 2001), Tyrolean Alps (Füreder et al., 2001; Schütz et al., 2001) and Central Swiss Alps (Lods-Crozet et al., 2001). In contrast, these studies generally addressed non-insects only at a coarse taxonomic level, making comparisons difficult.

3. LONGITUDINAL PATTERNS

Despite a well documented interest in high mountain streams and their fauna at the beginning of the 20th century (e.g., Dorier, 1937; Steinmann, 1907; Thienemann, 1912), studies examining longitudinal distribution patterns of alpine benthic assemblages have typically been of limited duration or intensity .

Many studies on the ecology of glacial streams relate longitudinal patterns in zoobenthic assemblages to temperature gradients (e.g., Ward, 1994). In their conceptual model on glacial-fed streams, Milner et al. (2001) relate longitudinal changes in the benthic fauna to rising water temperatures and increasing channel stability. However, this model does not account for the spatio-temporal heterogeneity resulting from seasonal changes and hydro-geomorphological features.

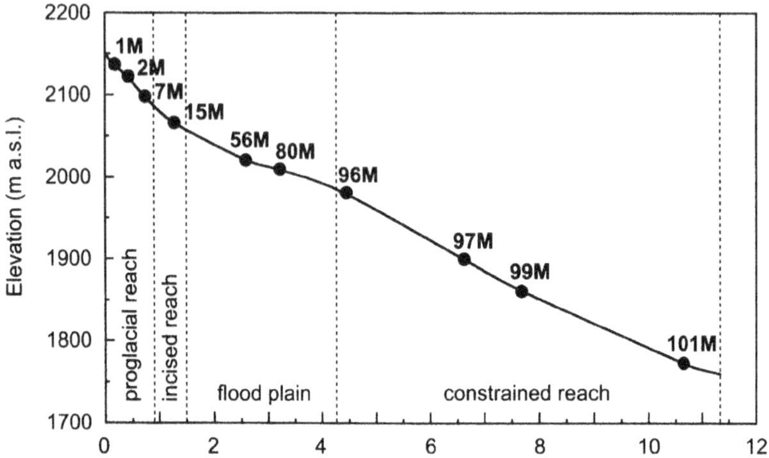

Figure 2. Location of zoobenthic sampling sites along the main channel (M) of the Roseg River. Note the four geomorphic reaches delineated with dashed lines. Modified from Burgherr & Ward (2001), Fig. 3a, p. 1708.

In a study of longitudinal patterns at main channel sites, samples were taken from ten sampling locations that traversed four geomorphic reaches (Fig. 2). Sampling was conducted from just below terminus of the Tschierva glacier (1M) down to the river mouth (101M) Further details on sampling sites are available in Appendix I. The temporal dimension was considered by collecting samples in June (increasing discharge), August (high discharge), October (decreasing discharge), and November 1997 (low and constant discharge) (compare Fig. 10 in Chapter 1).

3.1. Zoobenthic distribution patterns

Total zoobenthic density, all sites combined, was lowest in August (1472 m^{-2}), intermediate in June (4526 m^{-2}) and November (8676 m^{-2}), and highest in October (13,398 m^{-2}). Site averages for total density, taxon richness and Simpson's index of diversity showed similar trends. Values of all three measures tended to increase with distance from the glacier, attaining maxima in the constrained reach, as indicated by the highly significant ($p \leq 0.001$) logarithmic regression lines fitted to the data (Fig. 3). However, seasonal variability of all three measures was high in some cases, with coefficients of variation up to 150%. These findings suggest that changes in density and diversity of benthic assemblages reflect (1) spatial phenomena, i.e., reduced harshness with increasing distance from the glacier, and (2) temporal phenomena, i.e., more favorable periods in spring and late autumn/early winter than in summer.

In general, Chironomidae, especially the genus *Diamesa*, were a predominant constituent of the zoobenthos in the glacial stream of Val Roseg, as reported for many glacier-fed streams in different areas (e.g., Bretschko, 1969; Elgmork & Sæther, 1970). Common *Diamesa*-groups found, were *D. steinboecki*-gr., *D. latitarsis*-gr., *D. bertrami*-gr., *D. cinerella-zernyi*-gr., and *D.* cf. *incallida*. Close to the glacier terminus only *Diamesa* spp. was found, whereas the chironomid subfamily Orthocladiinae first appeared at site 7M, then progressively increased in relative density at the expense of *Diamesa* spp. Other taxa such as the Ephemeroptera *Baetis alpinus* and *Rhithrogena* spp., the Plecoptera *Leuctra* spp. and *Protonemura* spp., as well as Simuliidae and Oligochaeta also made substantial contributions to benthic assemblages, especially at lower sites during periods of favorable environmental conditions. Other taxa contributed on average less than 1% of the total assemblage.

Baetis alpinus was ca. 2.5x more abundant than *Rhithrogena* species. Within the genus *Baetis* only *B. alpinus* was found. The absence of *B. melanonyx* Pictet, another common species of this region, is attributed to the high altitude of the Val Roseg and the crystalline substratum (Sartori & Landolt, 1999). Within the genus *Rhithrogena*, *R. loyolaea* and *R. nivata* were the two most abundant species at

Figure 3. (next page). Mean values and trendlines for density, taxon richness and Simpson's index (SI) of the macroinvertebrate assemblages at the ten sampling sites. SI values were converted to -ln SI such that values increase as the number of species increases, independently of sample size. Dashed lines delineate the geomorphic reaches shown in Figure 2.

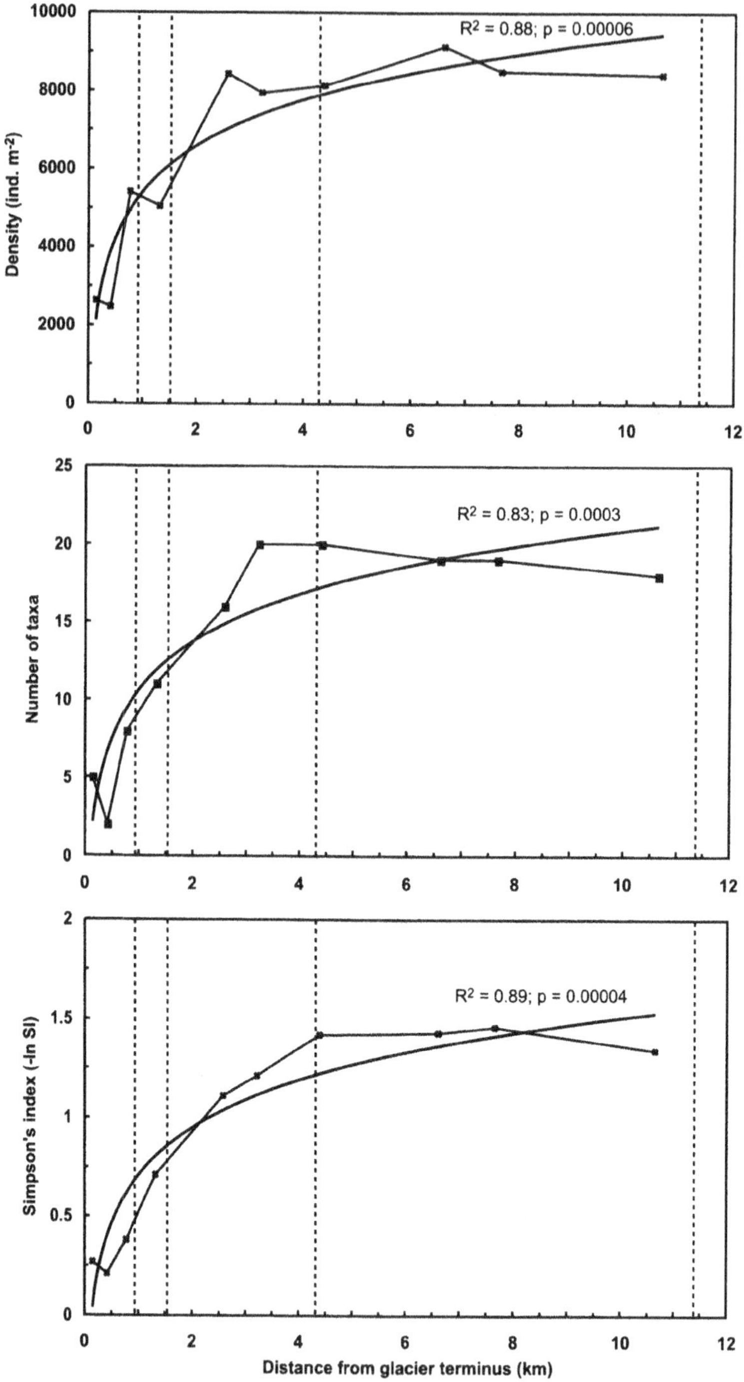
R² = 0.88; p = 0.00006
Density (ind. m-2)
10000
8000
6000
4000
2000
0
0 2 4 6 8 10 12
R² = 0.83; p = 0.0003
Number of taxa
25
20
15
10
5
0
0 2 4 6 8 10 12
R² = 0.89; p = 0.00004
Simpson's index (-ln SI)
2
1.5
1
0.5
0
0 2 4 6 8 10 12
Distance from glacier terminus (km)

most sites, corresponding to their expected altitudinal optima between 1700 and 2200 m a.s.l. (Sartori & Landolt, 1999). In contrast, *R. alpestris* and *R. degrangei* were only abundant at the more downstream sites because they have their altitudinal optima in the lower alpine zone and in the upper subalpine zone (Sartori & Landolt, 1999).

3.2. Relationships between fauna and environment

We further examined if there was concordance among biotic and environmental gradients. Because the relationship between fauna and environment may vary in space and time, we focused on changes in the spatial configuration of sampling sites among dates. This was analyzed by means of a co-inertia analysis[1], the results of which are visualized in Figure 4. Figure 4A demonstrates how environmental gradients were related to downstream changes in glacial influence (i.e., opposing arrows for turbidity and specific conductance). Turbidity and specific conductance also reflected seasonal shifts in sources and flow paths of water in the Val Roseg catchment as shown by Figures 4A and 4C (Malard et al., 1999). Water temperature represented the typical sequence of hypokryal, metakryal and glacio-rhithral habitats that is expected along the continuum of a glacial-fed stream (Füreder, 1999; Ward, 1994). For a detailed description of the physico-chemical templet of the Val Roseg refer to Chapters 3-6 of this book.

Figure 4. (next page) Within-date co-inertia analysis (spatial typology). The existence of a significant co-structure was confirmed using a Monte-Carlo permutation test (p=0), with factor 1 (90.6%) explaining most of the observed co-structure and factor 2 contributing only marginally (4.8%). A) Co-inertia scores of the environmental parameters (BOM = benthic organic matter, DOC = dissolved organic carbon, NO_3-N = nitrate, PN = particulate nitrogen, PP particulate phosphorus, SRP = soluble reactive phosphorus, SS = suspended solids, TIC = total inorganic carbon) and B) of the taxa on the F1xF2 factorial plane. In 4A the longer an arrow and the closer to an axis the stronger is the relationship to this factor. Abbreviations of taxa: Cre = Crenobia alpina, *Nmt = Nematoda, Oli = Oligochaeta, Hyd = Hydrachnellae, Cru = Crustacea, Bae =* Baetis alpinus, *Rhi =* Rhithrogena *spp., Rha =* Rhabdiopteryx alpina, *Nem =* Nemoura *spp., Pro =* Protonemura *spp., Leu =* Leuctra *spp., Cap =* Capnia vidua, *Iso =* Isoperla *spp., Per =* Perlodes intricata, *Lim = Limnephilidae, Wie =* Wiedemannia *spp., Dic =* Dicranota *spp., Rhy =* Rhypholophus *spp., Rhm =* Rhabdomastix *sp., Sim =* Simulium *spp. , Prs =* Prosimulium *spp., Dia =* Diamesa *spp. , Ort = Orthocladiinae, Tan = Tanytarsini. C) Standardized co-inertia scores of sampling sites of the environmental (arrowbase) and faunistic (arrowhead) data sets onto the F1xF2 factorial plane for each sampling date. Arrows measure the strength of the co-structure between the fauna and the environment, so the shorter the arrow the better the agreement between the two structures. Modified from Burgherr & Ward (2001), Fig. 6, p. 1715.*

[1] Multivariate statistical techniques are well suited to analyze and illustrate the relationships between the environmental and faunistic data. Two-table ordination methods, such as co-inertia analysis (CIA) simultaneously examine the structure in both data sets and identify the corresponding variations (i.e., co-structure) of these structures (Dolédec & Chessel, 1994). The temporal and spatial components of variation in the two data sets can be explicitly addressed using within-class CIA (Franquet & Chessel, 1994; Franquet et al., 1995).

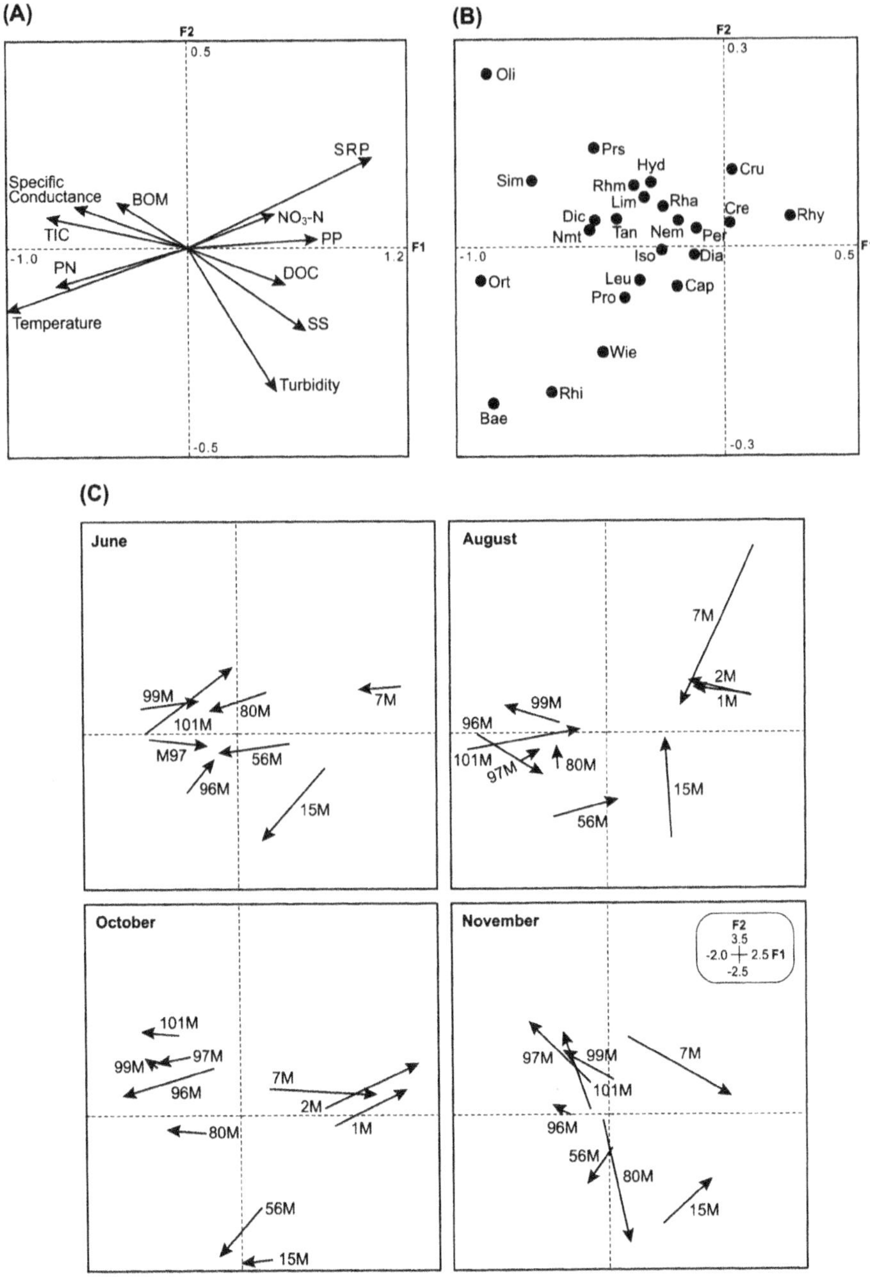
(A)
F2
0.5
-0.5
F1
-1.0
1.2
SRP
Specific Conductance
BOM
TIC
NO3-N
PP
PN
DOC
Temperature
SS
Turbidity
(B)
F2
0.3
-0.3
F1
-1.0
0.5
Oli
Prs
Hyd
Sim
Rhm
Cru
Lim
Rha
Dic
Cre
Rhy
Nmt
Tan
Nem
Per
Iso
Dia
Ort
Leu
Cap
Pro
Wie
Rhi
Bae
(C)
June
99M
101M
80M
7M
M97
56M
96M
15M
August
7M
2M
1M
99M
96M
101M
97M
80M
15M
56M
October
101M
97M
99M
96M
7M
2M
1M
80M
56M
15M
November
F2
3.5
-2.0
2.5 F1
-2.5
97M
99M
7M
101M
96M
56M
80M
15M

Baetis alpinus, Orthocladiinae, *Simulium* spp. , Tanytarsini and *Isoperla* spp. contributed most to the definition of axis F1, whereas axis F2 was best described by Oligochaeta, *Prosimulium* spp., *Rhithrogena* spp., *Protonemura* spp. and *Leuctra* spp. (Fig. 4B). Seasonal shifts in taxonomic composition were most evident at sites in the flood plain and in the constrained reach (Figs. 4B and 4C).

Although longitudinal patterns differed among dates, three groups of sites were observed (Fig. 4C). Sites 1M, 2M and 7M in the proglacial reach, and site 80M in the lower flood plain together with sites 96M, 97M, 99M, 101M in the constrained reach formed two distinct groups. The group formed by sites 15M and 56M reflected the transition between them. This pattern was clearest in August and October, but less distinct in June, and even weaker in November, indicating conditions were most homogenous on this last date. In summary, co-inertia analysis clearly demonstrated a strong linkage between environmental conditions and benthic assemblage structure along the longitudinal gradient and among seasons.

3.3. Seasonality of longitudinal gradients

Spatio-temporal changes in zoobenthic gradients are not solely related to temperature and channel stability. Seasonal shifts in sources and pathways of water (i.e., extent of glacial influence), and periods of favorable environmental conditions (in spring and late autumn/early winter) also strongly influenced zoobenthic distributions. Although longitudinal response patterns of zoobenthos varied across seasons, individual sites exhibited similar temporal patterns (see also Burgherr & Ward, 2001). During periods of favorable environmental conditions or "ecological windows" (spring and late autumn/early winter) invertebrate density and taxon richness peak, and cold-adapted taxa other than just *Diamesa* spp. can occur rather close to the glacier terminus. Similar evidence of macroinvertebrate assemblages reflecting seasonal changes in harshness of environmental conditions was recently reported in several other studies of glacial streams (e.g., Brittain et al., 2001; Füreder et al., 2001; Robinson et al., 2001; Schütz et al., 2001; Snook & Milner, 2001).

Overall, Milner & Petts' (1994) original conceptual model and their modified version (Milner et al., 2001) predicted macroinvertebrate distribution patterns accurately during glacial melt in summer, but did not account for seasonal changes in assemblage patterns. The modified model of Figure 5 focuses on these temporal changes along the upper course of a glacier-fed stream. Minimum and maximum curves for zoobenthic performance indicators reflect temporal changes in glacial influence, which is determined by discharge, and the relative contributions of glacial melt water and ground water. These factors in turn determine channel stability, and, together with air temperature, maximum water temperature at the next lower hierarchical level.

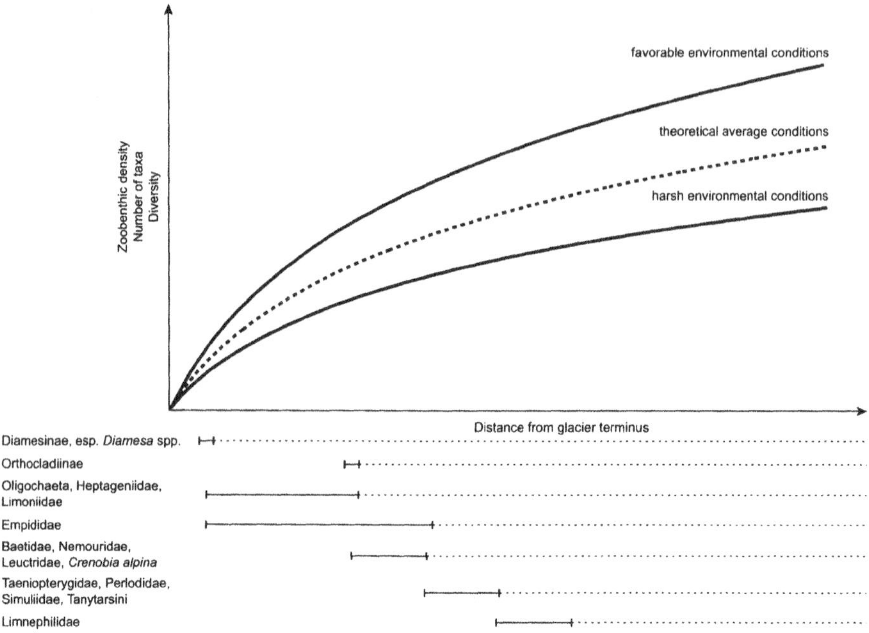

Figure 5. Conceptual view of changes in zoobenthic summary statistics with increasing distance from glacier terminus. Solid curves represent endpoints when environmental conditions are most harsh and favorable, respectively. The dotted curve is indicative of a theoretical average over an annual cycle. In the lower half, the solid lines indicate the first appearance of taxa during favorable and harsh environmental conditions. Many additional taxa are added with further distance downstream.

4. HETEROGENEITY OF ZOOBENTHOS ACROSS HABITAT GRADIENTS

4.1. Habitat heterogeneity and stability

Compared to longitudinal distribution patterns, spatio-temporal dynamics of stream networks within glacial flood plains have received even less attention and virtually nothing is known regarding distribution patterns of benthic macroinvertebrates in these systems. In river flood plains at low elevations, the pulsing of discharge (flood pulse) is a key factor structuring biotic assemblages (Junk et al., 1989). In the glacial flood plain of the Val Roseg, seasonal changes in discharge, water sources, and flow paths create a shifting habitat mosaic (Malard et al., 1999; Tockner et al., 1997). Therefore, the effects of habitat stability, spatio-temporal habitat heterogeneity and hydrological connectivity on the abundance and composition of benthic macroinvertebrates was examined in streams of the alluvial flood plain of Val Roseg.

We compared the zoobenthic assemblages of three different habitat (channel) types – the main channel, intermittently-connected channels and groundwater channels – that represented a gradient of increasing channel stability. Within the groundwater channel type we looked at alluvial groundwater streams and groundwater-fed springbrooks fed by hillslope ground water. Springbrooks are characterized by several times higher concentrations of particulate organic carbon (POC) and particulate nitrogen (PN) than alluvial groundwater channels or any other channel type, and contain the highest concentrations of nitrate (NO_3-N) and soluble reactive phosphorus (PO_4-P) within the flood plain (Klein & Tockner, 2001). Thus, they may be considered a distinct subtype within groundwater-fed channels. The longitudinal perspective was incorporated by sampling in the proglacial reach (7M, 8G)[2], in the upper part of the flood plain (56M, 46I, 64G, 38SP, 63SP) and in the lower part of the flood plain (80M, 73I, 81G, 69SP, 74SP). For exact locations of sampling sites see Appendix I. Bimonthly sampling intervals from June to December 1997 accounted for seasonal changes in discharge, water sources and flow paths.

Channel types can be distinguished, based on temperature regime, turbidity, and specific conductance as indicators of hydrological connectivity and water source (Table 1).

Table 1. Summary statistics for specific conductance, turbidity and water temperature. Measures are reported as means and standard deviations except water temperature, for which daily mean and range in parentheses were calculated from continuous logger data in 1997. M = Main channel, I = Intermittently-connected channels, G = Groundwater channels, SP = Springbrooks.

Site	*Specific conductance (μS/cm)*	*Turbidity (NTU)*	*Water temperature (°C)*
7M	49.2 ± 31.7	44.5 ± 29.5	0.7 (0.0 – 1.9)
56M	52.7 ± 23.6	59.0 ± 41.1	1.8 (0.0 – 4.8)
80M	51.5 ± 18.2	51.6 ± 35.5	2.6 (0.0 – 6.1)
46I	55.9 ± 20.3	28.8 ± 45.1	4.5 (0.0 – 7.1)
73I	64.0 ± 6.3	3.7 ± 2.7	6.0 (0.2 – 11.0)
8G	56.2 ± 27.8	4.9 ± 3.3	2.9 (0.3 – 6.2)
64G	62.6 ± 19.2	15.8 ± 28.5	4.0 (0.1 – 8.1)
81G	88.1 ± 4.5	1.0 ± 0.7	5.4 (0.5 – 9.4)
38SP	98.5 ± 21.5	0.6 ± 0.3	2.9 (0.6 – 5.0)
63SP	61.8 ± 23.0	9.0 ± 8.2	4.4 (1.5 – 7.4)
69SP	62.1 ± 3.7	0.5 ± 0.2	5.2 (3.2 – 7.3)
74SP	68.1 ± 8.0	0.4 ± 0.2	5.9 (0.9 – 10.4)

Mean water temperature and range in water temperature were lowest in the main channel sites that were influenced most directly by glacial meltwater. Furthermore,

[2] Intermittently-connected channels and springbrooks did not occur in the proglacial reach.

the mean temperature and temperature range increased with increasing distance from the glacier terminus. Specific conductance and turbidity were inversely related. Specific conductance generally increased from the main channel towards the groundwater channels, whereas turbidity showed the opposite pattern. These two parameters also showed temporal differences. With decreasing discharge from summer to winter, turbidity in the main channel decreased whereas specific conductance increased. This temporal trend was less apparent in intermittently-connected and groundwater channels.

4.2. Macroinvertebrate distribution patterns across habitat gradients

Repeated measures analysis of variance was used to test for differences in mean density and Simpson's index as a measure of diversity. Significant differences were found among habitat types, longitudinal position (reach), and sampling dates (Fig. 6). Groundwater channels and springbrooks, which exhibited low variability in environmental conditions over time due to minimal glacial influence, sustained the highest density and diversity. Similarly, glacial influence decreased with increased distance from the glacier terminus as indicated in maximum values for both measures in the lower floodplain section. Finally differences among sampling dates suggest that density patterns reflect transitions between high and low flow periods in the annual hydrograph, whereas diversity patterns revealed that peaks were associated with periods of favorable environmental conditions during late autumn/early winter. Higher zoobenthic density and species numbers in reaches strongly influenced by ground water is also known from other alpine and arctic streams (Craig & McCart, 1975; Füreder et al., 2001; Kownacki, 1991).

The distribution of zoobenthic taxa along the habitat stability gradient was examined using OMI-analysis[3]. The first two axes of OMI-analysis explained 85.8% and 10.3% of the variability; therefore only the first axis was used for graphical representation of ordination results (Fig. 7). The first axis arranged the sampling sites along a habitat gradient based on glacial influence (e.g., turbidity) versus groundwater contributions (e.g., specific conductance) as indicated by the separation of environmental variables into two distinct groups (Figs. 7A and 7B). Main channel sites were clearly separated, whereas the other habitat types showed some overlap. However, the behavior of individual channels may be explained by factors such as isolation from the channel network during low flow periods or upwelling ground water (Burgherr et al., 2002; Malard et al., 2001). Although individually stable, groundwater channels and springbrooks collectively encompassed greater spatial heterogeneity than other channel types (Ward et al., 1999).

[3] The Outlying Mean Index (OMI), or species marginality, measures the distance between mean habitat conditions used by a taxon (taxon centroid) and mean habitat conditions in the benthic layer of the stream Thus, the position of species depends on their niche deviation from a reference that represents a theoretical ubiquitous species uniformly distributed among habitat conditions. For more details refer to Chessel & Gimaret (1996) and Dolédec et al. (2000).

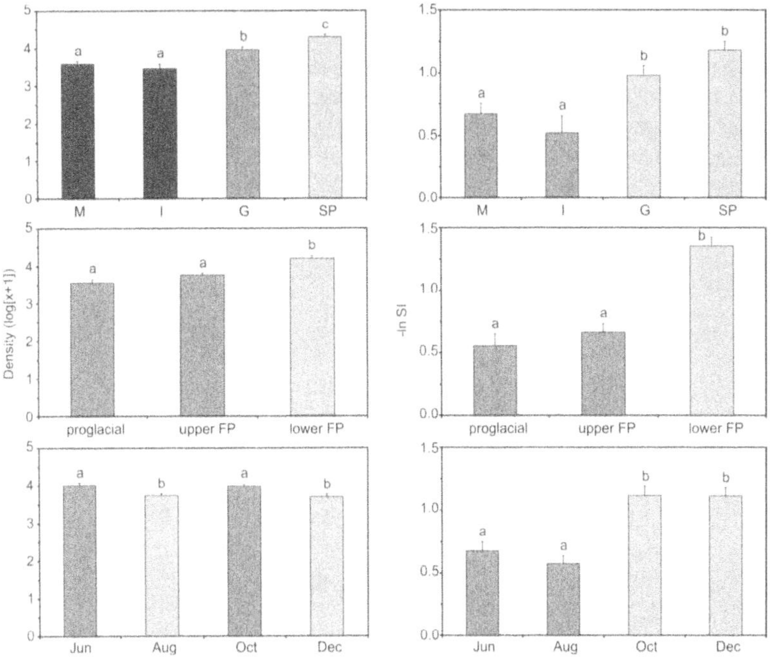

Figure 6. Comparison of mean density and Simpson's index among habitat types, reach and season, based on repeated measures analysis of variance. Simpson's index (SI) was expressed as -ln SI so that values increase as the number of species increases independently of sample size. Error bars represent +1 SE. Bars with the same letters are not significantly different (p > 0.05). M = Main channel, I = Intermittently-connected channels, G = Groundwater channels, SP = Springbrooks, FP = Flood plain.

A global permutation test on the outlying mean index was statistically significant ($p = 0$), justifying the plot of species positions on the ordination diagram. The niche of 13 out of 22 taxa deviated significantly ($p < 0.1$) from a uniform distribution along this habitat gradient (Fig. 7C). In general, taxa were arranged along a krenal to kryal/glacio-rhithral gradient. Ostracoda, Harpacticoida, Hydrachnellae, Nematoda, Oligochaeta, Limnephilidae and *Crenobia alpina* were predominant in groundwater channels and springbrooks, but rarely abundant in the main channel. This indicates a strong dependence on environmental conditions of these taxa, and thus a rather narrow niche breadth. In contrast, taxa such as *Rhithrogena* spp., *Baetis* spp., *Leuctra* spp. or *Rhypholophus* spp. were more indicative of glacial influenced channels but also occurred in lower density in the other habitat types, indicating they were more generalist taxa. This suggests that groundwater and intermittently connected channels, and to a lesser extent springbrooks, may serve as important refugia during high discharge in summer, thus also providing a potential recolonization pathway (besides the hyporheos and oviposition by winged adults) for the main channel where environmental conditions were unsuitable for most taxa. Towards winter the system becomes groundwater-controlled resulting in more homogenous physico-chemical conditions among channel types. During these periods of favorable environmental conditions the main channel can also sustain higher density and diversity, thus becoming more similar to the other habitat types. In conclusion, spatial variability in assemblage distribution patterns followed changes in discharge and associated degrees of hydrological connectivity of the channel network.

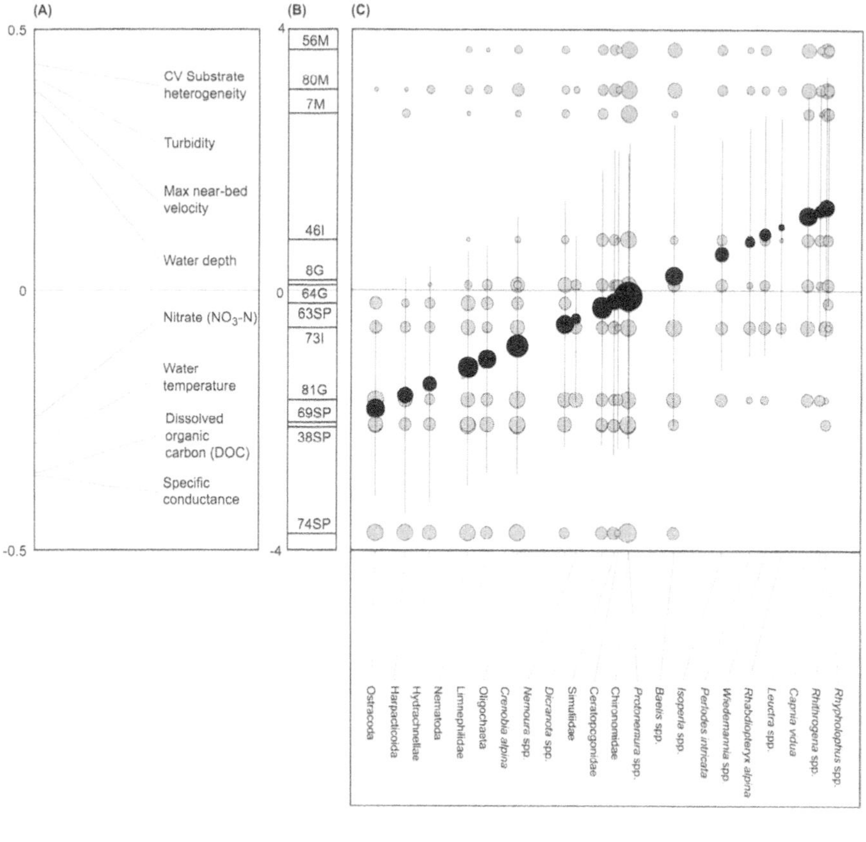

Figure 7. OMI-analysis of zoobenthos using a one-axis representation. A) Canonical weights of environmental variables. B) Scores of sampling sites. C) Distribution of taxa across the habitat gradient according to their weighted average position along site scores. The sizes of black circles are proportional to the total frequencies of taxa. Vertical lines correspond to ± 2 standard deviations. The sizes of grey circles are proportional to the densities of taxa. Dashed lines indicate taxa whose distribution significantly deviates from a uniform distribution across the habitat gradient.

5. HABITAT TEMPLET FOR ZOOBENTHOS IN THE ROSEG RIVER

The previous sections revealed some distinct patterns within and among different habitat types. Here we present how zoobenthic distribution patterns in different habitat types can be summarized using a two-dimensional habitat templet. Our classification includes the three basic habitat types, i.e., the main channel, intermittently-connected channels, groundwater channels, and additionally the outlet stream of Lake Roseg. We abstained from a separation of the subtypes alluvial groundwater channels and springbrooks because of their great spatial heterogeneity resulting in partial overlap. The outlet stream of Lake Roseg was included because interruption of the river course by lakes generally results in the amelioration of physical conditions further downstream allowing increased faunal diversity (Milner et al., 2001). This phenomenon was also observed in the Val Roseg, where a more diverse community in the kryal lake outlet stream was found than in the adjacent kryal stream originating from Tschierva glacier (Burgherr & Ward, 2000). Similar evidence has been reported from other sites in the Swiss Alps (Hieber et al., 2002). First, a centered principal components analysis was performed using all benthic samples collected during 1997 to examine differences in assemblage composition among habitat types (Fig. 8). The centroids of the different habitat types were clearly separated along the first axis, although this pattern was less apparent for the main channel and the lake outlet stream (Fig. 8A). Figure 8B illustrates how taxonomic composition responds to the kryal/rhithral–krenal gradient from the main channel towards groundwater channels. Addionally, Figure 8A also indicates that assemblages of different channels types may be nested to a certain degree.

Pronounced nestedness is a feature typically found in assemblage distribution patterns within naturally fragmented habitat (Atmar & Patterson, 1993). The Val Roseg flood plain can be put in this context due to the distinct environmental conditions in different channel types (Tockner et al., 1997; Tockner et al., 2002) and the seasonal expansion/contraction cycle of the channel network (Malard et al., 1999). To determine the degree of nestedness in the Val Roseg, we used the Temperature Calculator of Atmar & Patterson (1993). The calculated system "temperature" (T) reflects the degree of order present in presence-absence matrices and ranges from $T = 0^{\circ}$ (perfectly nested) to $T = 100^{\circ}$ (random) [4]. The calculation of T resulted in a matrix temperature of 25.7°, which was significantly lower than the Monte Carlo simulated random temperature (70.5°, $p < 0.0001$), indicating assemblages were nested. In conclusion, this finding provides additional support to the concept that certain channel types play an important role as refugia for zoobenthos during periods of harsh environmental conditions.

[4] The software calculates the metric by packing the data into rows (sampling sites) and columns (taxa) such that nestedness is maximized. A Monte Carlo randomization procedure (100 permutations in our case) is used to test whether the calculated temperature T from packing the matrix is significantly lower (more nested) than if the matrix is packed randomly (Atmar & Patterson, 1995).

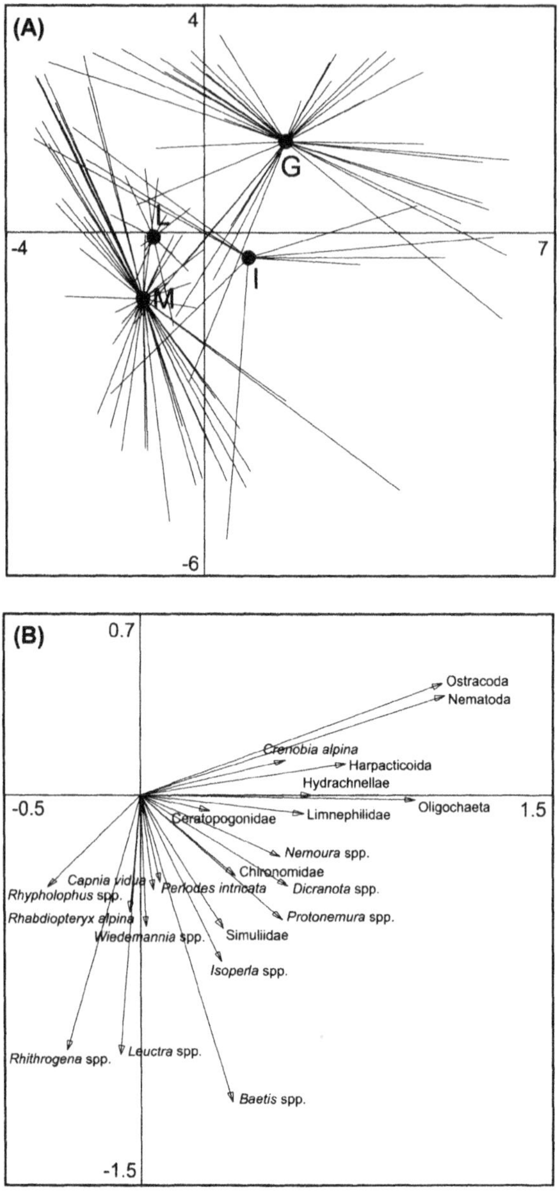

Figure 8. Differences in zoobenthic assemblage structure among four habitat types. The first two axes of the centered principal components analysis explained 31.0% and 27.4%, respectively. A) Plot of factorial scores of samples. Black circles correspond to the centres of gravity (centroids) and lines to individual samples of the four habitat types. M = Main channel, L = Lake outlet stream, I = Intermittently-connected channels, G = Groundwater channels. B) Plot of factorial scores of taxa.

Based on these analyses and those from the previous sections, it appears that habitat stability, spatio-temporal habitat heterogeneity and hydrological connectivity constitute the major factors in determining the habitat templet for the benthic assemblages of different channel types, as well as individual channels, in the glacial flood plain of Val Roseg. Following the rationale of the intermediate disturbance hypothesis (Connell, 1978) and its adaptation to lotic freshwaters (Ward & Stanford, 1983), alpine flood plains are positioned on the descending limb of the harshness-diversity curve (Fig. 9A). However, temporal dynamics of the complex channel network in the Val Roseg flood plain enhanced overall ecosystem stability, which is not the case for glacial streams lacking flood plains.

Based on seasonal changes in spatial habitat heterogeneity (Malard et al., 1999; Tockner et al., 1997; Ward et al., 1999), a two-dimensional habitat templet can be obtained for the three different channel types investigated (Fig. 9B). From this

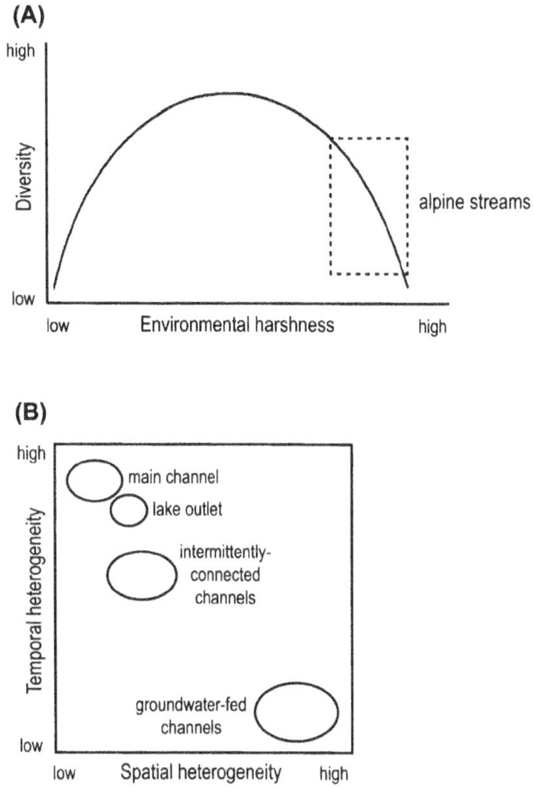

Figure 9. A) Alpine streams are positioned on the descending limb of the harshness-diversity curve. B) Within the glacial flood plain of Val Roseg, groundwater channels are characterized by a high spatial but low temporal heterogeneity. In contrast, the main channel and intermittently-connected channels exhibit a high temporal but low spatial heterogeneity. The diameter of the ellipses represents within channel type spatio-temporal variability.

templet, scale-specific predictions about distribution patterns of their benthic assemblages can be made.

- Whole flood plain: high habitat heterogeneity increases overall ecosystem stability by providing numerous refugia (after Tockner et al., 1997).
- Among channel types: more stable habitats (i.e., groundwater channels) exhibit low fluctuations in density and assemblage structure, and high levels of species diversity (Burgherr et al., 2002).
- Within channel types: channel types encompassing a greater spatial heterogeneity, such as groundwater channels (Ward et al., 1999), also show distinct differences in their benthic assemblage structure.

6. THREATS TO BIODIVERSITY

Alpine flood plains, such as the Val Roseg, exhibit biodiversity levels similar to the lowlands, and thus are considered regional "hotspots" of biodiversity. However, alpine stream ecosystems are among the most fragile and endangered with respect to the effects of climate change and diverse human impacts (McGregor et al., 1995; Messerli & Ives, 1997). These multiple impacts represent a threat to zoobenthic biodiversity of alpine streams.

In the Alps about 90% of all streams and rivers are moderately or strongly affected by human developments such as generation of hydrolelectric power, tourism or habitat degradation/loss (FOWG, 2002; Weingartner & Aschwanden, 1994). These human activities result in environmental changes that often affect species abundance and distribution more than they determine presence or absence of taxa. Additionally, they promote fragmentation of natural and often species-rich habitats such as alpine flood plains. In a recent study, Monaghan et al. (2002) showed that habitat fragmentation can reduce dispersal of alpine stream insects, leading to genetic differences among populations. Overall, reduced dispersal and habitat fragmentation potentially reduce species richness and may lead to species extinction in the long run (inbreeding depression).

Like land use developments, climate change is a dominant driver of biodiversity change in high altitude streams (Sala et al., 2000). Climate change scenarios predict that up to 95% of Alpine glacier mass could disappear by 2100 (Watson et al., 1997). Furthermore, a global increase in average temperature will lead to an intensified hydrological cycle and heavier precipitation in Switzerland, particularly during the winter half year (Beniston, 2003; Frei et al., 1998). For alpine glacial streams, such as the Roseg River, this mechanism is expected to result in a gradual shift from a glacial-nival to a nival-pluvial regime (SAEFL, 2001). This would affect downstream extent of glacial water influence, and as a consequence modify the distribution range of species. Taxa that are currently restricted to the lower reaches and/or to other channel types than the main glacial channel of the Roseg River are expected to colonize more upstream sites. Cold adapted organisms are likely to become locally extinct because upstream migration to more suitable habitats is strongly limited (summit trap phenomenon). In conclusion, the data presented here and the derived conceptual insights can serve as a basis for developing a predictive model of changes induced by the retreat of glaciers.

7. SUMMARY

Alpine glacial flood plains are highly dynamic systems with respect to both their environmental conditions and zoobenthic assemblages. Seasonal changes in the balance between glacial meltwater and groundwater contributions result in shifting patterns of spatial heterogeneity. These changes in spatial heterogeneity of the complex channel network patterns in the Val Roseg flood plain enhanced overall biodiversity by (1) providing numerous habitats for taxa not able to colonize the main glacial channel at all or during most of the year, and (2) the provision of refugia during maximum glacial melting. From an organismic view, plasticity in life history traits (e.g., feeding behavior) may play an important role in sustaining aquatic invertebrate populations in alpine glacial streams. Finally, the sensitivity of zoobenthos may be used as a potentially useful indicator to determine ecosystem health and integrity in respect to potential climate change effects and human developments.

8. REFERENCES

Atmar, W., & Patterson, B. D. (1993). The measure of order and disorder in the distribution of species in fragmented habitat. *Oecologia, 96*, 373-382.

Atmar, W., & Patterson, B. D. (1995). The nestedness temperature calculator: a visual basic program, including 294 presence-absence matrices. Chicago: AICS Research, University Park, and The Field Museum.

Beniston, M. (2003). Climatic change in mountain regions: a review of possible impacts. *Climatic Change, 59*, 5-31.

Bretschko, G. (1969). Zur Hydrobiologie zentralalpiner Gletscherabflüsse. *Verhandlungen der Deutschen Zoologischen Gesellschaft, 1968*, 741-750.

Brittain, J. E., Saltveit, S. J., Castella, E., Bogen, J., Bonsnes, T. E., Blakar, I., Bremnes, T., Haug, I., & Velle, G. (2001). The macroinvertebrate communities of two contrasting Norwegian glacial rivers in relation to environmental variables. *Freshwater Biology, 46*, 1723-1736.

Burgherr, P., & Ward, J. V. (2000). Zoobenthos of kryal and lake outlet biotopes in a glacial flood plain. *Verhandlungen der Internationalen Vereinigung für Theoretische und Angewandte Limnologie, 27*, 1587-1590.

Burgherr, P., & Ward, J. V. (2001). Longitudinal and seasonal distribution patterns of the benthic fauna of an alpine glacial stream (Val Roseg, Swiss Alps). *Freshwater Biology, 46*, 1705-1721.

Burgherr, P., Ward, J. V., & Glatthaar, R. (2001). Diversity, distribution and seasonality of the Simuliidae fauna in a glacial stream system in the Swiss Alps. *Archiv für Hydrobiologie, 152*, 19-37.

Burgherr, P., Ward, J. V., & Robinson, C. T. (2002). Seasonal variation in zoobenthos across habitat gradients in an alpine glacial flood plain (Val Roseg, Swiss Alps). *Journal of the North American Benthological Society, 21*, 561-575.

Chessel, D., & Gimaret, C. (1996). Thema 47Fr: Niches écologiques et couplages de tableaux. In D. Chessel & S. Dolédec (Eds.), *ADE Version 4: Hypercard Stacks and Quick Basic Microsoft programme library for the analysis of environmental data. User's Manual.* Lyon: ESA CNRS 5023, Ecologie des Eaux Douces et des Grands Fleuves, Université Lyon 1.

Connell, J. H. (1978). Diversity in tropical rain forests and coral reefs. *Science, 199*, 1302-1310.

Craig, P. C., & McCart, P. J. (1975). Classification of stream types in Beaufort Sea drainages between Prudhoe Bay, Alaska, and the Mackenzie Delta, N.W.T., Canada. *Arctic and Alpine Research, 7*, 183-198.

Dolédec, S., & Chessel, D. (1994). Co-inertia analysis: an alternative method for studying species-environment relationships. *Freshwater Biology, 31*, 277-294.

Dolédec, S., Chessel, D., & Gimaret-Carpentier, C. (2000). Niche separation in community analysis: a new method. *Ecology, 81*, 2914-2927.

Dorier, A. (1937). La faune des eaux courantes alpines. *Verhandlungen der Internationalen Vereinigung für Theoretische und Angewandte Limnologie, 8*, 33-41.

Elgmork, K., & Sæther, O. R. (1970). Distribution of invertebrates in a high mountain brook in the Colorado Rocky Mountains. *University of Colorado Studies. Series in Biology, 31*, 1-55.

FOWG. (2002). *Hydrologischer Atlas der Schweiz. 1. - 6. Lieferung*. Berne, Switzerland: Federal Office for Water and Geology.

Franquet, E., & Chessel, D. (1994). Approache statistique des composantes spatiales et temporelles de la relation faune-milieu. *Sciences de la vie, 317*, 202-206.

Franquet, E., Dolédec, S., & Chessel, D. (1995). Using multivariate analyses for separating spatial and temporal effects within species-environmental relationships. *Hydrobiologia, 300/301*, 425-431.

Frei, C., Schär, C., Lüthi, D., & Davies, H. C. (1998). Heavy precipitation processes in a warmer climate. *Geophysical Research Letters, 25*, 1431-1434.

Füreder, L. (1999). High alpine streams: cold habitats for insect larvae. In R. Margesin & F. Schinner (Eds.), *Cold adapted organisms. Ecology, physiology, enzymology and molecular biology* (pp. 181-196). Berlin: Springer.

Füreder, L., Schütz, C., Wallinger, M., & Burger, R. (2001). Physico-chemistry and aquatic insects of a glacier-fed and a spring-fed alpine stream. *Freshwater Biology, 46*, 1673-1690.

Hieber, M., Robinson, C. T., Uehlinger, U., & Ward, J. V. (2002). Are alpine lake outlets less harsh than other alpine streams? *Archiv für Hydrobiologie, 154*, 199-223.

Irons III, J. G., Miller, K., & Oswood, M. W. (1993). Ecological adaptations of aquatic macroinvertebrates to overwintering in interior Alaska (U.S.A) subarctic streams. *Canadian Journal of Zoology, 71*, 98-108.

Junk, W. J., Bayley, P. B., & Sparks, R. E. (1989). The flood-pulse concept in river floodplain systems. *Canadian Special Publication of Fisheries and Aquatic Sciences, 106*, 110-127.

Klein, B., & Tockner, K. (2001). Biodiversity in springbrooks of a glacial flood plain (Val Roseg, Switzerland). *Verhandlungen der Internationalen Vereinigung für Theoretische und Angewandte Limnologie, 27*, 704-710.

Koshima, S. (1984). A novel cold-tolerant insect found in a Himalayan glacier. *Nature, 310*, 225-227.

Kownacka, M., & Kownacki, A. (1972). Vertical distribution of zoocenoses in the streams of the Tatra, Caucasus and Balkan Mts. *Verhandlungen der Internationalen Vereinigung für Theoretische und Angewandte Limnologie, 18*, 742-750.

Kownacki, A. (1991). Zonal distribution and classification of the invertebrate communities in high mountain streams in South Tirol (Italy). *Verhandlungen der Internationalen Vereinigung für Theoretische und Angewandte Limnologie, 24*, 2010-2014.

Kownacki, A., & Kownacka, M. (1973). The distribution of the bottom fauna in several streams in the Middle Balkan in the summer period. *Acta Hydrobiologia, 15*, 295-310.

Lavandier, P. (1981). Cycle biologique, croissance et production de *Rhithrogena loyolaea* Navas (Ephemeroptera) dans un torrent pyrénéen de haute montagne. *Annales de Limnologie, 17*, 163-179.

Lods-Crozet, B., Castella, E., Cambin, D., Ilg, C., Knispel, S., & Mayor-Simeant, H. (2001). Macroinvertebrate community structure in relation to environmental variables in a Swiss glacial stream. *Freshwater Biology, 46*, 1641-1661.

Maiolini, B., & Lencioni, V. (2001). Longitudinal distribution of macroinvertebrate assemblages in a glacially influenced stream system in the Italian Alps. *Freshwater Biology, 46*, 1625-1639.

Malard, F., Lafont, M., Burgherr, P., & Ward, J. V. (2001). A comparison of longitudinal patterns in hyporheic and benthic oligochaete assemblages in a glacial river. *Arctic, Antarctic and Alpine Research, 33*, 457-466.

Malard, F., Tockner, K., & Ward, J. V. (1999). Shifting dominance of subcatchment water sources and flow paths in a glacial floodplain, Val Roseg, Switzerland. *Arctic, Antarctic and Alpine Reseach, 31*, 135-150.

McGregor, G., Petts, G. E., Gurnell, A. M., & Milner, A. M. (1995). Sensitivity of alpine stream ecosystems to climate change and human impacts. *Aquatic Conservation: Marine And Freshwater Ecosystems, 5*, 233-247.

Messerli, B., & Ives, J. D. (1997). *Mountains of the world: a global priority*. London: The Parthenon Publishing Group.

Milner, A. M., Brittain, J. E., Castella, E., & Petts, G. E. (2001). Trends of macroinvertebrate community structure in glacier-fed rivers in relation to environmental conditions: a synthesis. *Freshwater Biology, 46*, 1833-1847.

Milner, A. M., & Petts, G. E. (1994). Glacial rivers: physical habitat and ecology. *Freshwater Biology, 32*, 295-307.

Monaghan, M. T., Spaak, P., Robinson, C. T., & Ward, J. V. (2002). Population genetic structure of 3 alpine stream insects: influences of gene flow, demographics, and habitat fragmentation. *Journal of the North American Benthological Society, 21*, 114-131.

Olsson, T. I. (1981). Overwintering of benthic macroinvertebrates in ice and frozen sediment in a North Swedish river. *Holarctic Ecology, 4*, 161-166.

Oswood, M. W., Miller, L. K., & Irons, J. G. I. (1991). Overwintering of freshwater benthic macroinvertebrates. In R. E. Lee jr. & D. L. Denlinger (Eds.), *Insects at low temperature* (pp. 360-375). New York: Chapman & Hall.

Robinson, C. T., Uehlinger, U., & Hieber, M. (2001). Spatio-temporal variation in macroinvertebrate assemblages of glacial streams in the Swiss Alps. *Freshwater Biology, 46*, 1663-1672.

Rossaro, B., & Bettinetti, R. (2000). Chironomid distribution in north-western Italian glacial streams and cold springs. *Verhandlungen der Internationalen Vereinigung für Theoretische und Angewandte Limnologie, 27*, 2388-2391.

SAEFL. (2001). *UN Framework Convention on Climate Change: third national communication of Switzerland 2001*. Berne, Switzerland: Swiss Agency for the Environment, Forests and Landscape.

Sæther, O. A. (1968). Chironomids of the Finse area, Norway, with special reference to their distribution in a glacier brook. *Archiv für Hydrobiologie, 64*, 426-483.

Sala, O. E., Chapin III, F. S., Armesto, J. J., Berlow, E., Bloomfield, J., Dirzo, R., Huber-Sannwald, E., Huenneke, L. F., Jackson, R. B., Kinzig, A., Leemans, R., Lodge, D. M., Mooney, H. A., Oesterheld, M., Poff, N. L., Sykes, M. T., Walker, B. H., Walker, M., & Wall, D. H. (2000). Biodiversity - Global biodiversity scenarios for the year 2100. *Science, 287*, 1770-1774.

Sartori, M., & Landolt, P. (1999). *Atlas de distribution des éphémères de Suisse (Insecta, Ephemeroptera)*. Neuchâtel, Switzerland: CSCF/SES.

Schütz, C., Wallinger, M., Burger, R., & Füreder, L. (2001). Effects of snow cover on the benthic fauna in a glacier-fed stream. *Freshwater Biology, 46*, 1691-1704.

Snook, D. L., & Milner, A. M. (2001). The influence of glacial runoff on stream macroinvertebrate communities in the Taillon catchment, French Pyrenees. *Freshwater Biology, 46*, 1609-1623.

Steffan, A. W. (1971). Chironomid (Diptera) biocoenoses in Scandinavian glacier brooks. *The Canadian Entomologist, 103*, 477-486.

Steinmann, P. (1907). Die Tierwelt der Gletscherbäche. Eine faunistisch-biologische Studie. *Annales de Biologie Lacustre, 2*, 30-150.

Thienemann, A. (1912). Der Bergbach des Sauerland. *Internationale Revue der gesamten Hydrobiologie Supplement, 4*, 1-25.

Tockner, K., Malard, F., Burgherr, P., Robinson, C. T., Uehlinger, U., Zah, R., & Ward, J. V. (1997). Physico-chemical characterization of channel types in a glacial floodplain ecosystem (Val Roseg, Switzerland). *Archiv für Hydrobiologie, 140*, 433-463.

Tockner, K., Malard, F., Uehlinger, U., & Ward, J. V. (2002). Nutrients and organic matter in a glacier-fed river-floodplain system (Val Roseg, Switzerland). *Limnology and Oceanography, 47*, 266-277.

Ward, J. V. (1992). *Aquatic insect ecology: 1. biology and habitat*. New York: Wiley & Sons.

Ward, J. V. (1994). Ecology of alpine streams. *Freshwater Biology, 32*, 277-294.

Ward, J. V., Malard, F., Tockner, K., & Uehlinger, U. (1999). Influence of ground water on surface water conditions in a glacial flood plain of the Swiss Alps. *Hydrological Processes, 13*, 277-293.

Ward, J. V., & Stanford, J. A. (1983). The intermediate-disturbance hypothesis: an explanation for biotic diversity patterns in lotic ecosystems. In T. D. Fontaine III & S. M. Bartell (Eds.), *Dynamics of lotic ecosystems* (pp. 347-356). Ann Arbor, Michigan: Ann Arbor Science.

Watson, R. T., Zinyowera, M. C., & Moss, R. H. (1997). *The regional impacts of climate change : an assessment of vulnerability. A special report of IPCC Working Group II.* Cambridge: Cambridge University Press.

Weingartner, R., & Aschwanden, H. (1994). Quantification des débits des cours d'eau des Alpes suisses et des influences anthropiques qui les affectent. *Revue de Géographie Alpine, 82*, 45-57.

Zah, R., Burgherr, P., Bernasconi, S. M., & Uehlinger, U. (2001). Stable isotope analysis of macroinvertebrates and their food sources in a glacier stream. *Freshwater Biology, 46*, 871-882.

CHAPTER 11

F. MALARD

INTERSTITIAL FAUNA

1. INTRODUCTION

Global climate warming and the resulting retreat of many European glaciers have stimulated a number of recent studies on the fauna of glacial streams (Füreder, 1999; Milner et al., 2000, 2001; Lods-Crozet et al. 2001). Ecological research conducted on 7 streams distributed across Europe, from the French Pyrenees to the arctic archipelago of Svalbard, provided insight into the factors affecting the composition and distribution of benthic invertebrate assemblages in glacial streams (Brittain & Milner, 2001; Milner et al. 2001). Castella et al. (2001) emphasized the strong deterministic nature of kryal zoobenthic communities and demonstrated that their distribution pattern was driven by longitudinal changes in water temperature and substrate stability. However, despite the recognition of the hyporheic zone as a key component of stream ecosystems (Brunke & Gonser, 1997; Boulton et al., 1998; Ward et al., 2000; Malard et al., 2002), most studies have focussed on the ecology of the surface benthos. Because large amounts of sediments are transported and deposited by glacial water along Alpine valleys, the hyporheic corridor may extend meters vertically beneath the channel and hundreds of meters away from the channel.

Water-saturated interstitial habitats beneath and along glacial streams are physically more stable than benthic habitats and may provide more favorable thermal conditions (Malard et al., 2001a, 2001b). Thus, they may contribute significantly to the diversity of invertebrate assemblages in glacial streams. Insect larvae, especially Chironomidae, are the predominant members of the benthos in kryal stream segments (Ward, 1994). In contrast, sampling carried out in the hyporheic zone of the Roseg River showed that interstitial invertebrate assemblages were dominated by permanent aquatic taxa, including Oligochaeta, Cyclopoida, Harpacticoida, and Ostracoda. Several species of Harpacticoida are typical inhabitants of interstitial groundwater biotopes of high alpine and boreal regions, some of them (e.g., *Maraenobiotus insignipes insignipes* Lilljeborg 1902) being strictly bound to the border zone of glaciers (Husmann, 1975; Tilzer, 1968).

Few definitive data are available on the main factors affecting the distribution of invertebrates in interstitial habitats of glacial streams (Tilzer, 1968). One would expect the distribution patterns of benthic and shallow hyporheic invertebrate assemblages to share some similarities because advection of surface water into the bed sediments strongly influences vertical gradients of temperature, nutrients and

J.V. Ward & U. Uehlinger (eds.), Ecology of a Glacial Flood Plain, 175-198
© 2003 *Kluwer Academic Publishers. Printed in the Netherlands.*

organic matter. However, environmental conditions within the hyporheic zone of glacial streams are also controlled by the upwelling of ground water from subsurface reservoirs of the catchment. Ecological studies conducted in low elevation streams have repeatedly demonstrated the prominent role of groundwater – surface water hydrological exchanges on the distribution pattern of interstitial invertebrates (Dole-Olivier et al., 1994; Brunke & Gonser, 1999; Boulton, 2000). The influence of ground water on hyporheic biodiversity should be all the more accentuated in glacial streams because of the contrasting abiotic conditions prevailing in ground waters and surface waters. In the Val Roseg, upwelling of ground water in kryal stream segments increases hyporheic temperature, buffers environmental variations, and ensures a permanent flow of water (Malard et al., 1999, 2001a).

This chapter summarizes results of research that examined the distribution patterns of the fauna in shallow hyporheic sediments of the Roseg River. Emphasis is put on the influence of groundwater – surface water hydrological exchanges because groundwater inputs play a major role in structuring ecological conditions in the hyporheic zone of the Roseg River. First, we examine the distribution of taxa along a gradient of increasing temperature and groundwater influence as a function of the distance to the glacier terminus. Second, we compare the longitudinal patterns of taxa in the benthic and hyporheic layers of the main channel and address the role of the hyporheic zone as a refuge and as a migration pathway. Third, we test the influence of the large-scale upwelling of ground water on the density and taxonomic composition of shallow hyporheic assemblages. Finally, we propose a two-dimension habitat templet which accounts for most of the variation in the taxonomic composition and density of interstitial invertebrate assemblages among floodplain habitats.

2. COMPOSITION OF THE HYPORHEOS

Hyporheic samples were collected by driving a mobile pipe (0.025 m internal diameter, with 0.005-m holes at the base) to a depth of 0.3 m below the streambed. Ten liters of interstitial water were immediately extracted using a Bou-Rouch pump (Bou & Rouch, 1967) and filtered through a 100-μm-mesh net. Then, a 1-L volume of hyporheic water for physico-chemical analysis was withdrawn using a peristaltic pump (WAB type).

A total of 72 taxa were collected from the hyporheic zone of the Roseg River (Appendix V). However, total species richness was strongly underestimated because insect larvae could only be identified to the family level. Following the classification by Gibert et al. (1994), we distinguished between the occasional hyporheos, the permanent hyporheos and stygobites (Tables 1 and 2).

Table 1. Average density (± standard deviation), relative density, and frequency of occurrence of taxa collected in the hyporheic zone of the main channel (n=118 samples). Abundance is expressed as number of individuals per 10 L of pumped water. OH = occasional hyporheos, PH = permanent hyporheos, ST = stygobite. See Appendix V for authorities. .

	Code	Ecological category	Density Mean	Density Range	Relative density (%)	Frequency (%)
Troglochaetus cf. *beranecki*	TROG	ST	123.5 ±212.2	0 - 1163	44.4	61.9
Nematoda	NEMA	PH	58.8 ±109.0	0 - 667	21.1	89.0
Other Orthocladiinae	ORTH	OH	18.6 ±40.5	0 - 327	6.7	80.5
Parastenocaris glacialis	PAGL	ST	15.3 ±43.5	0 - 272	5.5	54.2
Leuctridae	LEUC	OH	12.9 ±18.0	0 - 120	4.6	88.1
Heptageniidae	HEPT	OH	9.9 ±16.7	0 - 107	3.6	86.4
Propappus volki	PPVO	PH	8.6 ±17.6	0 - 157	3.1	46.6
Diacyclops bisetosus	DIBI	PH	4.2 ±25.8	0 - 260	1.5	13.6
Diamesinae	DIAM	OH	4.2 ±8.0	0 - 45	1.5	66.1
Cernosvitoviella atrata	CEAT	PH	3.6 ±14.3	0 - 113	1.3	26.3
Ostracoda	OSTR	PH	3.3 ±9.1	0 - 60	1.2	30.5
Diacyclops cf. *disjunctus*	DIDI	PH	2.0 ±10.9	0 - 89	0.7	7.6
Maraenobiotus insignipes insignipes	MAIN	PH	1.7 ±5.1	0 - 46	0.6	37.3
Corynoneura spp.	CORY	OH	1.4 ±4.0	0 - 31	0.5	33.1
Nemouridae	NEMO	OH	1.3 ±10.6	0 - 115	0.5	21.2
Dorydrilus michaelseni	DOMI	PH	1.1 ±3.1	0 - 27	0.4	39.8
Baetidae	BAET	OH	1.1 ±8.6	0 - 93	0.4	11.0
Bathynellidae	BATH	ST	1.1 ±3.1	0 - 22	0.4	21.2
Hydrachnellae	HYDR	PH	0.9 ±1.4	0 - 8	0.3	43.2
Bryocamptus (A.) *cuspidatus*	BRCU	PH	0.9 ±6.1	0 - 66	0.3	15.3
Rhypholophus sp.	RHYP	OH	0.6 ±1.2	0 - 9	0.2	29.7
Acanthocyclops robustus	ACRO	PH	0.4 ±2.2	0 - 20	0.2	10.2
Dicranota sp.	DICR	OH	0.4 ±0.9	0 - 5	0.1	28.0
Nais communis	NACO	PH	0.4 ±2.8	0 - 29	0.1	3.4
Cernosvitoviella carpatica	CEOO	PH	0.4 ±1.9	0 - 16	0.1	9.3
Wiedemannia sp.	WIED	OH	0.3 ±0.9	0 - 6	0.1	17.8
Chloroperlidae	CHLO	OH	0.3 ±0.7	0 - 4	0.1	21.2
Bryocamptus (L.) *echinatus*	BREC	PH	0.3 ±1.4	0 - 13	0.1	8.5
Tanytarsini	TANY	OH	0.3 ±1.0	0 - 7	0.1	11.9
Ceratopogonidae	CERA	OH	0.2 ±1.1	0 - 8	0.1	10.2
Diacyclops languidus	DILA	PH	0.2 ±0.7	0 - 5	0.1	6.8
Rhabdomastix sp.	RHAB	OH	0.1 ±0.3	0 - 2	< 0.1	5.9
Limnephilidae	LIMN	OH	0.1 ±0.4	0 - 3	< 0.1	4.2
Haplotaxis gordioides	HAPG	PH	0.1 ±0.2	0 - 1	< 0.1	5.1
Moraria (M.) *alpina*	MOAL	PH	< 0.1 ±0.2	0 - 2	< 0.1	3.4
Bryocamptus (A.) *raethicus*	BRRA	PH	< 0.1 ±0.3	0 - 3	< 0.1	2.5
Simuliidae	SIMU	OH	< 0.1 ±0.2	0 - 1	< 0.1	4.2
Enchytraeidae g1. sp. 1	HENC	PH	< 0.1 ±0.2	0 - 1	< 0.1	3.4
Bryocamptus (R.) *zschokkei*	BRSZ	PH	< 0.1 ±0.2	0 - 2	< 0.1	2.5
Crenobia alpina	CRAL	OH	< 0.1 ±0.2	0 - 1	< 0.1	2.5
Cognettia glandulosa	COGL	PH	< 0.1 ±0.1	0 - 1	< 0.1	0.8
Niphargus cf. *tatrensis*	NIPH	ST	< 0.1 ±0.1	0 - 1	< 0.1	0.8

The occasional hyporheos consists mainly of benthic insect larvae (e.g. Chironomidae, Plecoptera, and Ephemeroptera), the early instars of which can occasionally reside in the hyporheic zone. The permanent hyporheos consists mainly of Nematoda, Oligochaeta, Copepoda, and Ostracoda that can spend all their life cycle either in the hyporheic layer or in the benthic layer. Stygobites are specialised subterranean forms that complete their whole life cycle exclusively in subsurface water. In the Val Roseg, stygobites were represented by the Archiannelida *Troglochaetus* cf. *beranecki* (Delachaux), the Harpacticoida *Parastenocaris glacialis* (Noodt), one yet undetermined species of Bathynellidae and the Amphipoda *Niphargus* cf. *tatrensis* (Wrzesniowsky). The average density of hyporheic invertebrate assemblages equalled 243 ±245 individuals per 10 L of pumped water (range: 0-1508, n=232). Assemblages were dominated by taxa belonging to the permanent hyporheos and stygobites which collectively represented 81.4 and 66.9 % of all individuals collected in the main channel and within the flood plain, respectively (Tables 1 & 2). Chironomidae (mainly Orthocladiinae), Leuctridae, and Heptageniidae were by far the most abundant and frequent insect taxa.

3. DISTRIBUTION PATTERN ALONG THE RIVER CONTINUUM

3.1. Distribution of taxa along a gradient of increasing groundwater influence

Longitudinal changes in key environmental factors (e.g. temperature, organic matter content of bed sediments, stability of the river bed) with distance from the glacier terminus are driving forces for glacial stream invertebrates (Castella et al., 2001). In order to examine the relationship between the environmental gradient and the longitudinal pattern of the hyporheos, three hyporheic samples were collected at 11 sites located along the main channel of the Roseg River from 400 m below the terminus of the Tschierva glacier (site 2M) down to the end of the catchment (site 101M) (Fig. 1). Faunal sampling was carried out in June (increasing discharge; Q = 3.78 ±0.51 m^3s^{-1}), August (high discharge; Q = 7.85 ±0.49 m^3s^{-1}), September (decreasing discharge; Q = 3.19 ±0.13 m^3s^{-1}), and November 1997 (low discharge; Q = 0.65 ±0.10 m^3s^{-1}).

In summer, hyporheic temperature increased from 1.5 °C in the proglacial reach to 6.4 °C at the end of the constrained reach (Fig. 1). Specific conductance and silica concentration increased markedly in the lower part of the flood plain, which behaved as a major groundwater upwelling zone (see Chapter 3). Although the amount of LOM (i.e. the ash-free dry mass of loosely associated organic matter > 100 µm elutriated from 10-L hyporheic samples) in hyporheic samples was extremely low, it increased from the glacier terminus to the end of the catchment. Zah (2001) also found that the amount of LOM in hyporheic sediment collected with the freezing core method increased along the altitudinal gradient from the proglacial area into the zone of subalpine forest.

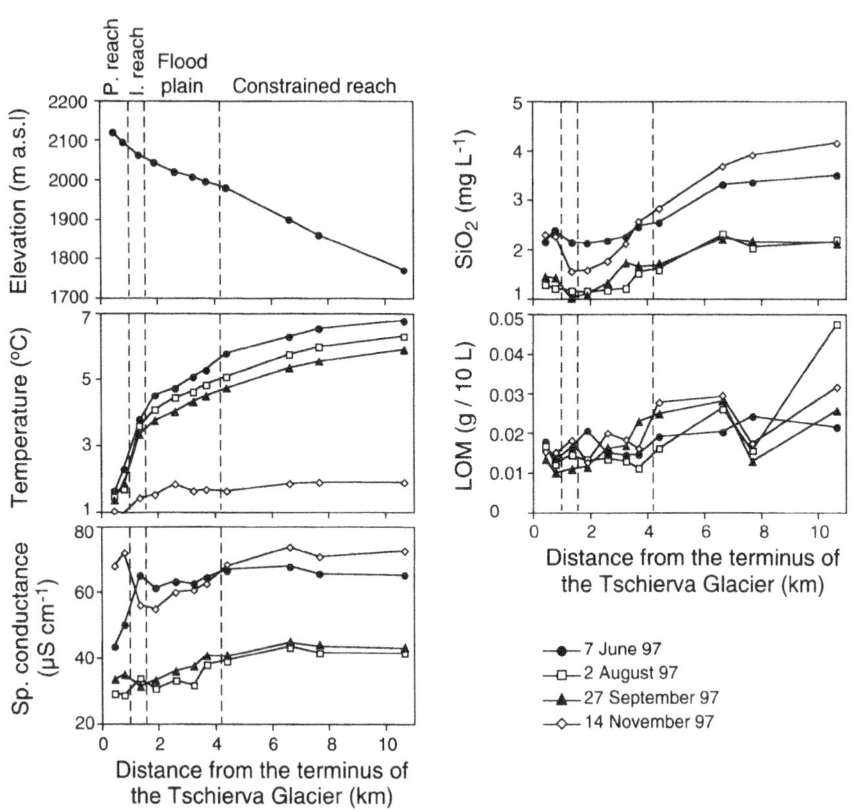

Figure 1. Location of sampling sites along the river corridor (upper left panel) and longitudinal patterns of temperature, specific conductance, silica concentration, and ash-free dry mass of loosely associated organic matter (LOM) in the hyporheic zone of the main channel.

Longitudinal patterns of biodiversity in glacial streams reflect the response of taxa to multiple environmental factors among which temperature has been ascribed a major role (Milner & Petts; 1994, Ward, 1994; Milner et al., 2001) (Fig. 2). In the Val Roseg, taxonomic richness was significantly correlated with average hyporheic water temperature ($r=0.76$, $n=11$, $p<0.05$). However, the longitudinal increase in taxonomic richness also paralleled the increase in specific conductance and concentrations of groundwater indicators including bicarbonates, base cations, and silica. Indeed, taxonomic richness exhibited a major discontinuity along the longitudinal dimension: the number of taxa increased markedly as the main channel entered the lower floodplain reach where ground water influence also became more important (Fig. 2). This suggests that surface water - groundwater linkages had a strong influence on hyporheic biodiversity along the river continuum (see chapter 3). Floodplain habitats originating from the upwelling of ground water could also act as a major source of species to the main channel.

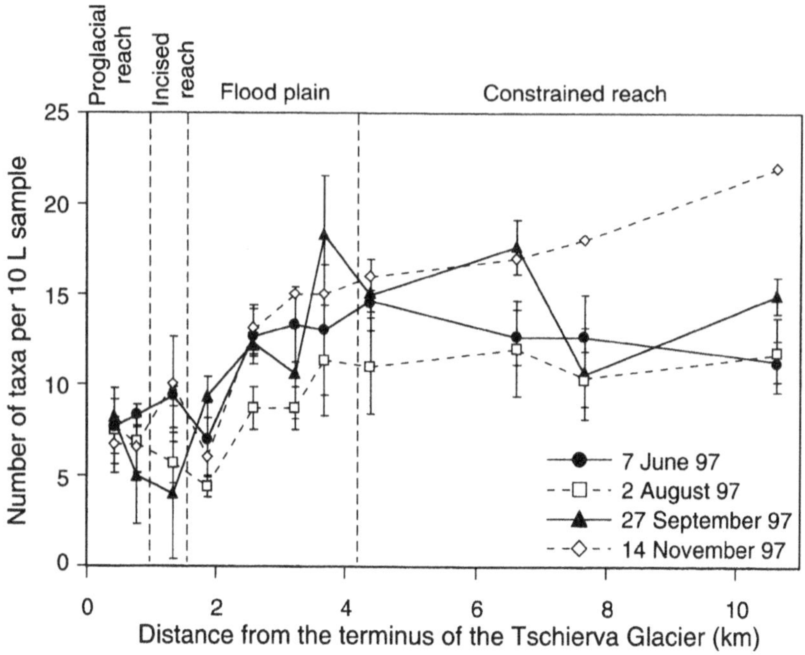

Figure 2. Longitudinal changes in the average taxonomic richness ± standard deviation (n = 3 replicates per site and date) of invertebrate assemblages in the hyporheic zone of the main channel.

A recent multivariate method, the outlying mean index (OMI; Dolédec et al., 2000) was used to examine the distribution of hyporheic taxa along a gradient of increasing groundwater influence with distance from the glacier terminus (Fig. 3). The OMI measures the distance between the mean hyporheic habitat conditions used by a taxon (taxon centroid) and the mean habitat conditions in the hyporheic zone of the river. The position of the taxa along the environmental gradient described by the first axis of the principal component analysis on physico-chemical variables (Fig. 3 A & B) depends on their niche deviation from a theoretical ubiquitous taxon that would be uniformly distributed among habitat conditions. The first axis of the principal component analysis arranged the sampling sites along the longitudinal gradient based on the increase in specific conductance, concentrations of silica, base cations and bicarbonates, temperature, and LOM (Fig. 3 A & B). The spatial niche of 20 taxa (out of 42 taxa) deviated significantly ($p<0.1$) from a uniform distribution along this gradient of increasing groundwater influence (Fig. 3C). Only 2 taxa, the Turbellaria *Crenobia alpina* (Dana) and the Harpacticoida *M. insignipes insignipes* colonized preferentially the proglacial reach. Most of the taxa present in the proglacial reach occurred over the entire longitudinal gradient.

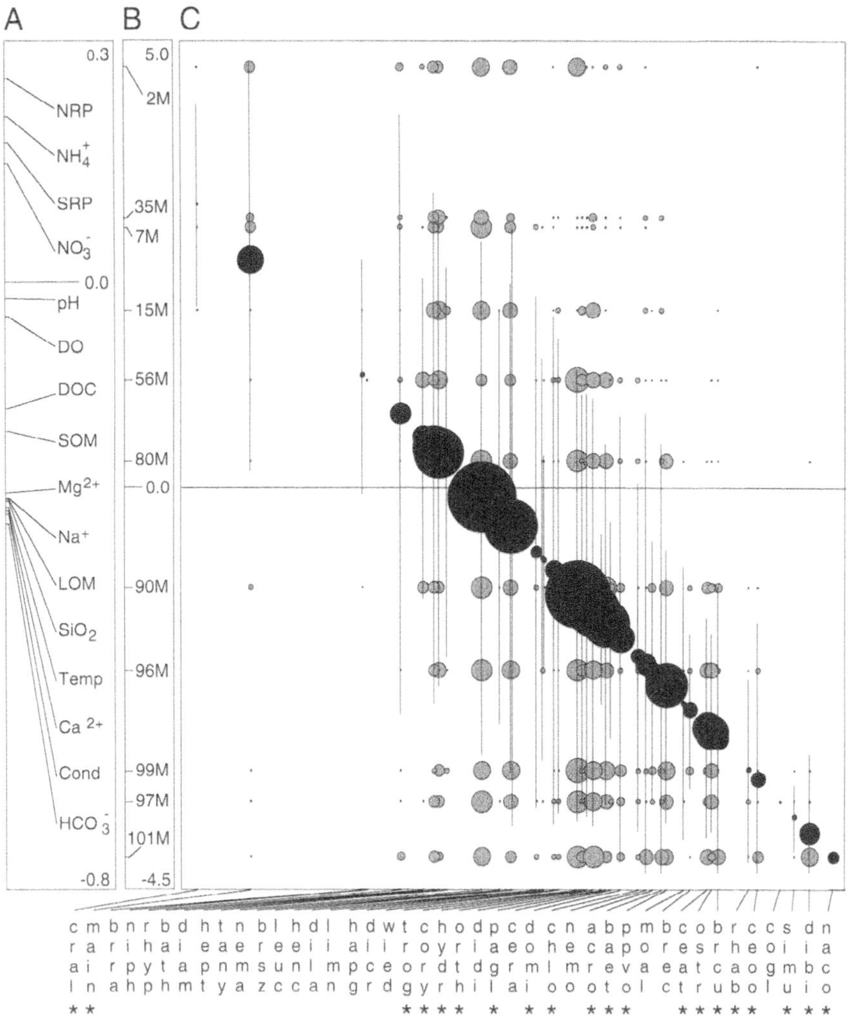

Figure 3. OMI analysis of interstitial fauna. Only the first axis, which represented 87 % of the explained variability, is used for graphical presentation. A) Canonical weights of environmental variables. NRP: non reactive phosphorus; SRP: soluble reactive phosphorus; DO: dissolved oxygen; DOC: dissolved organic carbon; SOM: strongly associated organic matter; LOM: loosely associated organic matter; Temp: temperature; Cond: specific conductance. B) Scores of sampling sites. C) Distribution of taxa along the environmental gradient as a function of their weighted average position along site scores. The sizes of black circles are proportional to the total frequencies of taxa. Vertical lines correspond to the standard deviations. The sizes of grey circles are proportional to the densities of taxa. Asterisks indicate taxa whose distribution deviates significantly from a uniform distribution along the environmental gradient. See Table 1 for the codes of taxa.

The location of the centroids showed that almost all taxa occurred preferentially downstream of the major groundwater upwelling zone (i.e. downstream of site 56 M). The increase in taxonomic richness in the lower floodplain reach was essentially due to the appearance of taxa belonging to the permanent hyporheos including several species of Oligochaeta, Harpacticoida, Cyclopoida, and Ostracoda. The sharp increase in the diversity and abundance of hyporheic invertebrate assemblages was closely linked to the convergence of ground water and surface glacial water, a process determined by the hydrogeomorphologic characteristics of the catchment. The longitudinal distribution of taxa in the hyporheic zone of the main channel varied little over the year despite strong temporal changes in the environmental gradient (Fig. 1). Overall, the distribution pattern of the hyporheos appeared to be more stable than that of the surface benthos. Temporal changes in total densities were also much more attenuated in the hyporheic layer than in the benthic layer. Total hyporheic densities, all sites combined, varied from 214 invertebrates per 10 L of pumped water in August to 360 in June, whereas total zoobenthic densities varied from 1472 invertebrates per m^2 in August to 13,398 invertebrates per m^2 in October (Burgherr & Ward, 2001).

3.2. The hyporheic zone as a migration pathway and refuge

The hyporheic zone may contribute significantly to the biodiversity of glacial streams by providing refugia against harsh environmental conditions prevailing in the benthic layer. The hyporheic zone can also be used as a nursery by benthic invertebrates, although this aspect is still conjectural due mainly to the lack of appropriate data. Most insect larvae collected in the hyporheic zone of the Roseg River were early instars whose body size did not exceed 5 mm. However, it is not yet clear whether the dominance of small size insect larvae is the result of a behavioral process (i.e. early instars preferentially migrate into the sediment) or more simply the result of a physical filtration process (i.e. small pore size only allows the occurrence of early instars).

There seemed to be strong differences in the colonization of the hyporheic layer between insect larvae. For example, Diamesinae constituted 79 % of the Chironomidae in the benthic layer, whereas hyporheic Chironomidae were dominated by Orthocladiinae (65 %). Differences in the use of the hyporheic zone by Diamesinae and Orthocladiinae might reflect the greater adaptation of the former to the harsh environmental conditions prevailing in the benthic layer of glacial streams. Baetidae (59 %) and Heptageniidae (41 %) occurred in approximately the same proportion in the benthic layer, whereas Heptageniidae was the dominant Ephemeroptera in hyporheic sediments (88 % versus 12 % for the Baetidae). Leuctridae was almost the only Plecoptera to colonize the hyporheic zone (99 % at most of the sites), whereas Nemouridae represented about one third of the Plecoptera in the benthic layer. Leuctridae occurred about 350 m farther upstream in the hyporheic layer of the proglacial reach (site 2M) than in the benthic layer (site 7M).

Several taxa belonging to the permanent hyporheos showed distinct longitudinal distribution patterns in the hyporheic and benthic layers of the main channel. Copepoda and Ostracoda occurred almost exclusively in the hyporheic layer of the main channel. Only 17 individuals of Copepoda and 4 of Ostracoda were found in 102 benthic samples collected in the main channel, whereas 1155 Copepoda (excluding the stygobite *P. glacialis*) and 387 Ostracoda were recovered from the 118 hyporheic samples. Although the composition of hyporheic and benthic oligochaete assemblages was generally similar, several species extended farther upstream in the hyporheic zone than in the surface stream (Malard et al., 2001b). *Dorydrilus michaelseni* (Piguet) colonized the hyporheic layer of the proglacial reach (site 2M), whereas in the benthic layer it occurred only at a distance of 6.6 km from the glacier terminus (site 97M). *Propappus volki* (Michaelsen) and *Cernosvitoviella atrata* (Bretscher) also developed abundant populations farther upstream in the hyporheic zone (site 90M) than in the surface stream (site 97M).

Differences between the longitudinal patterns of the benthos and hyporheos suggest that the hyporheic zone could act both as the main upstream migration pathway and as a refuge for several taxa belonging to the permanent hyporheos including the Harpacticoida, Cyclopoida, Ostracoda, and Oligochaeta. Several authors have emphasized the potential role of the hyporheic corridor as a main dispersal pathway for permanent aquatic taxa. In their presentation of the interstitial highway model, Ward & Palmer (1994) hypothesized that alluvial aquifers were the major dispersal route between several spatially-discontinuous hypogean habitats with limited interconnections. Henry & Magniez (1992) postulated that the spreading of sediments during glacial retreat in the Quaternary promoted the upstream dispersal of some stygobites including the Isopoda *Proasellus cavaticus* (Leydig) and *Stenasellus virei* (Dolfus). Permanent aquatic taxa that could complete their entire life cycle in either ground water or in the surface stream were probably able to migrate, not only in the shallow hyporheic zone, but also deeper in alluvial sediments. In the Val Roseg, the hyporheic distribution of the most invasive Oligochaeta, *D. michaelseni*, matched that of the stygobite Archiannelida *T.* cf. *beranecki* and the stygobite Harpacticoida *P. glaclialis* (Malard et al., 2001b). Recolonization of suitable surface aquatic habitats via subsurface pathways could produce discontinuous patterns of distribution within the catchment. For example, the surface stream at the outlet of Roseg Lake, characterized by relatively higher temperature and bed stability, harbored six species of Oligochaeta (*Nais communis* (Piguet), *C. atrata*, *Cernosvitoviella carpatica* (Nielsen & Christensen), *Henlea* sp., *Hemienchytraeus* sp., and *Fridericia* sp.), whereas Oligochaeta were almost totally absent from the benthic layer at other sites in the most upstream reaches of the Roseg River (sites 2M to 56M). Although the longitudinal pattern of permanent aquatic taxa in the hyporheic zone of the main channel was primarily determined by river - aquifer interactions, a catchment-scale feature, the differential distribution of hyporheic and benthic assemblages appeared to result from particular instream conditions (i.e. high flow velocity and instability of the streambed) that prevented the maintenance of permanent populations in the benthic layer.

4. DISTRIBUTION PATTERN WITHIN THE FLOOD PLAIN

A total of 28 sampling sites were selected to examine the distribution pattern of the hyporheos within the flood plain (Fig. 4). More particularly, the sampling design was made to test the effect of the large scale upwelling of ground water in the lower flood plain on the diversity and density of the hyporheos (Malard et al., 2003). Thirteen sites were located in the upper flood plain and 15 in the lower flood plain. Sites were located either in surface-connected channels or surface-disconnected channels.

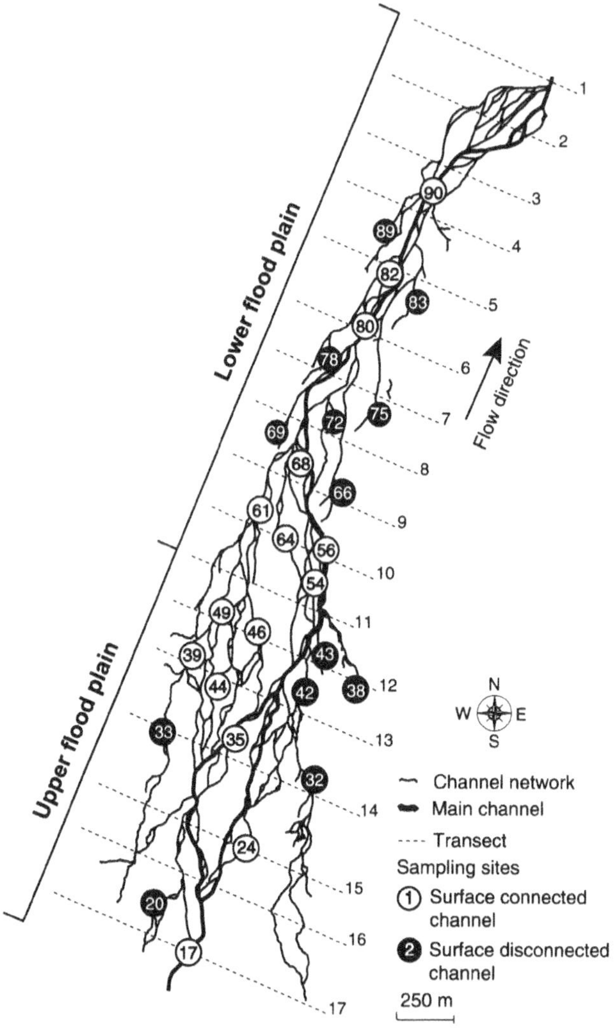

Figure 4: Detailed map of the flood plain showing the location of hyporheic sampling sites.

Surface-connected channels had an upstream connection with the main channel during the main ablation period. Surface-disconnected channels had no upstream connection with the main channel because they were fed by subsurface water. Surface-disconnected channels encompassed a diversity of channel types including hillslope groundwater-fed tributaries entering the flood plain as surface water channels (sites 20, 32, 33, 42, 78), floodplain spring brooks fed by upwelling of hillslope ground water (sites 38, 43, 66, 69, 75, 83, 89), and floodplain spring brooks originating from the resurfacing of main channel water (site 72). Sampling was carried out during the expansion phase (2-6 June 1997; Q = 3.78 ±0.51 m^3s^{-1}, total channel length: T = 22.5 km), the maximum expansion of the channel network (28 July - 2 Aug., Q = 7.85 ±0.49 m^3s^{-1}; T = 23.9 km), the contraction phase (22-27 Sept., Q = 3.19 ±0.13 m^3s^{-1}; T = 20.6 km), and 1 wk prior to the drying of most channels in the upper flood plain (10-14 Nov., Q = 0.65 ±0.10 m^3s^{-1}, T = 16.7 km).

A total of 37,000 invertebrates belonging to 20 higher taxa were collected from the hyporheic zone of the Roseg River flood plain (Table 2). The mean density was 219 ±272 invertebrates per 10 L of pumped water (range: 0-1458, n=169). Chironomidae was by far the most abundant and frequent taxon. However, 8 of the 11 most abundant taxa, representing 97 % of the total number of invertebrates collected in the flood plain, belong to the permanent hyporheos or stygobitic fauna.

Table 2. Average density (± standard deviation), relative density, and frequency of occurrence of taxa collected in the hyporheic zone of the flood plain (n=169 samples). Abundance is expressed as number of individuals per 10 L of pumped water. OH = occasional hyporheos, PH = permanent hyporheos, ST = stygobite.

			Density			
	Code	*Ecological category*	*Mean*	*Range*	*Relative density (%)*	*Frequency (%)*
Chironomidae	Chiro	OH	55.5 ±105.1	0 - 771	25.4	95.9
Troglochaetus cf. *beranecki*	Trogl	ST	42.4 ±102.1	0 - 683	19.4	63.9
Nematoda	Nemat	PH	33.7 ±75.0	0 - 667	15.4	88.2
Ostracoda	Ostra	PH	19.6 ±76.2	0 - 521	9.0	33.1
Oligochaeta	Oligo	PH	19.3 ±34.8	0 - 236	8.8	72.2
Parastenocaris glacialis	Paras	ST	14.6 ±47.9	0 - 358	6.7	49.1
Harpacticoida	Harpa	PH	8.4 ±42.1	0 - 392	3.8	51.5
Plecoptera	Pleco	OH	8.0 ±15.6	0 - 125	3.7	74.0
Heptageniidae	Hepta	OH	5.1 ±14.4	0 - 107	2.3	42.6
Cyclopoida	Cyclo	PH	5.1 ±15.7	0 - 105	2.3	34.3
Bathynellidae	Bathy	ST	2.3 ±3.9	0 - 22	1.0	50.3
Limoniidae	Limon	OH	1.0 ±2.3	0 - 13	0.5	36.1
Baetidae	Baeti	OH	1.0 ±5.7	0 - 71	0.4	20.1
Hydrachnellae	Hydr	PH	0.8 ±1.7	0 - 12	0.4	33.1
Limnephilidae	Limne	OH	0.7 ±3.3	0 - 33	0.3	17.2
Ceratopogonidae	Cerat	OH	0.6 ±1.7	0 - 16	0.3	24.9
Crenobia alpina	Creno	OH	0.3 ±1.6	0 - 17	0.1	11.2
Empididae	Empid	OH	0.2 ±0.8	0 - 7	0.1	8.9
Niphargus cf. *tatrensis*	Nipha	ST	0.1 ±0.7	0 - 7	0.1	8.3
Simuliidae	Simul	OH	0.1 ±0.4	0 - 4	< 0.1	5.3

Repeated measures analysis of variances were performed to test for differences in taxonomic richness and density of ecological categories between the upper and lower parts of the flood plain. The design was a 2 (longitudinal effect) by 2 (surface hydrological connectivity) by 4 (date) analysis of variance. Taxonomic richness and the densities of the occasional hyporheos, permanent hyporheos, and stygobitic fauna were significantly higher ($p<0.05$) in the lower flood plain where upwelling occurred on a large scale. (Fig. 5). The mean number of taxa averaged 10.2 ±2.8 in the lower part of the flood plain against 7.6 ±2.3 in the upper part. The downstream increase in taxonomic richness and densities occurred both in surface-connected channels and surface-disconnected channels of the flood plain.

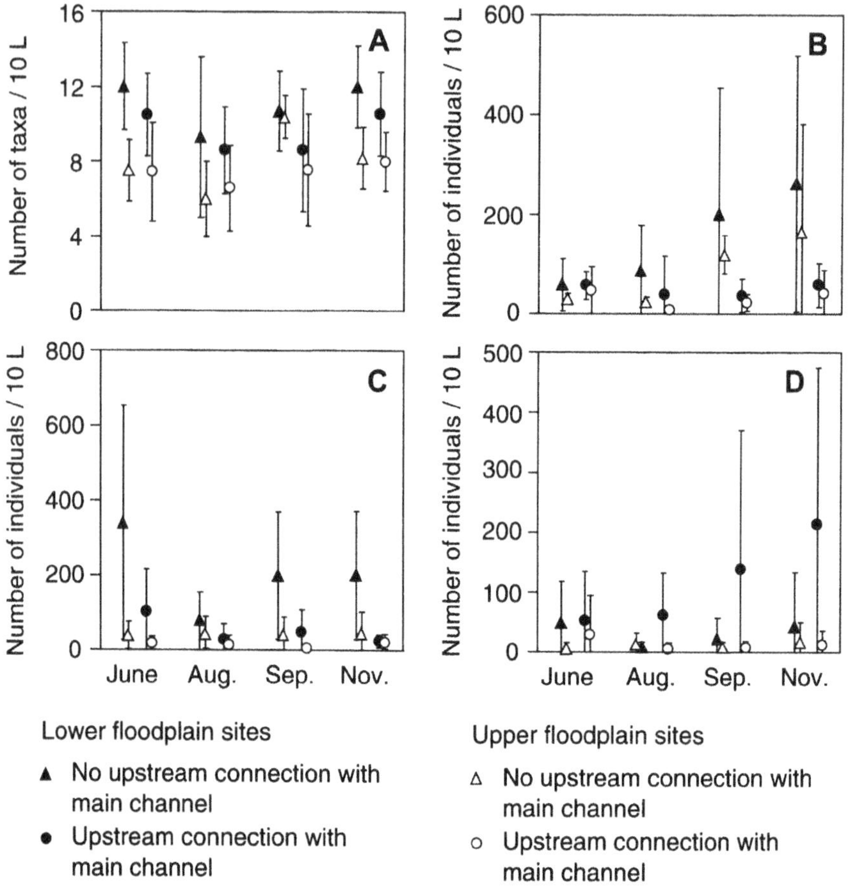

Figure 5. Differences in taxonomic richness (A) and density of occasional hyporheos (B), permanent hyporheos (C), and stybobites (D) between the upper and lower parts of the flood plain. Data are means ± standard deviation (n≥5 sampling sites).

We used the STATIS multivariate method (Lavit et al., 1994) to investigate the distribution patterns of hyporheic assemblages over the 4 different phases of the expansion-contraction cycle (Fig. 6). This method which is particularly well suited for analysing the stability of a spatial pattern aims at the simultaneous ordination of several data "tables" (4 site-by-taxa tables in the present study) by constructing a mean table of maximum inertia called the "compromise table". Then, the compromise table is analysed by a principal component analysis (PCA), thereby providing factorial axes and components as in usual multivariate methods. The mathematical principles of STATIS have recently been summarised by Blanc (2000) and Gaertner et al. (1998).

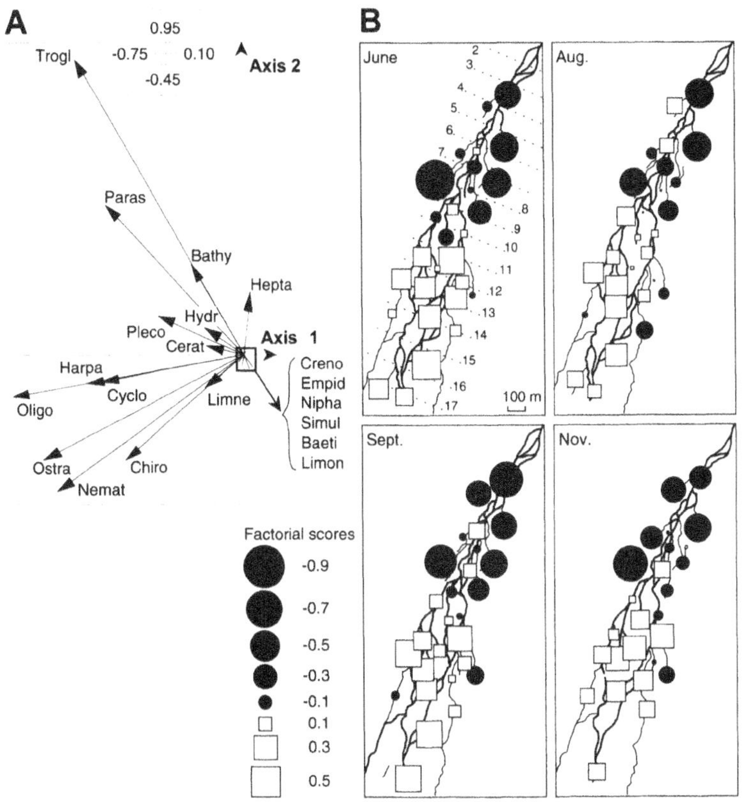

Figure 6. A) Plot of factorial scores of taxa along the first 2 axes of the principal component analysis of the STATIS compromise table. B) Factorial scores of sampling sites along the first axis are plotted directly on a floodplain map for each date. The surface areas of circles (negative scores) and squares (positive scores) are proportional to the factorial scores. Modified from Malard et al. (2003), Fig. 8, p. 104.

Taxa were distributed along the first axis of the principal component analysis of the compromise table following a density gradient (Fig. 6A). The most abundant taxa, including Oligochaeta, Cyclopoida, Harpacticoida, Ostracoda, the Archiannelida *T.* cf. *beranecki*, Nematoda, and Chironomidae were positioned on the left side of the axis, whereas insect taxa (except Chironomidae), Hydrachnellae, *C. alpina*, Bathynellidae and *N.* cf. *tatrensis* occurred on the right side of the axis. The factorial scores of sampling sites along the first axis of the principal component were plotted directly on a floodplain map using squares (positive scores) and circles (negative scores), the surface area of which is proportional to the absolute values of the factorial scores (Fig. 6B). The mapping of factorial scores revealed a strong longitudinal discontinuity in the composition of hyporheic assemblages by separating sites located in the lower flood plain (negative scores indicated by circles) from those located in the upper flood plain (positive scores indicated by squares). The lower floodplain sites harbored abundant and diversified hyporheic assemblages composed of taxa belonging to the permanent hyporheos, stygobitic fauna, and occasional hyporheos (Chironomidae), whereas upper floodplain sites contained fewer invertebrates, belonging mainly to the occasional hyporheos. The longitudinal shift in densities was particularly evident for the Archiannelida *T.* cf. *beranecki*, Oligochaeta, Cyclopoida, Harpacticoida, and Ostracoda, because these taxa were absent from many sites located in the upper flood plain (Fig. 7).

The longitudinal pattern of hyporheos remained remarkably stable over the 4 different phases of the expansion-contraction cycle despite major changes in the physico-chemistry of shallow hyporheic water caused by small scale downwelling-upwelling flow paths of glacial water at multiple locations within the flood plain. The stability of the spatial pattern of hyporheic assemblages stands in stark contrast with the high spatio-temporal dynamics of the surface benthos. The distribution pattern of benthic invertebrates was shown to vary in response to the pulsing of discharge and resulting longitudinal and lateral expansion of glacial water within the flood plain (Burgherr & Ward 2001; Burgherr et al. 2002).

Several studies have demonstrated the primary influence of downwelling / upwelling flow patterns on the composition and distribution of interstitial assemblages along small scale geomorphological features such as riffle-pool sequences (Sterba et al., 1992; Mermillod-Blondin et al., 2000), gravel bars (Dole-Olivier & Marmonier, 1992), gravel runs (Stanley & Boulton, 1993; Boulton & Stanley, 1995), and stream reaches (Creuzé des Châtelliers & Reygrobellet, 1990). In their formulation of the hyporheic corridor concept, Stanford & Ward (1993) predicted that the large-scale upwelling of ground water at the downstream end of flood plains might be a primary determinant of floodplain biodiversity and bioproduction. This prediction was largely supported by the distinctiveness of hyporheic assemblages between the upper and lower sections of the Roseg flood plain. The longitudinal shift in the composition of hyporheic assemblages was essentially due to an increase in the number and density of taxa belonging to the permanent hyporheos and stygobitic fauna. In the Val Roseg, faunal heterogeneity along the longitudinal dimension was even more pronounced than along the lateral dimension. In many river-floodplain systems, lateral changes in surface hydrological connection between floodplain water bodies and the main river channel are the main source of variation in the composition of hyporheic and benthic communities (Ward et al. 1998, 1999a).

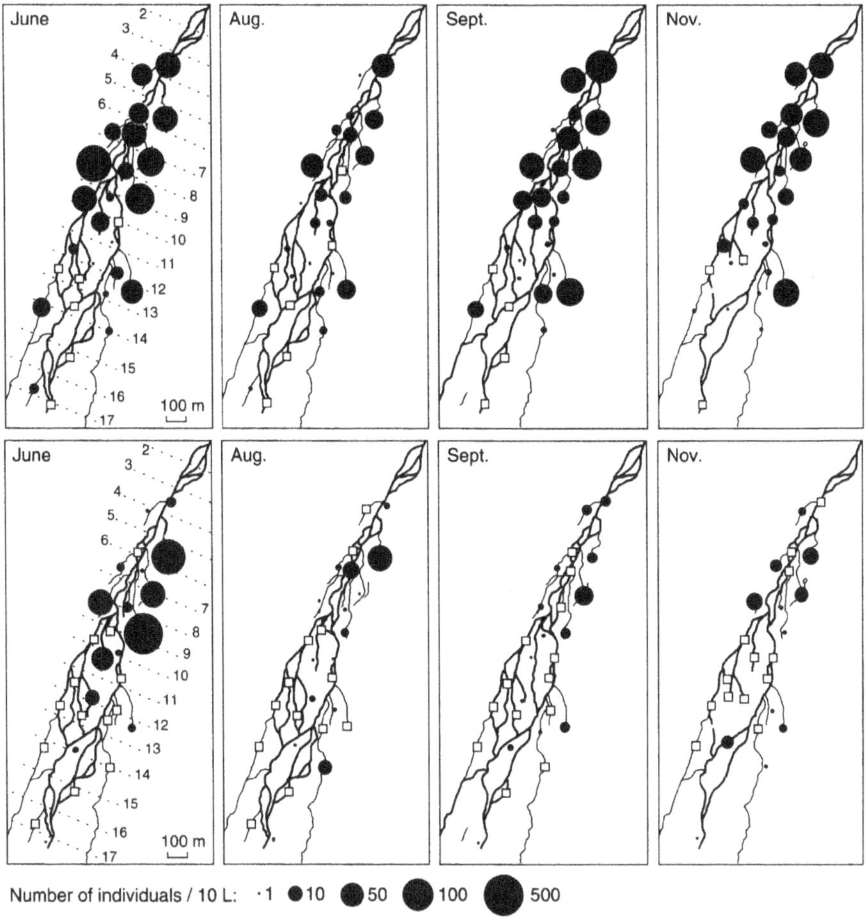

Figure 7. Spatial pattern of Oligochaeta (upper panels) and Harpacticoida (excluding Parastenocaris glacialis; *lower panels) in the flood plain. Diameters of circles are proportional to the number of individuals collected per 10 L of pumped water. Squares correspond to sites where no individuals were collected.*

The primary influence of groundwater upwelling was to maintain permanent flow conditions in the lower flood plain (Ward et al., 1999b; Malard et al., 1999). In the absence of pronounced differences in the physico-chemical characteristics of hyporheic water and sediment between the upper and lower parts of the flood plain (Malard et al., 2003), the longitudinal shift in invertebrate assemblages was more likely due to the low fluxes of organisms between newly created aquatic patches in the upper flood plain and the source communities established in the lower flood plain. Recolonization by permanent hyporheos and stygobites after rewetting of hyporheic sediments in the upper flood plain could occur via 3 major pathways: downstream drift, upstream migration in the benthic layer, and vertical migration from ground water. Additionally, some taxa could exhibit physiological adaptations that allowed them to survive during winter in dry and frozen bed sediment (Oswood et al. 1991; Irons et al. 1993). Because of the high physical instability of the streambed, non insect taxa occurred only occasionally at very low densities in the water column and surficial benthic layer of the main glacial channel (Burgherr & Ward 2001; Robinson et al. 2001). Moreover, vertical migration from ground water was probably limited in the upper flood plain because many channels had their bed separated from local ground water by an unsaturated layer of sediment (Malard et al., 2000). In contrast, vertical migration in the lower flood plain was facilitated by direct saturated linkages between the stream bed sediments and ground water. There was little difference in hyporheic assemblages among lower floodplain sites, even though some of them went dry in winter (e.g. sites 69, 72, 78). Thus, the longitudinal variation in groundwater – surface water linkages appeared as a major determinant of hyporheic biodiversity by controlling not only the degree of flow permanency but also the pathways of recolonization. The stability of the spatial pattern of hyporheic assemblages also suggested that the composition and density of shallow hyporheic assemblages was more influenced by the nature of surface – subsurface linkages (e.g. saturated *versus* unsaturated linkages) than by the direction of small-scale hydrological exchanges.

5. RIVER HABITAT TEMPLET

A habitat templet has been developed in an attempt to synthesize our current understanding of the factors that structure hyporheic assemblages in the Val Roseg. We have chosen the strength of saturated hydrological connection with ground water as one axis of the habitat templet. This axis also corresponds to a gradient of water permanency because channels that are in direct saturated hydrological connection with ground water tend to maintain a permanent flow of water. In contrast, channels that lack saturated connection to the ground water dry when the surface flow is too low to compensate for water infiltration into the bed sediments. Depending on their location within the flood plain, temporary channels maintain surface water flow during 4 to 8 months (Malard et al., 1999). The second axis of the habitat templet represents the degree of upstream surface hydrological connection with the main channel which carries most of the glacial water during the ice ablation period. Temporal heterogeneity increases along this second axis because surface connected channels fed by glacial water exhibit stronger fluctuations in discharge and associated environmental conditions (e.g., bedload transport) than surface disconnected channels fed by ground water. Using this habitat templet, we distinguished between 5 different types of habitat patches in the flood plain: I - permanent habitats that are never connected to the main channel at their upstream ends; II - permanent habitats that are connected to the main channel in summer; III - temporary habitats that are never connected to the main channel; IV - temporary habitats that are intermittently connected to the main channel; and V - temporary habitats that are always connected to the main channel. Table 3 shows the total length and relative proportions of these 5 habitat patches in the flood plain. The duration and timing of the period during which the different habitat patches carry clear water (i.e. turbidity < 9 NTU) correspond to the period of surface upstream disconnection with the main channel. Temporary channels represent 72.9 % of the total length of the channel network. Permanent channels that are never connected to the main channel represent a length of only about 2.4 km, which corresponds to less than 10 % of the total channel network.

Table 3. Permanency, life span as clear water, total length, and proportion of 5 types of habitat patches in the flood plain of the Roseg River (data are for year 1997).

	Floodplain habitat patch types				
	I	*II*	*III*	*IV*	*V*
Water permanency	yes	yes	no	no	no
Duration as clear water channel (month)	12.0	6.4 ±1.2	7.2 ±1.6	2.6 ±1.8	0.0
Period of clear water	All year	winter	Summer	Spring/ autumn	Never
Total length (m)	2382	4315	4352	6140	7550
Proportion of the total channel network %)	9.6	17.4	17.6	24.8	30.5

Our classification which is primarily based on the relative position of habitat patches along the axes of the habitat templet also takes into consideration the distribution of habitat patches along the longitudinal and lateral dimensions of the flood plain. Therefore, it includes both the compositional and configurational features of the riverscape. Among the permanent channels that are mostly restricted to the downstream part of the flood plain, surface disconnected channels (type I) are located at the floodplain margins (i.e. sites 38, 66, 69, 72, 75, 83, and 89), whereas surface connected channels (type II) are positioned centrally within the active flood plain (i.e. sites 54, 56, 78, 80, 82, 90) (Fig. 4). Most temporary channels (types III to V) are located in the upstream part of the flood plain. They comprise temporary surface disconnected channels (type III) that are located at the margins of the upper flood plain (i.e. sites 32, 33, 42, 43, 64), intermittently connected channels (type IV) which occur in the center of the upper flood plain between transects 9 and 13 (i.e. sites 39, 46, 49, 61, and 68) and permanently connected channels (type V) that are located in the center and uppermost part of the flood plain (transects 13 to 17; sites 17, 20, 24, 35, and 44).

There was a clear decreasing gradient of taxonomic richness and density of invertebrate assemblages among habitat patches (Fig. 8). Taxonomic richness and density were highest in permanent surface disconnected habitat patches located along the margins of the lower flood plain and were lowest in temporary kryal channels located in the upper most part of the flood plain.

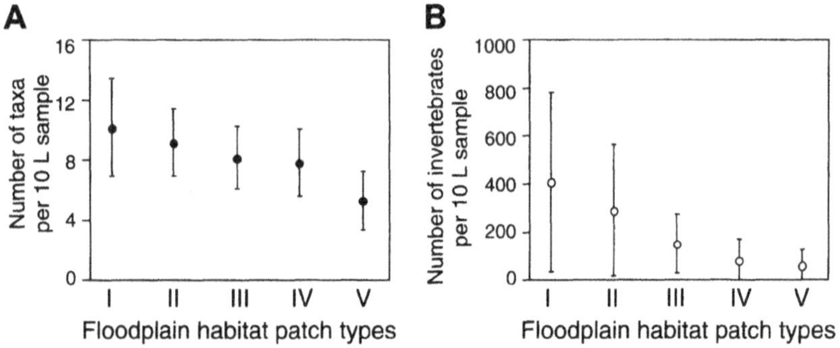

Figure 8. Average (± standard deviation) taxonomic richness (A) and density of interstitial invertebrate assemblages(B) in 5 distinct habitat patch types of the flood plain. The number of samples for each habitat patch varies from 26 to 43. See text and table 3 for the definition of floodplain habitat patch types.

Using all hyporheic replicate samples collected in the flood plain from June to November 1997 (n=169), we performed a centered principal component analysis (PCA) to examine differences in the composition of invertebrate assemblages among habitat patches (Fig. 9). The centroids of the different habitat patches were clearly separated along the first axis of the principal component analysis. Permanent habitat patches that were never connected to the main channel (type I) harbored abundant invertebrate assemblages composed mainly of taxa belonging to the permanent hyporheos (Ostracoda, Oligochaeta, Cyclopoida, and Harpacticoida) and occasional hyporheos (Chironomidae). Permanent channels that carry glacial water during summer (type II) were characterized by their higher densities of the cold stenothermal stygobites *T.* cf. *beranecki* and *P. glacialis*. Although there were some differences in the composition of hyporheic assemblages among temporary habitat patches (types III, IV, and V), they contained fewer invertebrates belonging mainly to the occasional hyporheos.

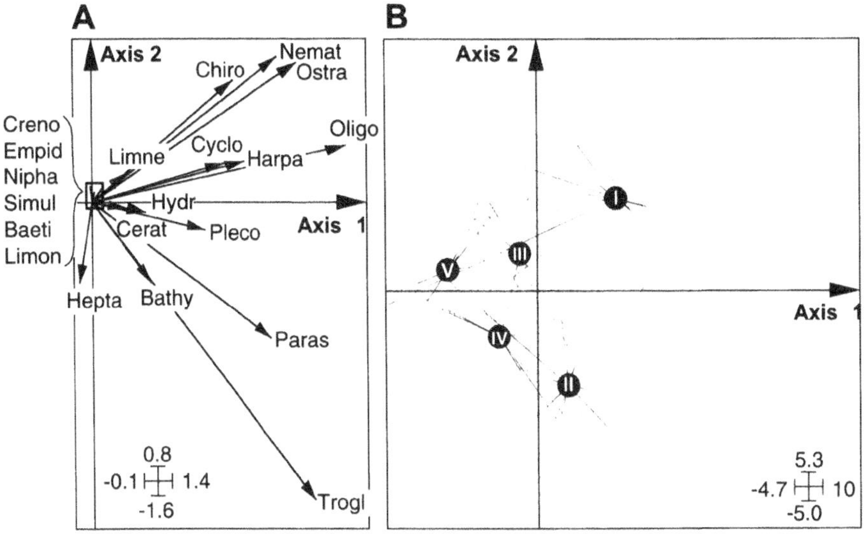

Figure 9. Differences in the composition of interstitial invertebrate assemblages among 5 floodplain habitat patch types. A) Plot of factorial scores of taxa along the first 2 axes of the principal component analysis. B) Plot of factorial scores of samples. Black circles correspond to the centers of gravity of the 5 floodplain habitat patch types. The number of samples for each habitat patch varies from 26 to 43. See text and table 3 for the definition of floodplain habitat patch types.

A few studies have examined the main factors affecting the distribution pattern of interstitial assemblages in river-floodplain systems (Ward et al., 1998). Dole-Olivier et al. (1993) demonstrated that spatial variation in the relative influence of ground water and advected surface water in interstitial habitats of the Rhône River flood plain, France, created repeated gradients in the composition of the stygobitic fauna along the vertical, lateral, and longitudinal dimensions. Ward et al. (1994) suggested that the patchy distribution of interstitial crustaceans in the Flathead River flood plain, Montana, reflected site-specific geomorphic and hydrogeological features induced by a subterranean latticework of alluvia-filled paleochannels of high hydraulic conductivity. The composition of interstitial invertebrate assemblages in the Roseg River flood plain appears to be driven mainly by the strength of saturated hydrological connection with ground water and to a lesser extent by the degree of upstream surface connection with the main channel (Fig. 10). The type of groundwater – surface water hydrological linkage largely determines the life span of aquatic habitats, whereas the strength of surface hydrological connection with the main channel strongly influences several key environmental features such as water temperature and streambed stability. Permanent habitats that are never connected to the main channel are hotspots of biodiversity in the Roseg River flood plain. The Ostracoda, Cyclopoida, and Harpacticoida (except *P. glacialis*) were almost restricted to a few hillslope groundwater-fed channels located in the lower flood plain (e.g. sites 66, 69, 75, 83, and 89).

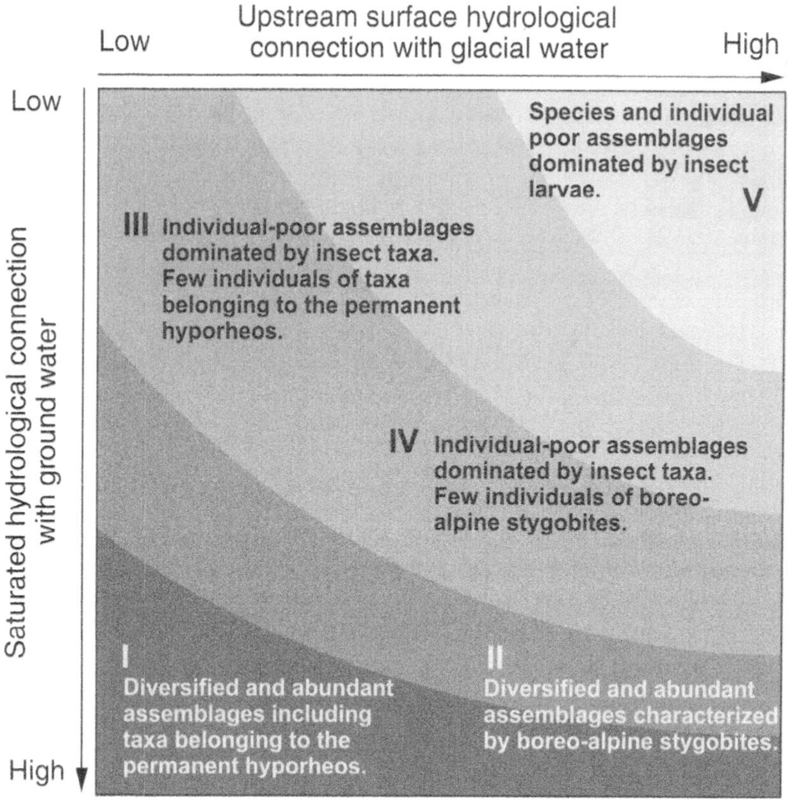

Figure 10. Variation in the composition and biodiversity of interstitial assemblages along the axes of a two-dimension habitat templet. The change from light to dark grey patterns corresponds to a gradient of increasing biodiversity. Numbers identify the position of the 5 floodplain habitat patch types along the axes of the templet (See text and table 3 for the definition of floodplain habitat patch types).

Permanent groundwater-fed channels may constitute source areas from which temporary habitat patches presenting suitable environmental conditions can be colonized. However, as previously discussed in section 4 of this chapter, upstream migration of permanent aquatic organisms between source areas and newly created patches presenting suitable environmental conditions (e.g. temporary surface disconnected channels in the upper flood plain) is probably limited. Thus, the distribution pattern of interstitial fauna within the flood plain may be due in part to the strength of ecological connectivity between habitat patches rather than to differences in environmental features among patches.

6. SUMMARY

Upwelling of ground water exerts a dominating influence on the taxonomic richness and density of interstitial invertebrate assemblages in the Roseg River. The distribution patterns of interstitial fauna along the river corridor and within the flood plain clearly demonstrate the significance of convergence zones between ground water and surface water in sustaining biodiversity in shallow hyporheic sediments of a glacial river system. The beneficial effect of groundwater inputs lies in the differences between the intrinsic features of the groundwater and glacial environments: the glacial surface water environment is highly variable and ephemeral, whereas the groundwater environment is typically more stable and permanent. Differences between the distribution patterns of the surface benthos and hyporheos along the river continuum suggests that the hyporheic corridor may facilitate the recolonization of recently deglaciated catchments by several permanent aquatic taxa. Saturated alluvia of the Val Roseg are probably used as a major migration pathway by several taxa belonging to the permanent hyporheos to colonize suitable benthic habitats separated by inhospitable kryal stream segments. Dole-Olivier et al. (1993) showed that surface water / groundwater interactions induced repeated gradients of interstitial fauna in subterranean landscapes of the Rhône River flood plain. Spatial changes in the relative influence of surface glacial water and ground water create repeated gradients of interstitial fauna along the longitudinal, lateral, and vertical dimensions of Val Roseg. Taxa that are sensitive to the extreme conditions of the glacial environment (e.g. Ostracoda and Cyclopoida) exhibit the most downstream distribution in the hyporheic layer of the main channel, colonize preferentially the interstitial habitats of surface disconnected channels that are never flooded by glacial water, and do not occur in the benthic layer of kryal stream segments. The diversity and density of the interstitial fauna are expected to vary widely as a function of the pore size and permeability of the sediment deposited by the glaciers. Although the hyporheic zone appears to contribute significantly to the biodiversity of the Roseg River, similar studies need to be conducted in other glacial streams before this result can be generalized. The retreat of the Tschierva and Roseg glaciers and resulting changes in the downstream extent of glacial water would modify the distribution range of species. Taxa that are actually restricted to the lower reaches of the Roseg River are expected to colonize more upstream sites if the Roseg and Tschierva glaciers continue to retreat. Conversely, the retreat of the glaciers may result in the loss of species rich groundwater-fed habitats if the floodplain channel network becomes more braided in response to increased sediment delivery. Data on the present-day distribution of interstitial invertebrates in the Val Roseg can serve as a basis for predicting and monitoring changes in biodiversity induced by the retreat of Alpine glaciers.

7. ACKNOWLEGEMENTS

Data included in this chapter were collected while the senior author was a member of the Department of Limnology at the EAWAG, Duebendorf, Switzerland. I thank C. Boesch and D. Ferreira for processing the hyporheic samples and S. Dolédec for his valuable help in using the multivariate methods. I am indebted to the following specialists for their precious help in the identification of taxa: P. Marmonier, (Ostracoda), D. Galassi (Copepoda), M. Lafont (Oligochaeta), and B. Lods-Crozet (Chironomidae).

8. REFERENCES

Blanc, L. (2000). *Données spatio-temporelles en écologie et analyses multitableaux: examen d'une relation.* PhD Thesis, University Lyon 1, Lyon, France, 244 pp.

Bou, C., & Rouch, R. (1967). Un nouveau champ de recherches sur la faune aquatique souterraine. *Comptes Rendus de l'Académie des Sciences de Paris, Sciences de la Vie*, 265, 369-370.

Boulton, A. J. (2000). The subsurface macrofauna. In J. B. Jones & P. J. Mulholland (Eds.), *Streams and Ground Waters* (pp. 337-361). San Diego: Academic Press.

Boulton, A. J., & Stanley, E. H. (1995). Hyporheic processes during flooding and drying in a Sonoran Desert stream. II. Faunal dynamics. *Archiv für Hydrobiologie*, 134, 27-52.

Boulton, A. J., Findlay, S., Marmonier, P., Stanley, E. H., & Valett, H. M. (1998). The functional significance of the hyporheic zone in streams and rivers. *Annual Review of Ecology and Systematics*, 29, 59-81.

Brittain, J. E., & Milner, A. M. (2001). Ecology of glacier-fed rivers: current status and concepts. *Freshwater Biology*, 46, 1571-1578.

Brunke, M., & Gonser, T. (1997). The ecological significance of exchange processes between rivers and groundwater. *Freshwater Biology*, 37, 1-33.

Brunke, M., & Gonser, T. (1999). Hyporheic invertebrates – the clinal nature of interstitial communities structured by hydrological exchange and environmental gradients. *Journal of the North American Benthological Society*, 18, 344-362.

Burgherr, P., & Ward, J. V. (2001). Longitudinal and seasonal distribution patterns of the benthic fauna of an alpine glacial stream (Val Roseg, Swiss Alps). *Freshwater Biology*, 46, 1705-1721.

Burgherr, P., Ward, J. V., & Robinson, C. T. (2002). Seasonal variation in zoobenthos across habitat gradients in an alpine glacial flood plain (Val Roseg, Swiss Alps). *Journal of the North American Benthological Society* , 21, 561-574.

Castella, E., Adalsteinssoon, H., Brittain, J. E., Gislason, G. M., Lehmann, A., Lencioni, V., Lods-Crozet, B., Maiolini, B., Milner, A. M., Olafsson, J. S., Saltveit, S. J., & Snook, D. L. (2001). Macrobenthic invertebrate richness and composition along a latitudinal gradient of European glacier-fed streams. *Freshwater Biology*, 46, 1811-1831.

Creuzé des Châtelliers, M., & Reygrobellet, J.-L. (1990). Interactions between geomorphological processes, benthic and hyporheic communities: first results on a by-passed canal of the French upper Rhône River. *Regulated Rivers: Research and Management*, 5, 139-158.

Dolédec , S., Chessel, D., & Gimaret-Carpentier, C. (2000). Niche separation in community analysis : a new method. *Ecology*, 81, 2914-2927.

Dole-Olivier, M.-J., & Marmonier, P. (1992). Patch distribution of interstitial communities: prevailing factors. *Freshwater Biology*, 27, 177-191.

Dole-Olivier, M.-J., Creuzé des Châtelliers, M., & Marmonier, P. (1993). Repeated gradients in subterranean landscape – Example of the stygofauna in the alluvial floodplain of the Rhône River (France). *Archiv für Hydrobiologie*, 127, 415-471.

Dole-Olivier, M.-J., Marmonier, P., Creuzé des Châtelliers, M., & Martin, D. (1994). Interstitial fauna associated with the alluvial floodplains of the Rhône River (France). In J. Gibert, D. L. Danielopol, & J. A. Stanford (Eds.), *Groundwater Ecology* (pp. 313-346). San Diego: Academic Press.

Füreder, L. (1999). High alpine streams: cold habitats for insect larvae. In R. Margesin & F. Schinner (Eds.), *Cold Adapted Organisms. Ecology, Physiology, Enzymology and Molecular Biology* (pp. 181-196). Berlin: Springer.

Gaertner, J.-C., Chessel, D., & Bertrand, J. (1998). Stability of spatial structures of demersal assemblages: a multitable approach. *Aquatic Living Resources*, 11, 75-85.

Gibert, J., Stanford, J. A., Dole-Olivier, M.-J., & Ward, J. V. (1994). Basic attributes of groundwater ecosystems and prospects for research. In J. Gibert, D. L. Danielopol, & J. A. Stanford (Eds.), *Groundwater Ecology* (pp. 7-40). San Diego: Academic Press.

Henry, J.-P., & Magniez, G. (1992). Evénements quaternaires en Europe et répartition des asellotes stygobies. *Mémoires de Biospéologie*, XIX, 121-126.

Husmann, S. (1975). The boreoalpine distribution of groundwater organisms in Europe. Verhandlungen der Internationalen Vereinigung für Theoretische und Angewandte Limnologie, 19, 2983-2988.

Irons, J. G., Miller, L. K., & Oswood, M. W. (1993). Ecological adaptations of aquatic macroinvertebrates to overwintering in interior Alaska (U.S.A.) subarctic streams. Canadian *Journal of Zoology*, 71, 98-108.

Lavit, C., Escoufier, Y., Sabatier, R., & Traissac, P. (1994). The ACT (STATIS method). *Computational Statistics and Data Analysis*, 18, 97-119.

Lods-Crozet , B., Lencioni, V., Olafsson, J. S., Snook, D. L., Velle, G., Brittain, J. E., Castella, E., & Rossaro, B. (2001). Chironomid (Diptera: Chironomidae) communities in six European glacier-fed streams. *Freshwater Biology*, 46, 1791-1809.

Malard, F., Tockner, K., & Ward, J. V. (1999). Shifting dominance of subcatchment water sources and flow paths in a glacial floodplain, Val Roseg, Switzerland. *Arctic Antarctic and Alpine Research*, 31, 135-150.

Malard, F., Ward, J. V., & Robinson, C. T. (2000). *An expanded perspective of the hyporheic zone. Verhandlungen der Internationalen Vereinigung für Theoretische und Angewandte Limnologie*, 27, 431-437.

Malard, F., Lafont, M., Burgherr, P., & Ward, J. V. (2001b). A comparison of longitudinal patterns in hyporheic and benthic oligochaete assemblages in a glacial river. *Arctic Antarctic and Alpine Research*, 33(4), 457-466.

Malard, F., Mangin, A., Uehlinger, U., & Ward, J. V. (2001a). Thermal heterogeneity in the hyporheic zone of a glacial floodplain. *Canadian Journal of Fisheries and Aquatic Sciences*, 58, 1319-1335.

Malard, F., Tockner, K., Dole-Olivier, M.-J., & Ward J. V. (2002). A landscape perspective of surface – subsurface hydrological exchanges in river corridors. *Freshwater Biology*, 47, 621-640.

Malard, F., Ferreira, D., Dolédec, S., & Ward, J.V. (2003). Influence of groundwater upwelling on the distribution of the hyporheos in a headwater river flood plain. *Archiv für Hydrobiologie*, 157, 89-116.

Mermillod-Blondin, F., Creuzé des Châtelliers, M., Marmonier, P., & Dole-Olivier, M.-J. (2000). Distribution of solutes, microbes and invertebrates in river sediments along a riffle-pool-riffle sequence. *Freshwater Biology*, 44, 255-269.

Milner, A. M., & Petts, G. E. (1994). Glacial rivers: physical habitat and ecology. *Freshwater Biology*, 32, 295-307.

Milner, A. M., Brittain, J. E., Castella, E., & Petts, G. E. (2001). Trends of macroinvertebrate community structure in glacier-fed rivers in relation to environmental conditions: a synthesis. *Freshwater Biology*, 46, 1833-1847.

Milner, A. M., Knudsen, E. E., Soiseth, C., Robertson, A. L., Schell, D., Phillips, I. T., & Magnusson, K. (2000). Colonization and development of stream communities across a 200-year gradient in Glacier Bay National Park, Alaska, U.S.A. *Canadian Journal of Fisheries and Aquatic Sciences* 57, 2319-2335.

Oswood, M. W., Miller, L. K., & Irons, J. G. (1991). Overwintering of freshwater benthic macroinvertebrates. In R. E. Lee & D. L. Denlinger (Eds.), *Insects at Low Temperature* (pp. 360-375). New York: Chapman and Hall.

Robinson, C. T., Tockner, K., & Burgherr, P. (2001). Seasonal patterns in macroinvertebrate drift and seston transport in streams of an alpine glacial flood plain. Freshwater Biology, 47, 985 – 993.

Stanford J. A., & Ward J. V. (1993). An ecosystem perspective of alluvial rivers: connectivity and the hyporheic corridor. *Journal of the North American Benthological Society*, 12, 48-60.

Stanley, E. H., & Boulton, A. J. (1993). Hydrology and the distribution of hyporheos: perspectives from a mesic river and a desert stream. *Journal of the North American Benthological Society*, 12, 79-83.

Sterba, O., Uvira, V., Mathur, P., & Rulik, M. (1992). Variations of the hyporheic zone through a riffle in the Morava, Czechoslovakia. *Regulated Rivers: Research and Management*, 7, 31-43.

Tilzer, M. (1968). Zur Ökologie und Besiedlung des hochalpinen hyporheischen Interstitials im Arlberggebiet (Osterreich). *Archiv für Hydrobiologie*, 65, 253-308.

Ward, J. V. (1994). Ecology of alpine streams. *Freshwater Biology*, 32, 277-294.

Ward, J. V., & Palmer, M. A. (1994). Distribution patterns of interstitial meiofauna over a range of spatial scales, with emphasis on alluvial river-aquifer systems. Hydrobiologia, 287, 147-156.

Ward, J. V., Bretschko, G., Brunke, M., Danielopol, D. L., Gibert, J., Gonser, T., & Hildrew, A. G. (1998). The boundaries of river systems: the metazoan perspective. *Freshwater Biology*, 40, 531-569.

Ward, J. V., Malard, F., Tockner, K., & Uehlinger, U. (1999b). Influence of ground water on surface water conditions in a glacial flood plain of the Swiss Alps. *Hydrological Processes*, 13, 277-293.

Ward, J. V., Stanford, J. A., & Voelz, N. J. (1994). Spatial distribution patterns of Crustacea in the floodplain aquifer of an alluvial river. *Hydrobiologia*, 287, 11-17.

Ward, J. V., Tockner, K., & Schiemer, F. (1999a). Biodiversity of floodplain river ecosystems: ecotones and connectivity. *Regulated Rivers: Research & Management*, 15, 125-139.

Ward, J. V., Malard, F., Stanford, J. A., & Gonser, T. (2000). Interstitial aquatic fauna of shallow unconsolidated sediments, particularly hyporheic biotopes. In H. Wilkens, D. C. Culver & W. F. Humphreys (Eds.), *Ecosystems of the World, Subterranean Ecosystems, vol. 30.* (pp. 41-58). Amsterdam: Elsevier.

Zah, R. (2001). *Patterns, pathways, and trophic transfer of organic matter in a glacial stream ecosystem in the Alps*. PhD thesis Nr. 13998. Swiss Federal Institute of Technology, Zürich, Switzerland.

CHAPTER 12

U. UEHLINGER AND R. ZAH

ORGANIC MATTER DYNAMICS

1. INTRODUCTION

Organic matter, which includes detritus and plant biomass, is the dominant energy base of lotic ecosystems. The pool of organic matter is fueled by autochthonous primary production in the wetted channel and allochthonous inputs of organic detritus produced on the flood plains and uplands. The relative importance of autochthonous versus allochthonous production varies significantly along the river continuum as well as across geographical regions and affects the structure of benthic fauna. This has been documented by numerous studies in different biomes, with a bias to low altitude sites in the temperate zone (e.g., Benfield, 1997; Bott et al., 1985; Fisher & Likens, 1973; Lamberti & Steinman, 1997; Minshall, 1978; Vannote et al., 1980). Corresponding studies in alpine streams and particularly in glacial streams are lacking. The regional climate controls organic matter dynamics, largely through its effects on terrestrial vegetation (Webster & Meyer, 1997) and flow regimes, which influence standing stocks of detritus and plant biomass. High mountain streams, in particular glacier-fed streams, differ from streams at lower elevations in some key features strongly influencing organic matter dynamics, including the density and composition of terrestrial vegetation, the discharge regime, channel stability, and water clarity.

This chapter provides an overview on the dynamics of organic matter in the Roseg flood plain. Consideration is given to both autochthonous and allochthonous sources, to pathways of organic matter input, and to the storage, transport, and loss of organic matter. A provisional organic matter budget of the flood plain is cautiously presented.

2. THE AUTOCHTHONOUS ENERGY BASE

Benthic algae form the autochthonous energy base along the corridor of the Roseg River. Higher aquatic plants and mosses are sparse or absent in the Val Roseg. Stable isotope analysis indicated that benthic algae are an important food source to macroinvertebrates throughout the flood plain and along the river continuum (Zah et al., 2001). Allochthonous inputs, in constrast, are important as a food source only in groundwater-fed channel along the flood plain margins. In the constrained reach, despite substantially enhanced allochthonous inputs, algae remained the primary food source of macroinvertebrates.

J.V. Ward & U. Uehlinger (eds.), Ecology of a Glacial Flood Plain, 199-215.
© 2003 *Kluwer Academic Publishers. Printed in the Netherlands.*

2.1. Physico-chemical habitat template of periphyton

The physical habitat of the main stem of the Roseg River is characterized by strong seasonal variation. High flow and high concentrations of suspended solids (glacial flour) prevail from June to September. Turbidity in the main channel ranges from 40 to >2000 NTU during this time. High turbidity strongly affects light availability; more than 95% of the incident light will be attenuated at a depth of 0.5 m if turbidity exceeds 250 NTU. High summer discharge due to intense ice melt transports coarse sediments and may cause lateral channel shifting; transport of coarse sediment, however, is lacking in the lake outlet stream. Daily mean temperatures during summer range from < 1°C below the terminus of the Tschierva Glacier to about 7°C at the end of the catchment (see also Chapter 5). More benign environmental conditions prevail in spring (April/May) and autumn (September/October) when discharge and turbidity are relatively low (no transport of coarse sediments, minor attenuation of incident light). However, in spring average light intensity is 40% higher than in autumn. Between December and April the lower part of lake outlet stream and the proglacial reach are covered by snow. At the bottom of a 0.6 m deep snow pack light intensity is less than 0.5% of that measured at the surface (U. Uehlinger, unpublished data; Kappen et al., 1995). The main channel in the lower flood plain, which is fed by clear and relatively warm ground water during winter, remains largely open, whereas in the constrained reach ice and snow cover about 95% of the wetted channel.

Floodplain channels with upstream surface connections to the main channel are turbid during summer, but turbidity varies widely depending on the relative contribution of glacial melt water. Daily mean temperatures during summer range from 3 to 8°C (see Chapter 5). Many of these channels are dry during winter. Transport of coarse sediments may occur in large surface-connected channels during summer high flow.

In floodplain channels lacking upstream surface connection to the main channel turbidity is usually low throughout the annual cycle. Daily mean temperatures may increase to 10°C from March until July or August (see Chapter 5) and discharge lacks the power to move bed sediments. However, in the upper flood plain more than 90% of these disconnected channels fall dry during winter (see Chapter 3).

Nitrate concentrations ranged from 173 to 432 $\mu g NO_3\text{-}N\ L^{-1}$ in different channels (Tockner et al., 2002), which is far beyond concentrations reported to limit periphyton growth (e.g., Bothwell, 1989; Grimm & Fisher, 1986; Newbold, 1992). Concentrations of soluble reactive phosphorus (SRP) ranged from 0 to 5 $\mu g\ L^{-1}$, which is relatively low; transition from nutrient limitation to nutrient saturation is in the range of 0.6 to 15 μg SRP L^{-1} (Chapters 4 and 14, Bothwell, 1989; Newbold, 1992).

2.2 Periphyton dynamics

Periphyton biomass was monitored at 25 sites along the Roseg River corridor (proglacial reach 4, lake outlet 2, flood plain 17, constrained reach 2) from August 1996 to January 1998. Most sites were sampled monthly although some locations upstream of the flood plain were not accessible during winter. At each site 5 to 10 rocks were sampled at random within a 20 – 30 m long reach. Periphyton biomass was determined as ash free dry mass as described by Uehlinger (1991).

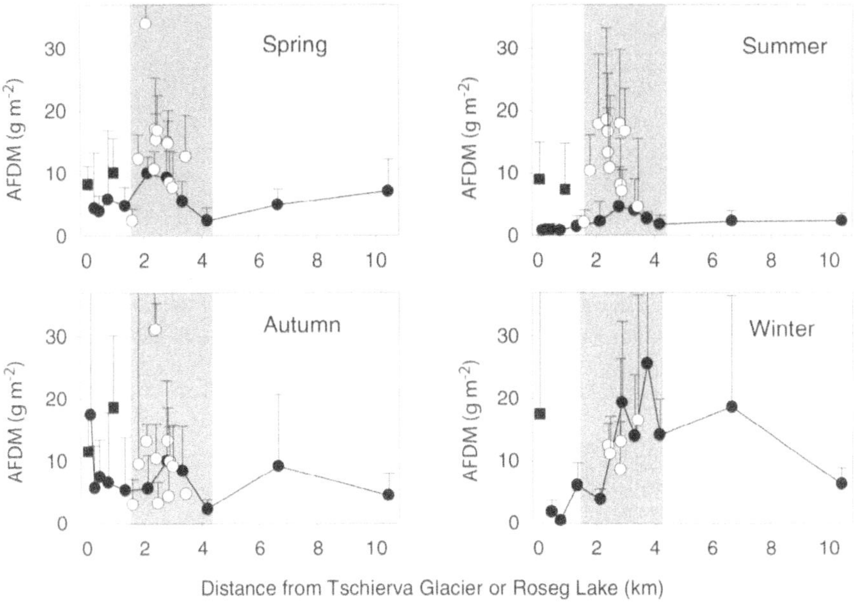

Figure 1. Average periphyton biomass as ash free dry mass (1996 – 1998) along the corridor of the Roseg River. Shaded area = flood plain. Squares = lake outlet sites, filled circles = main channel sites, open circles = floodplain sites excluding main channel sites.

During summer high flow (June to mid September) main channel periphyton biomass was low, particularly in the proglacial reach (0.9 ± 0.8 g m^{-2}; Fig. 1). The multiple channels (up to 11 in some transects) in the flood plain reach reduce main channel discharge and coincides with significantly higher biomass (3.2 ± 4.7 g m^{-2}) compared to the proglacial and constrained reaches. Biomass in the lake outlet exceeded that in the proglacial reach 2.5 to 9 times, but maximum biomass during summer (up to a seasonal average of 12.4 ± 9.6 g m^{-2}) accumulated in surface-disconnected floodplain channels. With the autumnal flow decline associated with reduced ice melt, periphyton biomass began to increase along the entire main channel. On 27 September 1996 biomass attained 53.3 ± 65.2 g m^{-2} only 150 m from the terminus of the Tschierva Glacier. Periphyton biomass declined when channels become covered by snow and ice (e.g. in the proglacial reach). However, environmental conditions at sites remaining open during winter (upper part of the lake outlet and sites fed by ground water) apparently impose less constraints on benthic algae than during the rest of the year; in the lower flood plain algae proliferated during winter at main channel sites (Fig. 2). Periphyton patterns in spring (April/May) were similar to those observed in autumn.

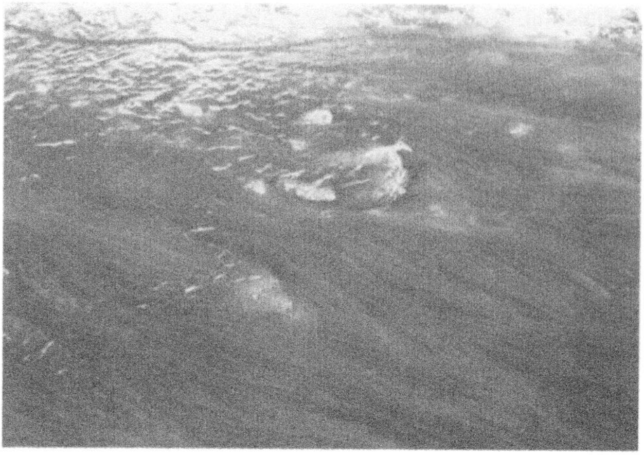

Figure 2. Dense algal mat formed by the chryosphyte Hydrurus foetidus *in the main channel (site 80) in January 1998.*

3. THE ALLOCHTHONOUS ENERGY BASE

3.1. Sources of allochthonous organic matter

Fig. 3 shows the spatial configuration of potential sources of particulate allochthonous organic matter along the floodplain reach of the Roseg River. Within the active flood plain, vegetation patches of different successional stages can be distinguished (see Chapter 7). The active channel area is characterized by gravel bars with initial stages of herbaceous pioneer plants such as *Epilobium fleischeri* (Hochst.), *Oxyria digyna* (L.) and *Linaria alpina* (L.). Less disturbed areas show denser plant communities (e.g. *Rumex scutatus-Poa nemoralis* community). Close to the floodplain border dense grassland with patches of *Juniperus communis* (L.) can be found and the hill slopes are covered with coniferous forests dominated by larch (*Larix decidua*), Arolla pine (*Pinus cembra*) and mountain pine (*Pinus mugo*). The larch, a deciduous conifer that sheds its needles in autumn, contributed about 55% to the needle litter produced in the subalpine forest.

Ground water, glacial melt water, and snow are sources of particulate and dissolved organic matter (Tockner et al., 2002). The glacier accumulates organic matter that is primarily blown in (aeolian drift, Teeri & Barrett, 1977) and is then released during the summer melt period. Floodplain soil is a source of particulate organic matter that enters the aquatic ecosystem by lateral bank erosion during high flow events.

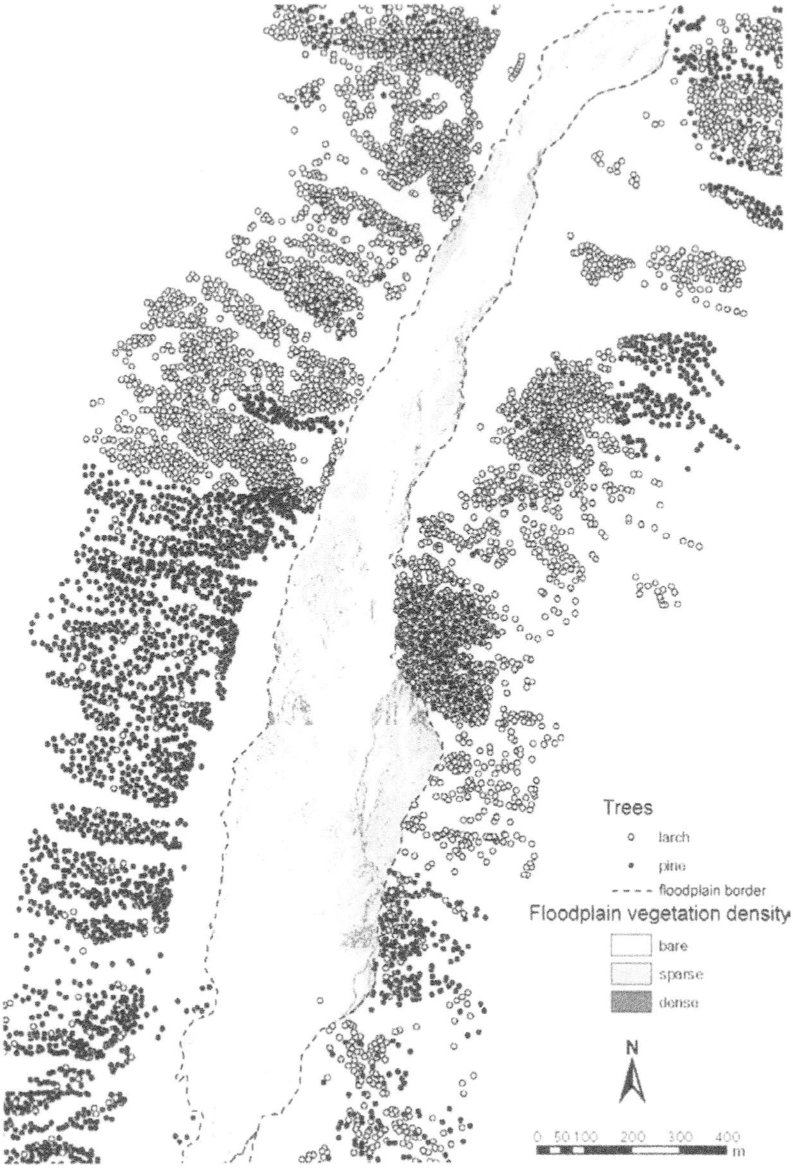

Figure 3. Distribution of individual trees on the side slopes along the floodplain reach and density of the floodplain vegetation.

In the Roseg catchment, input and transport of particulate organic matter from terrestrial vegetation (coniferous needles, twigs, cones etc.) was measured with litter traps (Zah & Uehlinger, 2001). Direct litter input was monitored with 70 traps (buckets with a cross section of 0.06 m^2) placed 20 cm above ground along a longitudinal gradient from the glacier terminus to the lower end of the catchment and along transects across the flood plain (Fig. 4; Zah & Uehlinger, 2001). Lateral transport of particulate organic matter > 0.25mm was determined with 26 lateral traps constructed according to Moser (1991). In the flood plain, four replicated lateral traps were placed at two different bank inclinations (flat: 10° ± 6°; steep: 33° ± 8°) in three riparian vegetation types: pioneer plants, dense grass cover and forest. In addition, three lateral traps were placed in the proglacial area and 2 traps at the lower end of the catchment. During the winter period, one survey was performed in February with a snow-corer in order to collect direct litterfall stored within the snowpack.

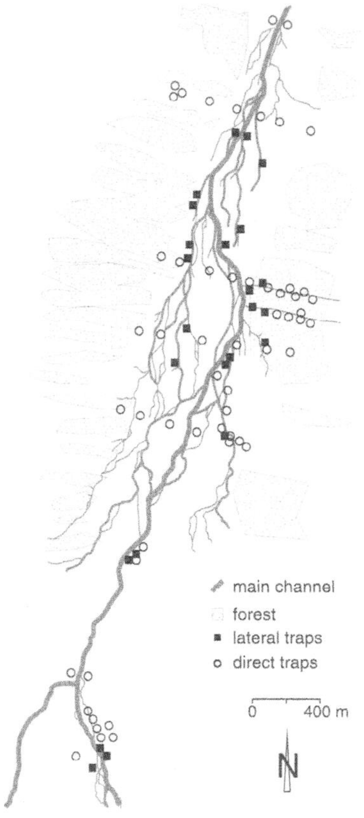

Figure 4. Part of the Roseg River corridor showing positions of traps for measuring direct and lateral inputs of allochthonous organic matter. Additional traps were located in the upper proglacial area and in the constrained reach downstream. From Zah & Uehlinger (2001), Fig. 3, p. 1601.

3.2. Pathways of organic matter input

3.2.1. Aerial and lateral transport

Amount and quality of input depend on terrestrial vegetation structure and the spatial configuration of vegetation patches. In areas covered by sparse pioneer vegetation, as for example along the proglacial reach below the Tschierva Glacier or along the braided corridor within the flood plain, annual input rates ranged from 1 to 10 g $m^{-2}y^{-1}$. The subalpine forest provided maximum amounts of particulate organic matter. Within the stands of larch and Arolla pines aerial input reached 280 g $m^{-2}y^{-1}$ (Zah & Uehlinger, 2001). However, direct inputs declined exponentially with distance from the forest (Fig. 5A). In the forest and at the forest edge, needles of larch and pine, and wood (twig fragments) were dominant (about 95%) but 30 to 50 m from the forest non-identifiable organic particles and grass became major components (≥90%).

Lateral transport was maximum in the forest and slowly declined with distance from the forest edge (Fig. 5B). A major part of the laterally transported organic matter consisted of non-identifiable particles. In contrast to direct inputs, larch and pine needles were still found in lateral traps more than 100 from the forest edge.

Direct and lateral transport of pine and in particular of larch needles peaked in October, but temporal variation of direct and lateral transport of wood, leaves, and grass was more stochastic (Zah & Uehlinger, 2001).

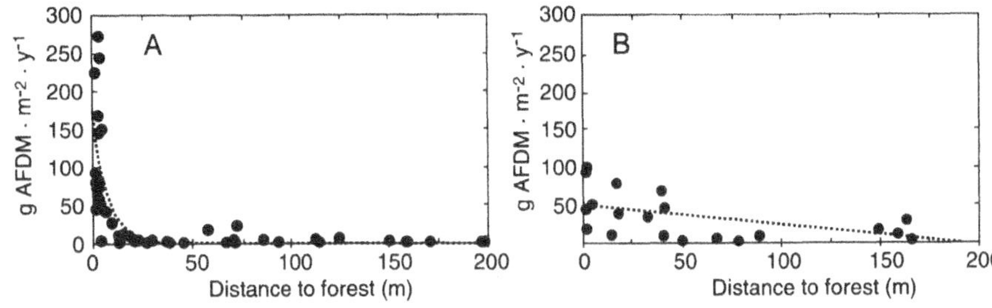

Figure 5. Allochthonous inputs across the flood plain: A) total direct input of organic matter and B) total lateral input of organic matter. Modified from Zah & Uehlinger (2001), Fig, 8, p. 1605.

3.2.2. Input to the main channel: longitudinal patterns

From the proglacial reach to the end of the catchment total annual input (direct and lateral) of particulate organic matter into the main channel increased from 1 to 35 g $m^{-2}y^{-1}$. Direct input prevailed in the constrained reach, where trees grew close to the channel margins (Fig. 6A). In the flood plain and reaches upstream, lateral transport was the dominant pathway of POM entering the main channel (Fig. 6B, Table 1).

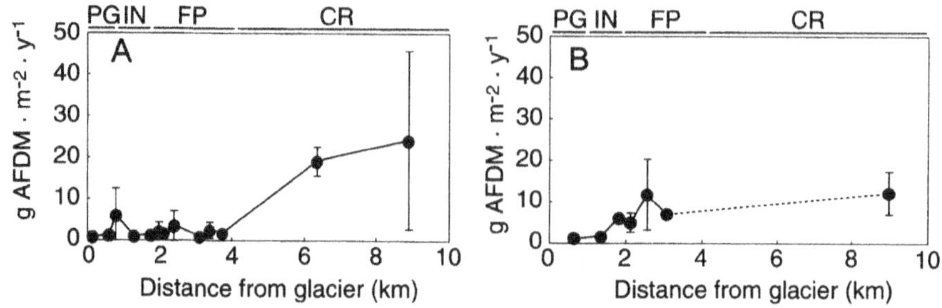

Figure 6. Longitudinal patterns of allochthonous organic matter inputs to the glacial main channel: A) total direct input of organic matter and B) total lateral input of organic matter. PG = proglacial reach, IN = incised reach, FP = flood plain, CR = constrained reach. Modified from Zah & Uehlinger (2001), Fig, 7. p. 1604.

Table 1. Direct, lateral, and total POM input into different reaches of the main channel of the Roseg River

Main channel reach	*Direct input* ($g\ m^{-2}y^{-1}$)	*Lateral input* ($g\ m^{-2}y^{-1}$)	*Total input* ($g\ m^{-2}y^{-1}$)
Proglacial reach	0.4	0.7	1.1
Flood plain	2.2	7.9	10.1
Lower constrained reach	22.3	12.5	34.8

3.2.3. Input across the flood plain

Assessments of direct or lateral inputs to single-thread channels are straightforward, because the input values can be measured along both shorelines and easily extrapolated to the stream area (e.g., Schade & Fisher, 1997). In flood plains characterized by complex channel networks, assessment of allochthonous inputs becomes more difficult. To account for the complexity of the Roseg flood plain, methods of landscape ecology were employed.

To estimate litter input to the Roseg flood plain a GIS-based model was applied. Using the trap data, litter input functions were derived that described direct litter input as function of distance from a single larch or pine tree. Lateral transport was calculated with a function that considered vegetation density, bank inclination and direct input. Aerial photographs (color and color infrared) of the Roseg flood plain were used to differentiate vegetation patterns and generate a digital elevation model (DEM) after rectification (Zah et al., 2000). More than 10,000 trees were digitized and stored as point coverage. The direct input functions were combined with the tree coverage and direct input was calculated for each grid point of the DEM within the flood plain. The lateral transport function was combined with the vegetation pattern coverage and DEM-derived data such as slopes or channel area, and lateral transport was calculated for each grid point.

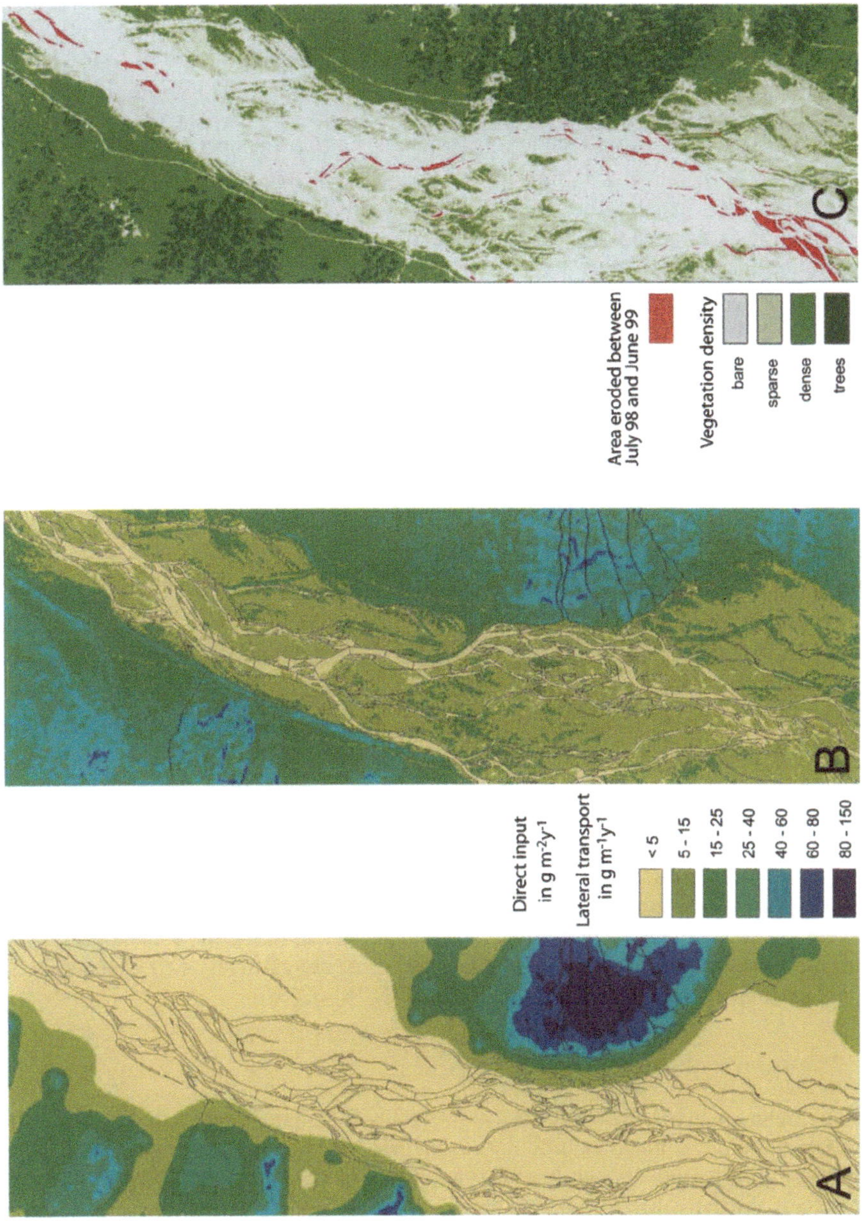

Figure 7. Results of the spatial modelling of the floodplain reach: A) annual mean of direct inputs of particulate organic matter, B) annual mean of lateral transport of particulate organic matter, and C) vegetation density and areas affected by bank erosion between July 1998 and June 1999.

Only channels flowing close to the forested edge of the flood plain received substantial direct POM inputs, with values ranging from 15 to 25 g AFDM $m^{-2}y^{-1}$ (Fig. 7A). Direct inputs to channels located more than 20m from the floodplain edge averaged <0.5g AFDM $m^{-2}y^{-1}$. Lateral transport is the dominating pathway of particulate organic matter in the flood plain (Figure 7B). Lateral inputs exceeded direct inputs 3 to 7 times (Table 2). In channels at the floodplain edge lateral input ranged from 10 to 19 g AFDM $m^{-2}y^{-1}$. Channels in the unstable braided corridor received 1 - 4 g AFDM $m^{-2}y^{-1}$, but in areas not affected by laterally shifting channels for several years lateral inputs were relatively high (about 9 g AFDM $m^{-2}y^{-1}$), even in the center of the flood plain.

Table 2. Direct, lateral, and total POM input into different channels of the Roseg River flood plain compared to streams in different biomes.

Stream	*Direct input* *($g\ m^{-2}y^{-1}$)*	*Lateral input* *($g\ m^{-2}y^{-1}$)*	*Total input* *($g\ m^{-2}y^{-1}$)*
Roseg flood plain (average of all channels)	0.5	3.5	4.0
Antarctic stream[1]	0	0	0
Arctic Tundra stream[1]	0 - 62	0 - 500	0 - 500
Desert streams (North America)[1,2]	3 - 242	3 - 18	3 - 242
Coniferous forest streams[1]	218 - 370	667 - 1111	218 - 1678
Deciduous forest streams[1]	37 - 700	19 - 344	37 - 761

[1]Benfield (1997), [2]first and second order streams

Compared to other systems, the input of particulate allochthonous organic matter to the Roseg River is small (Table 2), particularly in the proglacial reach The increase in POM input from the forefield of the Tschierva Glacier towards the end of the catchment reflects to some extent the altitudinal change in riparian vegetation. The input of terrestrial litter is expected to increase considerably if a stream crosses the boundary between the alpine zone and the subalpine forest belt. In the Val Roseg, however, the flood plain with its predominantly sparse vegetation, keeps riparian inputs at a low level. In the constrained reach, where the forest comes close to the banks, inputs are only moderate because of the relatively wide channel (e.g., wetted channel width during summer high flow: 10 - 23 m).

3.2.4. Input by soil erosion

Soil organic matter enters the streams of the flood plain by bank erosion. Overlaying orthophotographs taken in July 1998 and June 1999 provided a map of the area eroded during one annual cycle (Fig. 7C. This map was combined with the vegetation density map to estimate the proportions of different vegetation covers lost by erosion. Soil organic matter content (0 - 30 cm) was determined for different vegetation densities. Finally soil organic matter content was correlated with vegetation density and the amount of organic matter eroded over one year was calculated.

Organic matter content of the uppermost 0.1 m of the floodplain surface averaged 19 kg m^{-3} on bare gravel and 118 kg m^{-3} in densely vegetated areas; corresponding contents of the layer 0.1 – 0.3 m were 17 and 24 kg m^{-3}. The floodplain area subject to bank erosion in 1998/1999 was restricted to the braided corridor, where vegetation density is generally low. About 4500 m^2 of the eroded area was bare gravel and only 200 m^2 included densely vegetated banks. The annual input of soil organic matter was 23.4 t AFDM y^{-1}.

3.2.5. Hydrological import

Another pathway of organic matter imported to the floodplain channels includes inputs from upstream, from tributaries, and by ground water from the hillslope and valley aquifers. However, at this time we lack sufficient information with respect to discharge and organic matter concentrations to calculate hydrological import.

4. ORGANIC MATTER STORAGE

Organic matter stored in lotic ecosystems includes epilithic biofilms, mosses, higher aquatic plants, and detritus (dead organic matter). Detritus typically originating from the riparian vegetation and in-stream primary production is an energy source that sustains ecosystem functioning in case of the loss of primary production by disturbance (Reichle et al., 1975). The detritus pool is characterized by slow turnover and a mass which is large compared to primary producer biomass. Dead organic matter comprises woody debris, leaf litter and other parts of terrestrial plants usually over a wide range of decompositional stages. Large woody debris is known to influence channel stability and retention characteristics (Keller & Swanson, 1979).

In the Roseg River, large woody debris is of minor importance. Avalanches may transport log fragments from side slope forests to the floodplain margin but such events are infrequent. Input of trees by bank erosion occurs very rarely in the floodplain reach. Downstream of the flood plain large woody debris that enters the stream is removed by local authorities.

Benthic particulate organic matter (BPOM) (excluding periphyton) was collected with a Hess sampler (mesh size 0.1mm) at the channel surface to a depth of 0.1 m at 15 sites (7 floodplain sites) along the corridor of the Roseg River and determined as ash free dry mass (P. Burgherr, unpublished data). Average BPOM along the Roseg River corridor ranged from 0.7 g m^{-2} in the proglacial reach to 7.0. g m^{-2} at site 70, located in a surface-disconnected channel at the left floodplain margin. Benthic organic matter in floodplain channels, excluding the main channel, was significantly higher than in main channel and lake outlet sites (Table 5).

In autumn 1998, hyporheic sediments (0 – 0.5 m) were sampled with the freeze core method (Stocker & Williams, 1972) at nine 25 m-long reaches (3 replicates in each reach) between Tschierva Glacier and the end of the catchment. Two particulate organic matter fractions were distinguished: 1) loosely attached POM (LPOM), which was separated by elutriation, and 2) POM strongly attached to sediment grains (SPOM), presumably representing biofilms (polysaccharide matrix, microorganisms). Organic matter was determined as ash free dry mass. Total POM stored in the hyporheic zone between 0 and 0.5 m depth increased from the proglacial reach to the end of the catchment (Table 3). Within the flood plain, differences between main channel, surface-connected channels and surface-disconnected channels were not significant. SPOM was the dominated fraction, exceeding LPOM 3 to 9 times.

Table 3. Particulate organic matter in the upper hyporheic zone (0 – 0.5 m depth) of the Roseg River. Total POM = SPOM and LPOM. SPOM = strongly attached POM, LPOM = loosely attached POM.

Sites	*Total POM ($kg\ m^{-2}$)*	*LPOM ($kg\ m^{-2}$)*
Main channel		
Proglacial reach	1.0 ± 0.8	0.1 ± 0.2
Flood plain	1.3 ± 0.8	0.3 ± 0.7
Constrained reach	1.8 ± 0.6	0.5 ± 0.6
Channels excluding main channel		
Surface-connected	1.5 ± 0.8	0.4 ± 0.8
Surface-disconnected	1.3 ± 0.6	0.2 ± 0.3

In the Roseg River, the amount of BPOM is small compared to non alpine streams (Table 4); algal biomass even exceeds BPOM 2 – 6 times. This reflects small allochthonous inputs and the absence of retentive structures like large woody debris, both of which are known to control the size of the benthic detritus pool (e.g., Jones, 1997). In a glacial stream of the McMurdo Dry Valleys (Antarctica), where allochthonous inputs were virtually lacking, benthic organic matter almost exclusively consisted of dense algal mats (McKnight & Tate, 1997). Organic matter stored in the uppermost hyporheic sediments of the Roseg River exceeds by far BOM but it is 2 – 4 times lower than in streams draining forested catchments (e.g., Leichtfried, 1991; Metzler & Smock, 1990).

Table 4. Benthic particulate organic matter (algal biomass and detritus) along the Roseg River and some other streams and rivers.

Stream	*Algal biomass ($g\ m^{-2}$)*	*Detritus (BPOM) ($g\ m^{-2}$)*
Val Roseg		
Proglacial reach	4	1
Lake outlet stream	12	2
Main channel in flood plain	7	2
Main channel constrained reach	7	3
Floodplain channels excluding the main channel	13	5
Canada stream (glacier fed Antarctic stream)[1]	620	<1
Monument Creek (Subarctic stream)[2]	2	20
Kuparuk River (Subarctic stream)[3]	1	56
Sycamore Creek (desert stream)[4]	15	95
Bear Brook (deciduous forest stream)[5]	<1	663*
Satellite Branch (deciduous forest stream)[6]	<1	2000*

[1]McKnight & Tate (1997), [2]Irons & Oswood (1997), [3]Harvey et al. (1997), [4]Fisher & Grimm (1988), estimate of algal biomass based on chlorophyll-*a* using the ratio AFDM:chlorophyll-*a* = 130, [5]Findlay et al. (1997), [6]Wallace et al. (1997).
* wood not included.

5. LOSS OF ORGANIC MATTER

5.1. Sediment respiration

Respiration, as oxygen consumption of the uppermost bed sediments, was measured 4 times between March and October 1999 at 9 - 26 sites within the flood plain. The uppermost 0.2 m of sediment was collected (without the periphyton-coated top layer) with a shovel. The size fraction < 8mm was enclosed in respiration chambers as described by Jones (1995) and incubated in situ for 4 to 10 hours depending on respiratory activity. After incubation, sediments were analyzed for particulate organic matter, namely the SPOM and LPOM fractions. To calculate respiration rates per area or volume it was considered that only 16% of the bed sediments belong to the size fraction <8mm.

Average respiration rates ranged from 0.51 ± 0.41 g O_2 $m^{-3}d^{-1}$ in the main channel to 1.87 ± 1.49 g O_2 $m^{-3}d^{-1}$ in surface-disconnected channels (Table 5). In surface-connected channels respiration rates were 0.89 ± 0.71 g O_2 $m^{-3}d^{-1}$. These rates, which are very low compared to non-alpine streams, correlate with LPOM concentrations (R^2 = 0.32, n = 216, $p < 0.0001$) but not with SPOM. Sites of maximum metabolic activity are surface-disconnected channels, where periphyton biomass is high and distance to densely vegetated areas small.

Table 5. Respiration in the hyporheic sediments of the Roseg River and other streams. Results of measurements with respiration chambers.

Stream	*Depth (m)*	*Respiration ($g\ O_2\ m^{-3} d^{-1}$)*
Val Roseg (March – October)		
Main channel	0.1	0.5 ± 0.4
Surface-connected channels	0.1	0.9 ± 0.7
Surface-disconnected channels	0.1	1.9 ± 1.5
Steina (Blackforest mountain stream), annual mean[1]	0.3	7.9
River Necker (prealpine river) during summer[2]	0.3	6.3 ± 2.2
Sycamore Creek (desert stream)[3], annual mean in		
downwelling zones	0.1	26.8
upwelling zones	0.1	11.0
Hassayampa River (desert stream), autumn[4]	0.1	23.0 ± 2.6
Hassayampa River (desert stream), autumn[4]	0.3	13.2 ± 7.4

[1]Pusch & Schwoerbel (1994), [2]Naegeli & Uehlinger (1997), [3]Jones et al. (1995), [4]Uehlinger et al. (2002).

5.2. Hydrological export

Estimates of the export of organic matter are based on measurements of DOC and POC at the end of the flood plain (Tockner et al., 2002). Export equaled 24 t DOC and 27 t POC, which approximately corresponds to 48 and 54 t of ash free dry mass (see also chapter 6).

6. PROVISIONAL ORGANIC MATTER BUDGET

A complete organic matter budget for a stream segment would require independent measurements of all inputs, storage and outputs as well as the consideration of appropriate spatial and temporal scales (Cummins et al., 1983). The following preliminary organic matter budget for the Val Roseg flood plain (Table 6) is an attempt to integrate the information on organic matter dynamics gathered from 1996 to 1999. The lack of data on primary production and ecosystem respiration necessitate making certain simplifying assumptions. Gross primary production (GPP) was estimated using relationships between daily integrals of photosynthetic radiation and daily GPP normalized on chlorophyll a. These relationships are based on data from a prealpine river (Uehlinger & Naegeli, 1998; U. Uehlinger, unpublished data). Autotrophic respiration was assumed to account for 30% of GPP. Using these assumptions, a provisional organic matter budget was calculated for the Val Roseg flood plain (Table 6). The budget is not balanced; information on the hydrological import of organic matter and a possible change in organic matter stored is missing.

Table 6. Provisional organic matter budget for the Roseg Flood Plain channel network (ND = no data).

Variable	*t AFDM y^{-1}*	*g $m^{-2}y^{-1}$*
Inputs		
Gross primary production	15.5	318
Riparian POM	0.6	12
Floodplain soil erosion	23.4	-
Hydrological import	ND	-
Outputs		
Autotrophic respiration	5.2	107
Sediment respiration (0-0.2m depth)	5.6	120
Hydrological export (DOM, POM)	102	-
	t AFDM	*g m^{-2}*
Standing crops		
Benthic biofilms	0.6	12.3
Benthic particulate organic matter	0.2	4.1
Sediment organic matter (0-0.2m depth)	26.5	544.3

The estimate of GPP exceeds riparian inputs more than 20 times. The substantial input of soil organic matter is presumably of minor importance for the aquatic system because this material is only mobilized during high flow and the retention within the aquatic system is low. Low retention of particulate and dissolved organic matter characterizes the main channel, which transports most of the suspended and dissolved matter (Tockner et al., 2002). The GPP of 15.5 t organic matter and the loss by respiration of 10.8 t indicate that the metabolism of the floodplain system is autotrophic. This assumption is also supported by small standing crops of BPOM, which are 3 times lower than the standing crop of benthic biofilms.

7. SUMMARY AND CONCLUSIONS

Alpine glacial streams are presumably energy limited during summer when high flow and bed load transport suppress periphyton accrual. Periphyton biomass is usually high between autumn and spring, except when streams become covered by snow. In the flood plain of the Val Roseg, however, the extensive channel network provides habitats which are less affected by the harsh environmental conditions prevailing in the main channel during the period of glacial ablation and therefore, support relatively high periphyton biomass and primary production also during summer. Flow pathways that divert surface and subsurface water from the main channel into many small channels extend the photosynthetically active area and thus, substantially increase the autochthonous energy input.

Input of riparian organic matter to the floodplain system is very small. The subalpine forest, the richest source of organic matter within the catchment, is too far away from floodplain channels, considering the short transport distances of needle litter (Zah & Uehlinger, 2001). Moreover, the present fluvial dynamics of the flood plain and the climatic constraints result in a sparse pioneer vegetation covering major parts of the flood plain and providing minor amounts of organic matter (see Chapters 2 and 7, and Zah et al., 2001; Zah & Uehlinger, 2001). Particulate and dissolved organic matter imported from the glacier are presumably of minor importance because of low retention in the flood plain; more than 80% of the glacial melt water is transported through the main channel.

Storage of benthic organic matter reflects the density and structure of the adjacent riparian vegetation. Benthic organic matter is small along the entire river corridor particularly compared with periphyton biomass. Organic matter storage within the flood plain is small even if sediment organic matter is considered (Jones, 1997). This corroborates the notion that the energetics of the glacial flood plain of the Val Roseg, as presumably most alpine streams, strongly depend on autochthonous energy fixation.

8. REFERENCES

Benfield, E. F. (1997). Comparison of litterfall input to streams. *Journal of the North American Benthological Society, 16*, 104-108.

Bothwell, M. L. (1989). Phosphorus-limited growth dynamics of lotic periphytic diatom communities: areal biomass and cellular growth rate responses. *Canadian Journal of Fisheries and Aquatic Sciences, 46*, 1293-1301.

Bott, T. L., Brock, J. T., Dunn, C. S., Naiman, R. J., Ovink, R. W., & Petersen, R. C. (1985). Benthic community metabolism in 4 temperate stream systems: An inter-biome comparison and evaluation of the river continuum concept. *Hydrobiologia, 123*, 3-46.

Cummins, K. W., Sedell, J. R., Swanson, F. J., Minshall, G. W., Fisher, S. B., Cushing, C. E., Petersen, R. C., & Vannote, R. (1983). Organic matter budgets for stream ecosystems: problems in their evaluation. In J. R. Barnes & G. W. Minshall (Eds.), *Stream ecology. Application and testing of general ecological theory* (pp. 299-353). New York: Plenum Press.

Findlay, S., Likens, G. E., Hedin, L. O., Fisher, S. G., & McDowell, W. H. (1997). Organic matter dynamics in Bear Brook, Hubbard Brook Experimental Forest, New Hampshire, USA. *Journal of the North American Benthological Society, 16*, 43-46.

Fisher, S. G., & Grimm, N. B. (1988). Disturbance as a determinant of structure in a Sonoran Desert stream ecosystem. *Verhandlungen der Internationalen Vereinigung für Theoretische und Angewandte Limnologie, 23*, 1183-1189.

Fisher, S. G., & Likens, G. E. (1973). Energy flow in Bear Brook, New Hampshire: An integrative approach to stream ecosystem metabolism. *Ecological Monographs, 43*, 421-439.

Grimm, N. B., & Fisher, S. G. (1986). Nitrogen limitation in a Sonoran Desert (Arizona, USA) stream. *Journal of the North American Benthological Society, 5*, 2-15.

Harvey, C. J., Peterson, B. J., Bowden, W. B., Deegan, L. A., Finlay, J. C., Hershey, A. E., & Miller, M. C. (1997). Organic matter dynamics in the Kuparuk River, a tundra river in Alaska, USA. *Journal of the North American Benthological Society, 16*, 18-22.

Irons, J. G., & Oswood, M. W. (1997). Organic matter dynamics in 3 subarctic stream of interior Alaska, USA. *Journal of the North American Benthological Society, 16*, 23-28.

Jones, J. B. (1997). Benthic organic matter storage in streams: influence of detrital import and export, retention mechanisms, and climate. *Journal of the North American Benthological Society, 16*, 109-119.

Jones, J. B., Fisher, S. G., & Grimm, N. B. (1995). Vertical hydrologic exchange and ecosystem metabolism in a Sonoran Desert stream. *Ecology*, 942-952.

Kappen, L., Sommerkorn, M., & Schroeter, B. (1995). Carbon acquisition and water relations of lichens in polar regions - potentials and limitations. *Lichenologist, 27*, 531-545.

Keller, E. A., & Swanson, F. J. (1979). Effects of large organic material on channel form and fluvial processes. *Earth surface processes, 4*, 361-380.

Lamberti, G. A., & Steinman, A. D. (1997). A comparison of primary production in stream ecosystems. *Journal of the North American Benthological Society, 16*, 95-104.

Leichtfried, M. (1991). POM in bed sediments of a gravel stream (Ritrodat-Lunz study area, Austria). *Verhandlungen der Internationalen Vereinigung für Theoretische und Angewandte Limnologie, 24*, 1921-1925.

McKnight, D. M., & Tate, C. M. (1997). Canada stream: a glacial meltwater stream in Taylor Valley, South Victoria Land, Antarctica. *Journal of the North American Benthological Society, 16*, 14-17.

Metzler, G. M. & Smock, L.A. (1990). Storage and dynamics of subsurface detritus in a sand-bottomed stream. *Canadian Journal of Fisheries and Aquatic Sciences, 47*, 588-594.

Minshall, W. G. (1978). Autotrophy in stream ecosystems. *BioScience, 28*, 767-771.

Moser, H. (1991). Input of organic matter (OM) in a low order stream (Ritrodat-Lunz study area, Austria). *Verhandlungen der Internationalen Vereinigung für Theoretische und Angewandte Limnologie, 24*, 1913-1916.

Naegeli, M. W. & Uehlinger, U. (1997). Contribution of the hyporheic zone to ecosystem metabolism in a prealpine gravel-bed river. *Journal of the North American Benthological Society, 16*, 794-804.

Newbold, D. J. (1992). Cycles and spirals of nutrients. In P. Calow & G. E. Petts (Eds.), *The rivers handbook. Hydrological and ecological principles* (pp. 379-408). London: Blackwell.

Pusch, M., & Schwoerbel, J. (1994). Community respiration in hyporheic sediments of a mountain stream (Steina, Black Forest). *Archiv für Hydrobiologie, 130*, 35-52.

Reichle, D.E., O'Neill R.V. & Harris W.F. (1975). Principles of energy and material exchange in ecosystems. In W.H. Van Dobben. & R.H. Lowe-McConnell (Eds.), *Unifying concepts in ecology* (pp.27-43). The Hague: Junk

Schade, J. D., & Fisher, S. G. (1997). Leaf litter in a Sonoran desert stream ecosystem. *Journal of the North American Benthological Society, 16*, 612-626.

Stocker, Z. S. J., & Williams, D.D. (1972). A freeze core method for describing the vertical distribution of sediments in a streambed. *Limnology and Oceanography, 17*, 136-138.

Teeri, J. A., & Barrett, P. E. (1977). Detritus transport by wind in a high arctic terrestrial ecosystem. *Arctic and Alpine Research, 7*, 387-391.

Tockner, K., Malard, F., Uehlinger, U., & Ward, J. V. (2002). Nutrients and organic matter in a glacial river floodplain system (Val Roseg, Switzerland). *Limnology and Oceanography, 47*, 521-535.

Uehlinger, U. (1991). Spatial and temporal variability of the periphyton biomass in a prealpine river (Necker, Switzerland). *Archiv für Hydrobiologie, 123*, 219-237.

Uehlinger, U., & Naegeli, M. W. (1998). Ecosystem metabolism, disturbance, and stability in a prealpine gravel bed river. *Journal of the North American Benthological Society, 17*, 165-178.

Uehlinger, U., Naegeli, M. W., & Fisher, S. G. (2002). A heterotrophic desert stream? The role of sediment stability. *Western North American Naturalist, 62*, 466-473.

Vannote, R. L., Minshall, G. W., Cummins, K. W., Sedell, J. R., & Cushing, C. E. (1980). The river continuum concept. *Canadian Journal of Fisheries and Aquatic Sciences, 37*, 130-137.

Wallace, B. C., Cuffney, T. F., Eggert, S. L., & Whiles, M. R. (1997). Stream organic matter inputs, storage, and export for Satellite Branch at Coweeta Hydrologic Laboratory, North Carolina, U.S.A. *Journal of the North American Benthological Society, 16*, 67-74.

Webster, J. R., & Meyer, J. L. (1997). Organic matter budgets for streams: a synthesis. *Journal of the North American Benthological Society, 16*, 141-161.

Zah, R., Burgherr, P., Bernasconi, S. M., & Uehlinger, U. (2001). Stable isotope analysis of macroinvertebrates and their food sources in a glacier stream. *Freshwater Biology, 46*, 871-882.

Zah, R., Niederoest, M., & Uehlinger, U. (2000). Application of photogrammetry in freshwater ecology: analyzing the morphology of a high alpine floodplain. *International Archives of Photogrammetry and Remote Sensing, 33*, 1739-1746.

Zah, R., Niedröst, M., Rinderspacher, H., Uehlinger, U., & Ward, J. V. (2001). Long-term dynamics of the channel network in a glacial floodplain, Val Roseg, Switzerland. *Arctic, Antarctic, and Alpine Research, 33*, 440-446.

Zah, R., & Uehlinger, U. (2001). Particulate organic matter inputs to a glacial stream ecosystem in the Swiss Alps. *Freshwater Biology, 46*, 1597-1608.

CHAPTER 13

C. T. ROBINSON & M. O. GESSNER

LITTER DECOMPOSITION

1. INTRODUCTION

Leaf breakdown is considered a pivotal process in the metabolism of stream ecosystems (Webster et al., 1995; Wallace et al., 1997), providing important information on stream functioning (nutrient cycling and energy flow). Analyses of decomposing litter also inform us about the distribution and abundance (structure) of stream biota such as fungi, bacteria, invertebrates, and fish. Leaf breakdown is caused by the joint action of physical factors causing mechanical fragmentation, the activity of detritivorous macroinvertebrates (shredders), and the activity of micro-organisms such as aquatic fungi and bacteria (Hieber & Gessner, 2002). Although unimportant in the streams of the upper Roseg catchment, fish also can play an important functional role in the litter dynamics of low elevation streams, tropical streams in particular (Flecker, 1996). In concert with various chemical changes, micro-organisms cause transformations in litter quality (e.g., increases in nitrogen and phosphorus concentrations) that influence decomposition dynamics (Webster & Benfield, 1986; Gessner et al., 1999).

The upper Roseg catchment is mostly bare rock and glaciers (75%), but 25% is covered to some degree by vegetation of which ca 4% is woody. Important woody species include pine (*Pinus cembra* L. and *Pinus mugo* Miller) and larch (*Larix decidua* Mill.) on valley side-slopes, and alder (*Alnus viridis* (Chaix) Dc.) and willow (*Salix* spp.) in some riparian areas. Some rather dense stands of alder and willow shrubs occur at elevations above 2000 m a.s.l. However, most plant litter is transported to floodplain channels by wind or via forested tributaries (Zah & Uehlinger, 2001). Litter input into floodplain streams in the Roseg falls within the range of that entering arctic tundra and desert streams (Webster & Meyer, 1997). A further constraint in alpine streams, glacial streams in particular, is the absence of litter retention by large woody debris, thus posing the question as to whether these streams have the potential to function in a similar way as other lotic systems, specifically whether both macroinvertebrates and fungi play similar roles in streams of the Roseg flood plain as in forested streams at lower elevation. A secondary question is whether alder and larch, a deciduous conifer, share similar patterns in leaf breakdown. Although both species lose their leaves in autumn, we expected larch needles to be broken down more slowly than alder leaves, regardless of location in the flood plain, because of inherent differences in leaf quality.

J.V. Ward & U. Uehlinger (eds.), Ecology of a Glacial Flood Plain, 217-230.
© 2003 *Kluwer Academic Publishers. Printed in the Netherlands.*

A common perception is that harsh environmental conditions in alpine streams, especially glacial streams, impose major constraints on their functional potential. Due to glacial recession many kryal streams may be quite youthful with relatively low biological activity and different functional dynamics than glacial streams further downstream and upwelling springbrooks. Another, specific feature of the Roseg flood plain is the presence of a proglacial lake with the outlet stream having different habitat conditions and biotic properties than the respective kryal stream (Burgherr & Ward, 2000). The diversity of stream types in the flood plain (Tockner et al., 1997) suggests that functional attributes also may vary. Indeed, one could infer that the different stream types in the flood plain represent systems of differing successional ages, each possessing a distinctive habitat templet acting upon and dictating the kinds of organisms (e.g., fungi and invertebrates) inhabiting them. In essence, a better knowledge of the functional diversity of streams in the catchment allows an estimate of the overall productivity potential of this heterogenous flood plain. Consequently, we examined whether the differences in litter breakdown dynamics among stream types represented in the flood plain reflected differences in respective abiotic and biotic properties.

2. FIELD APPROACH

In an attempt to gain insight into the above issues, we conducted standard leaf breakdown experiments using freshly abscissed larch and alder leaves in a proglacial channel (kryal stream, site 7), the outlet of the proglacial lake (site 10), and 3 additional channel types typical of the Roseg flood plain (Chapter 4): a main channel site in the lower flood plain (site 60), a mixed channel (site 59), and a tributary fed by hillslope groundwater (site 22) (site locations in Appendix I). We selected these sites to express the range in physical conditions and inputs of terrestrially-derived organic matter possible in the flood plain. The kryal stream was the most physically harsh system with an unstable stream-bed and minimal inputs of terrestrial organic matter. It was located above treeline, the only woody vegetation being sparse low-lying willow shrubs growing along intermittent groundwater seeps. The glacial lake outlet, located at a similar elevation, also had minimal inputs of terrestrial organic matter, but its channel was much more stable than the kryal stream. The lower main channel site had greater inputs of terrestrial matter, mostly derived from side-slope tributaries, but it was still relatively unstable. This site contrasted with the mixed channel, which was more stable but with similar terrestrial inputs. The tributary was the most stable site and had direct inputs of terrestrial organic matter, including both alder leaves and larch needles.

To assess for differences between fungal and macroinvertebrate processing in litter breakdown dynamics, fresh leaves were enclosed in either coarse (5-mm mesh) nylon bags allowing colonization by fungi and macroinvertebrates or fine (0.5-mm mesh) nylon bags that excluded most macroinvertebrates (Gessner et al., 1998; Robinson et al., 1998, 2000). Bags initially contained either freshly abscissed alder leaves (7 g) or larch needles (5 g). Once submerged at each site, replicate bags were retrieved at pre-determined intervals over a 230-day period that encompassed autumn to early spring. Once the macroinvertebrates were removed from each pack, the fungal assemblages associated with leaf packs were assessed (see Chapter 8) and fungal biomass estimated using ergosterol (Gessner & Schmitt, 1996). Lastly, we determined the concentrations of nitrogen and phosphorus in the leaves during the breakdown period. To determine the distinctiveness of macroinvertebrate assemblages colonizing leaf packs from those naturally inhabiting each stream, we also collected benthic macroinvertebrates with a modified Hess sampler.

3. LEAF LITTER BREAKDOWN IN THE DIFFERENT CHANNEL TYPES

An exciting finding of our study was that leaves can be broken down quite quickly in alpine glacial streams, at rates comparable to those reported for other biomes (Gessner & Chauvet, 1994; Webster & Benfield, 1986), although being strongly dependent on leaf type, stream type, and the presence of shredders and microbes. We did find substantial fungal populations at all sites, and the sporulation rates from the immersed leaves resembled those observed in some forested streams (Weyers & Suberkropp, 1996). The nutrient results indicated immobilization of P and N by the decomposing litter, suggesting that aquatic fungi were important in the nutrient dynamics during litter breakdown. The assessment of litter breakdown in the different stream types in the Roseg flood plain suggested successional processes may be operating and that differential constraints imposed by the interaction between the physical habitat and resource inputs influenced the biota present, and consequently litter breakdown dynamics, within each stream type. In the following sections we provide additional details of our research findings to emphasize these highlights.

3.1. Breakdown rates of alder and larch leaves in the Alpine streams

The rates of breakdown were substantially faster for alder than larch in each of the stream types investigated (Table 1). Breakdown rate coefficients (k)[1] for alder in fine-mesh bags ranged from 0.0029 to 0.0036 d^{-1} with 73-77% of the leaf mass remaining after 45 days.

The patterns in alder breakdown followed our general expectations (within the first 45 days of exposure) with the unstable resource-limited proglacial and main channel sites showing the slowest rates, the more stable mixed channel being intermediate, and the most stable and resource-rich tributary being fastest (Fig. 1). The exception was the glacial lake outlet where less than 16% of the leaf material remained in coarse-mesh bags after 45 days. It was later determined that the extremely rapid processing rates at the outlet resulted from the feeding activity of a shredding caddisfly *(Acrophylax zerberus* Brauer) which colonized the packs in high numbers.

[1] The breakdown coefficient (k), describes the exponential loss of the leaf mass (M) with time (t) according to $M(t) = M(t_0)\, e^{-kt}$ (Wieder & Lang, 1982)

Table 1. Exponential breakdown coefficients (k ± standard error) for alder and larch from coarse-mesh bags in the different streams in the Roseg catchment.

Stream type	Alder		Larch	
	Mean k (d^{-1})	Standard error	Mean k (d^{-1})	Standard error
Kryal	0.0049	0.0002	0.0025	0.0012
Lake outlet	0.0305	0.0029	0.0017	0.0007
Main channel	0.0039	0.0005	0.0027	0.0006
Tributary	0.0113	0.0017	0.0055	0.0008
Mixed channel	0.0064	0.0006	0.0018	0.0007

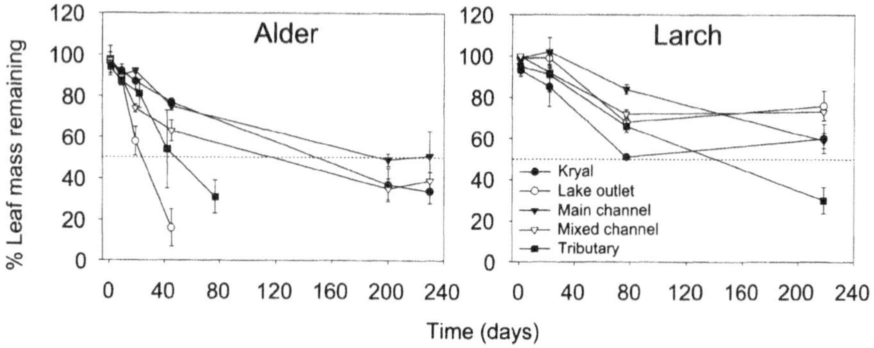

Figure 1. The percent (%) leaf mass remaining for alder and larch in five different stream types found in the Roseg catchment. Symbols represent the average of 4 samples with error bars as standard errors. The dotted line is at 50% leaf mass remaining. From Robinson et al.(1998), Fig. 1, p. 249, and Robinson et al. (2000), Fig. 1, p. 253.

The breakdown of larch needles was initially quite slow with no or very little mass loss occurring within the first 22 days of immersion (Fig. 1). However, significant mass loss had occurred by day 77 when needle mass remaining ranged from 54% at the proglacial site to 84% in the main channel. The tributary was the only site to have less than 50% needle mass remaining by the following spring, after 218 days of immersion. We suspect the low needle mass remaining in the proglacial site (ca 50%) was an artifact of needles being lost due to fragmentation. Regardless, needle breakdown was relatively slow but was in the range for needles in other streams (Iqbal, 1980; Rosset et al., 1982).

An additional 30-40% of the alder leaves were broken down over the winter in the glacial influenced sites (no leaf material remained in the tributary and lake outlet sites), whereas little additional breakdown of larch needles was evident (Fig. 1). Interestingly, substantial additional breakdown of larch needles was observed in the tributary during winter, suggesting that alder leaves (had any remained) would have been broken down even faster in this site. Clearly, the tributary was a site of higher biologically activity in the flood plain, especially in winter. The fact that >40% of the initial leaf mass remained after 200 days in the other sites suggests that leaf breakdown is quite slow in glacial influenced streams, and that the presence of tributaries, springbrooks and lake outlets enhances considerably the functional potential of alpine flood plains.

3.2. Fungal dynamics during the breakdown of alder and larch leaves

Fungal biomass, measured as the concentration of ergosterol per gram of leaf dry mass, attained values commonly observed from leaves decomposing in temperate forested streams. Indeed, some of these values were on the high end of values reported in the literature (Gessner 1997), suggesting substantial metabolic activity by fungi in these streams. Ergosterol concentrations increased over time for both alder and larch (Fig. 2), thus demonstrating the typical temporal pattern of colonization by micro-fungi on leaves in stream habitats (Gessner & Chauvet, 1994). However, values plateaued within 45 days on alder leaves and after 80 days on larch needles. The decrease in ergosterol concentrations over time at most sites is attributed partially to the reduced leaf mass remaining and leaf quality to support fungal growth. Surprisingly, we found values at some sites to be almost 1.5-2× greater for larch needles than for alder leaves, and ergosterol levels were higher on alder leaves in fine-mesh bags than on the leaves in coarse-mesh bags.

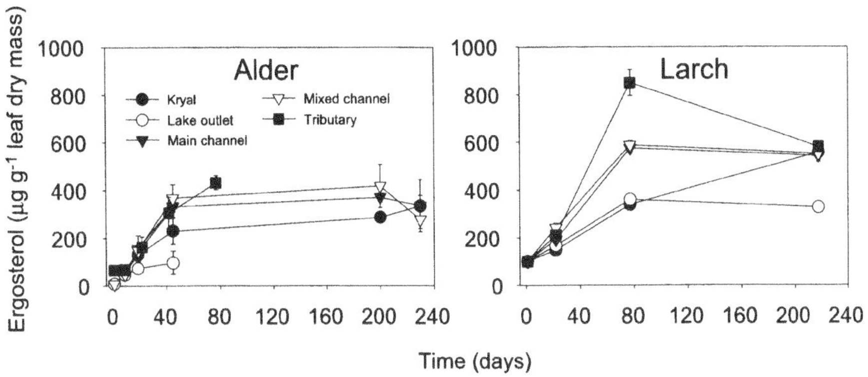

Figure 2. Ergosterol concentrations, as an indicator of fungal biomass, associated with alder leaves and larch needles collected from five different stream types in the Roseg catchment. Each symbol represents the average of 3 samples with error bars showing standard errors. From Gessner et al. (1998), Fig. 5, p. 410, and Robinson et al. (2000), Fig. 2, p. 254.

It is widely known that the microbes associated with leaves during decomposition provide a higher quality food resource for macroinvertebrates than leaves without microbes (Cummins & Klug, 1979; Suberkropp, 1992). The microbes are thought to provide essential nutrients and amino acids not available through the consumption of leaf material alone. Consequently, differences in the ergosterol patterns observed between needles and leaves (and between coarse- and fine-mesh bags of alder) may simply reflect differential feeding by macroinvertebrates. This would be the case between alder in fine and coarse-mesh bags, as the fine mesh excluded most invertebrates from access to the leaf material. This was most noticeable at the lake outlet after 40 days of immersion where leaves in coarse mesh packs had maximum ergosterol levels of ca 150 μg/g leaf dry mass (16% leaf mass remaining), whereas leaves in the fine mesh packs had ergosterol levels of over 500 μg/g leaf dry mass (ca 70% leaf mass remaining). This same explanation likely holds for larch, as needles of this species are known to be a poor-quality food resource, whereas alder is rapidly consumed by shredding invertebrates. The recalcitrant properties of larch needles, as of conifer needles generally, reduce consumption by shredders.

3.3. Nutrient dynamics during leaf breakdown

The assessment of nutrients during leaf breakdown provides important information on the function of microbes, fungal activity in particular, during leaf breakdown. For example, differential uptake of N and P may indicate potential nutrient limitation for heterotrophic processes, especially when viewed in parallel with the dynamics of fungal biomass. We observed somewhat different patterns in N and P on larch needles and alder leaves: larch needles had considerably lower concentrations of both N and P and maximum concentrations appeared to occur later (Fig. 3). Initial concentrations of P were less than 0.5 mg/g on needles but ca. 1.25 mg/g on alder;

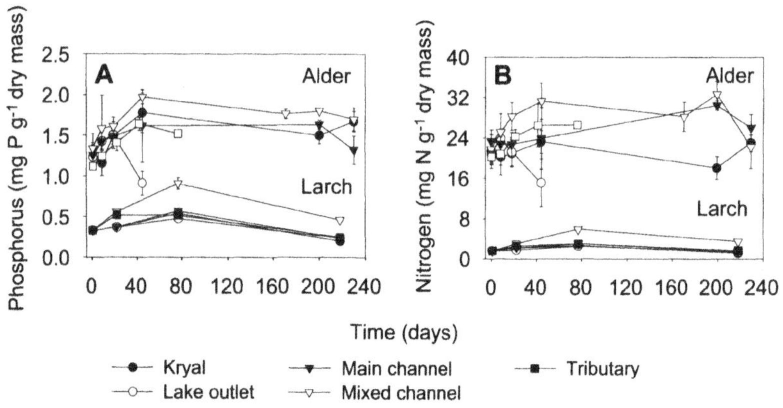

Figure 3. The amount of phosphorus (A) and nitrogen (B) associated with decomposing larch and alder leaves in five stream types in the Roseg catchment. Symbols represent the mean of 4 samples with error bars showing standard errors. From Gessner et al. (1998), Figs. 3, 4, p. 409, and Robinson et al. (2000), Fig. 3, p. 255.

initial concentrations of N were less than 2 mg/g on needles compared to over 20 mg/g on alder leaves. The large difference in nutrient concentrations between needles and alder leaves was expected, based on the findings by a number of previous studies (Iqbal et al., 1980; Rosset et al., 1982). However, what is interesting is that nutrient levels were higher on needles in the tributary than on needles from the glacially influenced sites, again suggesting that tributaries and springbrooks are more metabolically active sites in the flood plain.

Relative increases in phosphorus concentrations were greater on needles than increases in nitrogen concentrations but this pattern was not evident for alder leaves (Fig. 3). Trends in phosphorus concentrations generally followed patterns of leaf mass loss with concentrations decreasing once a substantial fraction of leaf material was broken down. This was especially evident in the lake outlet and tributary, where less than 50% leaf mass remained after about 40 days and leaves were essentially reduced to veins. Nitrogen concentrations increased on leaves from around 21 mg/g dry mass initially to ca. 33 mg/g in glacial sites of the lower flood plain, but not in the proglacial site. This is in contrast to phosphorus concentrations, which increased significantly over time in all sites, including the proglacial site. This difference in dynamics between phosphorus and nitrogen concentrations and the correspondence between phosphorus and fungal dynamics suggests that heterotrophic processes might have been limited by phosphorus availability. Fungi may have met their P requirements for growth in part by immobilizing P from the stream water, and since P levels in water were low, P acquisition via this pathway could have limited microbial decomposer activity.

3.4. Macroinvertebrate dynamics during the breakdown of alder and larch leaves

The use of different stream types to assess litter breakdown provided important information on the functioning of alpine streams, especially regarding the role of macroinvertebrates. Both the number and kinds of macroinvertebrates colonizing and leaves differed among the respective stream types, and this was reflected in the breakdown patterns observed in each site. The heterogeneous mosaic of stream types in glacial flood plains may enhance the overall functional potential of alpine environments, as leaf breakdown rates varied by an order of magnitude among sites. The number of macroinvertebrates colonizing leaf packs was highest in the more stable tributary and mixed-channel sites during the autumn collections, although

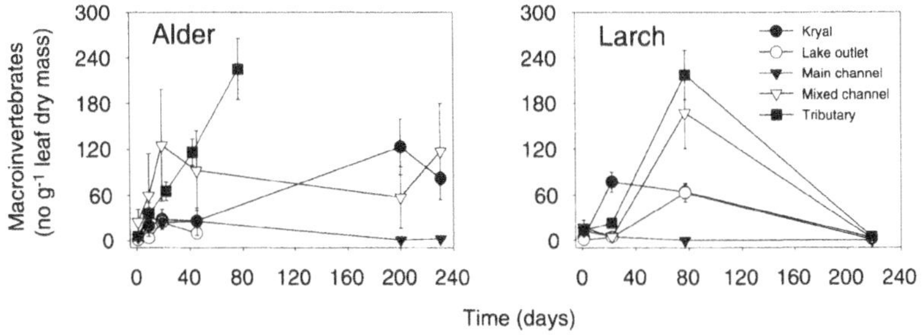

Figure 4. The number of macroinvertebrates associated with decomposing larch and alder leaves in five different stream types in the Roseg catchment. Symbols represent the average of 4 samples with error bars showing standard errors. From Robinson et al. (1998), Fig. 2, p. 220, and Robinson et al. (2000), Fig. 4, p. 258.

numbers increased at all sites over the initial breakdown period (Fig. 4). Typically, numbers peaked earlier in alder leaf packs (<40 days) than in larch needle packs (ca. 80 days), although numbers were comparable among packs for each stream type.

For needles, macroinvertebrate numbers generally increased up to day 80 followed by a decrease by spring (Fig. 4). For leaves, this general pattern was not observed at any site. The disparate patterns of colonization among channels reflected the rates of breakdown at each site. For example, macroinvertebrate numbers in alder packs peaked within the first 20 days at the lake outlet, and decreased by day 40 when little leaf material remained. This was a rather exciting finding, as the extremely fast breakdown rates of alder at this site were due to a shredding caddisfly, *Acrophylax zerberus*. We suspect that the leaf packs acted as resource islands, attracting this shredder to a higher quality food resource than usually found at this site. Although being glacially influenced, the outlet was a suitable habitat for the caddisfly, something not observed in the other glacial sites. Usually *A. zerberus* is found only in springbrooks or groundwater fed channels in glacial flood plains, suggesting that lake outlets in alpine environments may be more important in system function than previously thought. In addition, although the numbers of macroinvertebrates were lower in packs at the lake outlet, their actual biomass was 6× greater than in the other sites because of the large body size of individual caddis larvae (Fig. 5).

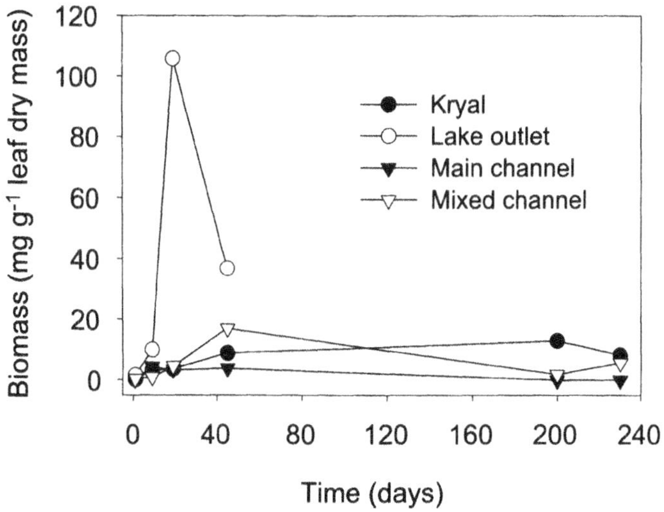

Figure 5. Macroinvertebrate biomass in alder leaf packs in four of the stream types in the Roseg catchment. Symbols represent the average of 4 samples, error bars not shown. From Robinson et al. (1998), Fig. 2, p. 220.

The other extreme was observed at the proglacial and main channel sites, both having a primary glacial influence. The proglacial site was dominated by chironomids that essentially used the packs as habitat, perhaps feeding on fine organic particles retained by the packs (Short et al., 1980). Their numbers in packs were actually higher in spring in the proglacial and mixed channel as numerous early instar chironomids colonized the packs; leaf mass remaining was still above 50% at these sites in spring. The main channel had low numbers throughout the study because this channel usually goes dry over winter and must be recolonized annually. Colonization and leaf breakdown patterns appeared to be similar for alder leaves and larch needles in the glacial influenced sites, although rates were always faster for alder than larch.

Our results suggest that heterotrophic processes (for which leaf breakdown was used as a surrogate) in streams of alpine glacial flood plains are constrained by the interaction of a physically harsh environment and limited resource availability. Landscape characteristics partly determine the kinds and amounts of organic matter as well as the hydrologic and temperature regimes of each stream type. The resulting habitat templet influences the composition of benthic assemblages (macroinvertebrates and microbes) and the interplay between the biota and available resources. For example, the presence of shredders in leaf packs was related to the specific stream site examined (Fig. 6). In accordance with these ideas, the springbrook displayed the greatest functional potential and the proglacial site showed the least. An interesting exception was the lake outlet, in which resource availability was low but the lower harshness of the physical environment provided adequate habitat stability for the persistence of a shredding caddisfly, *A. zerberus*. This caddisfly species is, however, a facultative shredder with highly plastic feeding behavior (Robinson et al., unpublished data), thus explaining why it remains in an environment normally lacking leaf litter.

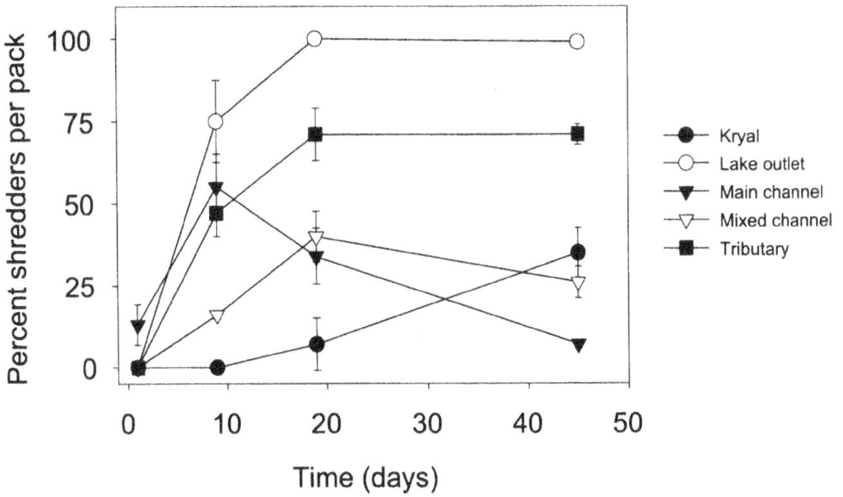

Figure 6. The percentage of shredders in the overall macroinvertebrate community associated with decomposing alder leaves in five stream types in the Roseg catchment. Symbols represent the average of 4 samples with error bars showing standard errors.

Another interesting aspect of using a variety of stream types in a glacial flood plain that has experienced glacial retreat relates to the interplay between successional processes and stream function. Because of glacial retreat, the proglacial stream is relatively young, unstable, and lacks substantial inputs of terrestrially derived organic matter. Based on our earlier assumptions, these factors constrain the functional potential of this stream, as evidenced by the slow breakdown rates of needles and leaves. In contrast, the tributary is a geologically older system with substantial inputs of terrestrially derived organic matter. This system clearly provides a more suitable environment for enhanced stream function, as demonstrated by leaf breakdown rates similar to those of temperate forested streams. This comparison suggests that as alpine streams mature and incorporate additional sources of organic matter, their functional capacity also increases. Consequently, it is the mosaic of stream types, from highly unstable proglacial streams with minimal organic inputs to stable tributaries and springbrooks with substantial terrrestrial inputs, that expresses the full range in the functional capacity of alpine glacial flood plains and optimizes their biocomplexity.

4. THE OTHER DIMENSION: LEAF BREAKDOWN IN THE TERRESTRIAL ENVIRONMENT

The kind of terrestrial environment on which a leaf falls and the length of time it remains there determines the extent to which leaf litter is preconditioned before entering the aquatic environment (Mayack et al., 1989). Terrestrial environments in the Roseg consist of conifer forest on valley side-slopes, grass covered areas in different parts of the active flood plain, and glacial alluvium lacking vegetative cover. To assess the influence of the terrestrial landscape on leaf breakdown, we examined larch needle breakdown in four different terrestrial areas in the flood plain over a 3-year period: the side-slope forest, open grassland in the middle of the flood plain, bare glacial alluvium in the middle flood plain, and open grassland near the lake outlet. After the initial placement of needle packs in autumn 1998, needles were collected each spring and autumn and analyzed for nutrients (N and P) and ergosterol (as a measure of fungal biomass). To assess the general physical environment at each site, we installed data-loggers at groundlevel for recording air temperature and percent relative humidity.

As is evident in Table 2, the environmental conditions among terrestrial habitat types in the flood plain varied predictably. Un-vegetated glacial alluvium in the middle of the flood plain (rock) appeared as the harshest environment, showing the greatest degree of desiccation (resulting from high temperatures and low relative humidity). In contrast, the side-slope forest (forest) appeared to be the least harsh environment (warm and wet) conducive for fungal colonization and growth. The grass area in the middle of the flood plain (grass) was a more benign environment than the grass area at the glacial lake outlet (outlet) because of its higher relative humidity. As a consequence, we predicted needles to decompose the fastest in the forest and the slowest in the un-vegetated glacial alluvium (rock).

Table 2. Temperature and relative humidity (mid-May to mid-September) at the four sites used to examine needle breakdown in different terrestrial environments of the alpine glacial flood plain. CV is the coefficient of variation.

Parameter	Site	Mean	CV (%)	Minimum	Maximum
Temperature (°C)	*Rock*	11.5	56	-3.8	31.9
	Grass	10.4	65	-5.3	29.9
	Outlet	9.2	42	-3.4	19.8
	Forest	9.4	48	-2.0	24.4
Relative Humidity (%)	*Rock*	74	34	10	100
	Grass	80	28	17	100
	Outlet	73	24	27	100
	Forest	84	20	22	100

Needles displayed a range in breakdown dynamics in the different terrestrial environments (Fig. 7). As expected, breakdown was fastest in the forest and slowest on the glacial alluvium (rock). The grass-vegetated sites (grass and outlet) were similar to each other and intermediate to rock and forest. In fact, needles in the forest were basically broken down by the second year, whereas they appeared to be essentially unchanged at the rock site. Needles at the grass sites showed obvious changes in color, although needle form remained unchanged. Nitrogen continued to accumulate in needles at all sites, and at concentrations somewhat higher than those in the aquatic environments. Phosphorus levels remained low in needles in the rock environment, increased slightly over time in the grass sites, and peaked in concert with ergosterol in the forest. Phosphorus values were similar to those found in the streams, except at the forest site where values actually were higher and more similar to values of alder leaves in the stream sites. This is quite interesting as it further indicates microbial uptake of phosphorus in both the aquatic and terrestrial environments, and that the temporal patterns in ergosterol are a good indicator of litter breakdown. These data from the terrestrial study of litter breakdown suggest that where a needle lands prior to entering the stream environment could play an

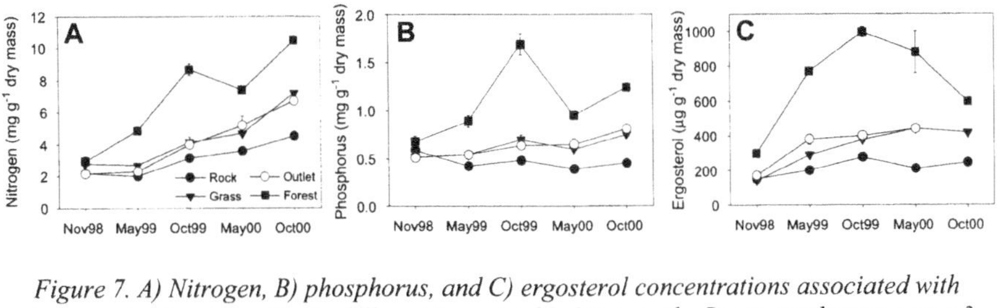

Figure 7. A) Nitrogen, B) phosphorus, and C) ergosterol concentrations associated with decomposing larch needles in different terrestrial habitats in the Roseg catchment over a 3 year period. Symbols represent the average of 4 samples with error bars showing standard errors.

important role in the breakdown of that needle in the stream environment. In effect, needles that spend some time in the forest environment could be broken down faster upon entering the aquatic environment than needles that spend some time in un-vegetated glacial alluvium.

5. SUMMARY AND CONCLUSIONS

Our results demonstrate that alpine streams, even glacial streams, have a functional potential comparable to their temperate stream counterparts. This finding was somewhat surprising considering the low water temperatures, unstable habitats, and low inputs of terrestrially derived organic matter at some sites. This suggests that functional differences between alpine glacial streams and forested lowland streams are more gradual rather than fundamental; the 'functional machinery' is present in alpine glacial streams, although the processes may be slower in the harsher stream types. The functional potential of alpine glacial streams appears constrained by the interplay of an extreme environment and limited resource availability. This indicates that alpine streams can functionally compensate for expected landscape transformations, such as increases in terrestrial vegetation resulting from climate change.

The contrast between needle and broad-leaf litter breakdown suggests complementarity between microbes and macroinvertebrates. Microbes were more important in needle breakdown, as the quality of needles as a food source for aquatic insects is much less than that of broad-leaf species such as alder. Microbes also were important in the breakdown of alder, but their association with the alder leaves enhanced the overall leaf quality for consumption by shredders. This finding was especially evident in these alpine systems, as leaves were a scarce commodity readily consumed in the presence of facultative shredders (e.g., at the lake outlet and tributary). Consumption of leaves by shredders may also have resulted in the lower ergosterol concentrations on leaves relative to needles, as macroinvertebrates probably fed preferentially on the microbes (fungi) associated with alder leaves, thereby reducing fungal biomass.

A comparison of leaf breakdown in the terrestrial environment with that in the streams suggests a land-water linkage is operating in alpine flood plains. The different results derived from different terrestrial environments also indicate similar successional processes are evident in both systems. The more recently derived environments, i.e., glacial alluvium or proglacial streams, showed the least functional potential, whereas the older more stable environments, i.e., the side-slope forests and springbrooks, showed the greatest. These successionally older sites, whether terrestrial or aquatic, had sufficient resources available and a greater habitat stability that allowed the development and persistence of a functionally competent biotic assemblage. These data indicate that as the proportions of habitat change in the flood plain, both aquatic and terrestrial, the functional capacity of the flood plain also will change, ultimately dictating functional patterns in floodplain biocomplexity.

6. ACKNOWLEDGEMENTS

Numerous individuals have assisted in the field and laboratory towards the completion of the various studies presented. We are particularly grateful to S. Meyns, R. Illi and B. Ribi for analytical chemistry, D. M. Anderson, P. Burgherr and C. Jolidon for help in the field, and K. Callies for identification of aquatic hyphomycetes. We thank J.V. Ward for initially bringing the topic of litter breakdown in these Alpine glacial streams to our attention.

7. REFERENCES

Bärlocher, F. (1992). Effects of drying and freezing autumn leaves on leaching and colonization by aquatic hyphomycetes. *Freshwater Biology, 28,* 1-7.

Burgherr, P., & Ward, J. V. (2000). Zoobenthos of kryal and lake outlet biotopes in a glacial flood plain. *Verhandlungen der Internationalen Vereinigung für Theoretische und Angewandte Limnologie, 27,* 1587-1590.

Cummins, K. W., & Klug, M. J. (1979). Feeding ecology of stream invertebrates. *Annual Review of Ecology and Systematics, 10,* 147-172.

Flecker, A. S. (1996). Ecosystem engineering by a dominant detritivore in a diverse tropical stream. *Ecology, 77,* 1845-1854.

Gessner, M.O. (1997). Fungal biomass, production and sporulation associated with particulate organic matter in streams. *Limnetica, 13,* 33-44.

Gessner, M. O., & Chauvet, E. (1994). Importance of stream microfungi in controlling breakdown rates of leaf litter. *Ecology, 75,* 1807-1817.

Gessner, M. O., Dobson, M., & Chauvet, E. (1999). A perspective on leaf litter breakdown in streams. *Oikos, 85,* 377-384.

Gessner, M. O., Robinson, C. T., & Ward, J. V. (1998). Leaf breakdown in streams of an alpine glacial floodplain: dynamics of fungi and nutrients. *Journal of the North American Benthological Society, 17,* 403-419.

Gessner, M. O., & Schmitt, A. L. (1996). Use of solid-phase extraction to determine ergosterol concentrations in plant tissue colonized by fungi. *Applied and Environmental Microbiology, 62,* 415-419.

Hieber, M., & Gessner, M. O. (2002). Contribution of stream detritivores, fungi, and bacteria to leaf breakdown based on biomass estimates. *Ecology, 83,* 1026-1038.

Iqbal, S. H., Bhatty, S.F., & Malik, K. S. (1980). Freshwater hyphomycetes on submerged decaying pine needles in Pakistan. *Transactions of the Mycological Society of Japan, 21,* 321-327.

Mayack, D. T., Thorp, J. H., & Cothran, M. 1989. Effects of burial and floodplain retention on stream processing of allochthonous litter. *Oikos, 54,* 378-388.

Robinson, C. T., Gessner, M. O., & Ward, J. V. (1998). Leaf breakdown and associated macroinvertebrates in alpine glacial streams. *Freshwater Biology, 40,* 215-228.

Robinson, C. T., Gessner, M. O., Callies, K. A., Jolidon, C., & Ward, J. V. (2000). Larch needle breakdown in contrasting streams of an alpine glacial floodplain. *Journal of the North American Benthological Society, 19,* 250-262.

Rosset, J., Bärlocher, F., & Oertli, J. J. (1982). Decomposition of conifer needles and deciduous leaves in two Black Forest and two Swiss Jura streams. *Internationale Revue der gesamten Hydrobiologie, 67,* 695-711.

Short, R. A., Canton, S. P., & Ward, J. V. (1980). Detrital processing and associated macroinvertebrates in a Colorado mountain stream. *Ecology, 61,* 727-732.

Suberkropp, K. (1992). Interactions with invertebrates. In F. Bärlocher (Ed.). *The ecology of aquatic hyphomycetes* (pp. 120-143). Berlin: Springer.

Tockner, K., Malard, F., Burgherr, P., Robinson, C. T., Uehlinger, U., Zah, R., & Ward, J. V. (1997). Physico-chemical characterization of channel types in glacial floodplain ecosystem (Val Roseg, Switzerland). *Archiv für Hydrobiologie, 140,* 433-463.

Wallace, J. B., Eggert, S. L., Meyer, J. L., & Webster, J. R. (1997). Multiple trophic levels for a forested stream linked to terrestrial litter inputs. *Science, 277,* 102-104.

Webster, J. R., & Benfield, E. F. (1986). Vascular plant breakdown in freshwater ecosystems. *Annual Review of Ecology and Systematics, 17,* 567-594.

Webster, J. R., & Meyer, J. L. (eds). (1997). Stream organic matter budgets. *Journal of the North American Benthological Society, 16,* 3-161.

Webster, J. R., Wallace, J. B., & Benfield, E. F. (1995). Organic processes in streams of the eastern United States. In C. E. Cushing, G. W. Minshall & K. W. Cummins (Eds.). *Ecosystems of the world 22: River and Stream Ecosystems* (pp. 117-187). Amsterdam: Elsevier.

Weyers, H. S., & Suberkropp, K. (1996). Fungal and bacterial production during the breakdown of yellow poplar leaves in 2 streams. *Journal of the North American Benthological Society, 15,* 408-420.

Wieder, R. K., & Lang, G. E. (1982). A critique of the analytical methods used in examining decomposition data obtained from litter bags. *Ecology, 63,* 1636-1642.

Zah, R., & Uehlinger, U. (2001). Particulate organic matter inputs to a glacial stream ecosystem in the Swiss Alps. *Freshwater Biology, 46,* 1597-1608.

CHAPTER 14

C. T. ROBINSON, U. UEHLINGER & M. O. GESSNER

NUTRIENT LIMITATION

1. INTRODUCTION

Primary production (algal growth) fuels the energy base of most streams in the Roseg flood plain, especially those streams having a glacial influence (see Chapter 12). Some side-slope tributaries do have substantial inputs of terrestrially derived organic matter, but these represent a small percentage of the total channel area in the flood plain (Zah & Uehlinger, 2001). Constraints on primary production vary by stream type and time of year. Channel instability, low transparency, and scour by glacial sediments limit primary producers in glacial channels during peak ice melt in summer, whereas snow-cover and low nutrients limit primary production in most stream types in winter (Robinson et al., 2002). Nutrient dynamics also change longitudinally along the main channel because of the increase in tributary and groundwater inputs as one moves downstream (Malard et al., 1999; Tockner et al., 2002).

In general, nitrogen concentrations in streams of the Roseg are high, whereas phosphorus levels are low (Chapter 6). The relatively high nitrogen levels result from high atmospheric deposition (Malard et al., 1999; Tockner et al., 2002). For instance, nitrogen levels increase in most channels during spring snowmelt as the accumulation of nitrogen deposited over the winter is released (Fig. 1). In contrast, soluble phosphorus is derived from geologic sources and tends to be low in these alpine streams because of the slow weathering rates. However, particulate phosphorus can be high in glacial streams (Tockner et al., 1997), but this form is not readily taken up by primary producers. Regardless, phosphorus levels tend to be higher in summer during the major period of glacial ablation (Fig. 1), although other factors mentioned above most probably limit algal growth. Silica is another important element that can limit the production of algae, diatoms in particular, but measured concentrations in Roseg streams were always above limitation levels.

We examined the role of seasonal nutrient dynamics among the different stream types in the Val Roseg using standard bioassay techniques. Four different avenues of investigation were used: 1) enrichment of nitrogen and phosphorus using nutrient diffusing substrata; 2) nutrient enrichment by whole stream fertilization; 3) assessment of nutrient kinetics by determination of uptake and retention; and 4) evaluation of the effects of nutrient enrichment on leaf litter breakdown. This complementary approach (nutrient enrichment and uptake) provided information on nutrient limitation and resulted in a better mechanistic understanding of nutrient

J.V. Ward & U. Uehlinger (eds.), Ecology of a Glacial Flood Plain, 231-241.
© 2003 *Kluwer Academic Publishers. Printed in the Netherlands*

dynamics in streams of the flood plain. The leaf breakdown experiment extended our findings on nutrient limitation from autotrophic to heterotrophic processes in streams of the Roseg flood plain.

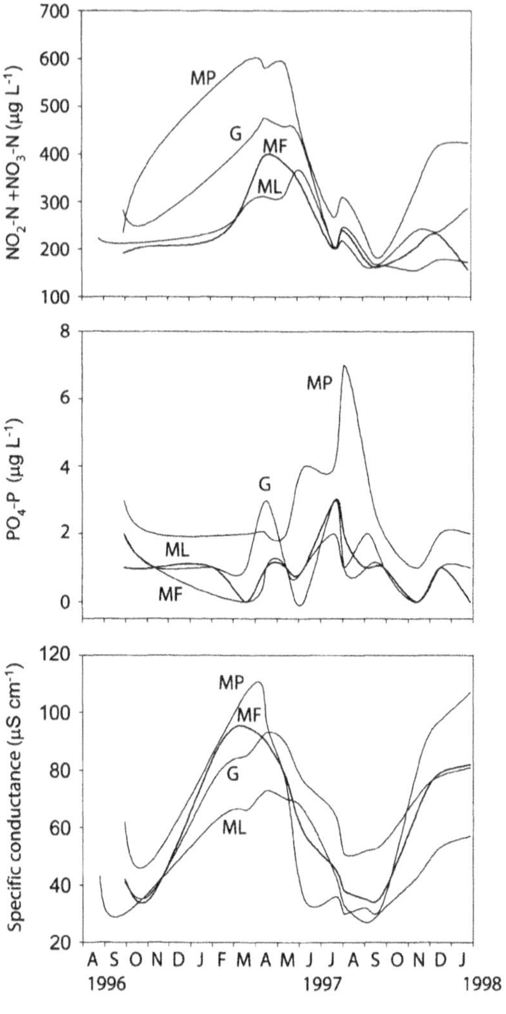

Figure 1. Seasonality in concentrations of nitrate-nitrogen, soluble reactive phosphorus, and specific conductance in four different channel types in the Roseg flood plain. MP is the proglacial stream (site 6), MF is the main glacial channel in the lower flood plain (site 80), ML is the lake outlet (site 10), and G is a groundwater fed channel in the lower floodplain area (site 75). Note that chemical fluctuations are more extreme in the proglacial reach, and less so in the lake outlet and groundwater channels. The lake acts as a sink for nutrients and buffers extreme changes in flow; the groundwater site is affected by snowmelt runoff but is not influenced by glacial runoff. Modified from Robinson et al. (2002), Fig. 1, p. 265.

2. NUTRIENT LIMITATION IN DIFFERENT STREAM TYPES

We assessed nutrient limitation using standard nutrient diffusing substrata of four different treatments: control (2% agar), N enriched, P enriched, and both N and P enriched (following Francoeur et al., 1999). Agar with the various treatments was poured into 30 ml glass vials and sealed with either 30 μm nitex netting or porous silicate ceramic that served as a surface for periphyton colonization. Diffusers were allowed to incubate in each study stream for about 30 days. Once collected, the nitex nets or ceramics were extracted in ethanol and analyzed for periphyton biomass as chlorophyll *a*.

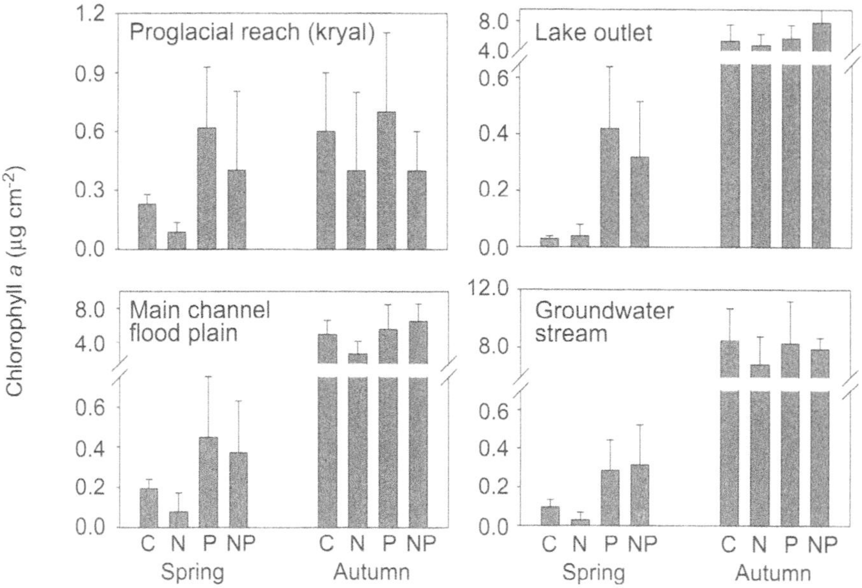

Figure 2. Results of nutrient diffuser experiments conducted in spring and autumn in the different stream types. Results are expressed in terms of chlorophyll a *as a measure of algal biomass, where C = control, N = nitrogen addition, P = phosphorus addition, and NP = nitrogen and phosphorus addition. Bars are an average of 5 replicates (+1 SD). Modified from Robinson et al. (2002), Fig. 2, p. 268.*

The results showed a significant effect of phosphorus enrichment in spring but not in summer (not shown) or autumn (Fig. 2). However, the groundwater sites showed little or no response to enrichment in spring. For all sites, vials enriched with phosphorus, or with N and P combined, had higher algal biomass than control or nitrogen enriched vials. The enrichment effect was more pronounced at the lake outlet and proglacial (kryal) sites than the lower main channel site, suggesting a longitudinal response to nutrient enrichment. Periphyton biomass on diffusers was an order of magnitude higher in autumn than in spring or summer at all sites except the proglacial site, regardless of treatment. The high biomass at all sites in autumn probably over-shadowed any effect of nutrient limitation, indicating nutrients were not limiting primary production at this time of year. The lack of a nutrient effect in glacial channels in summer may simply be related to other factors, e.g., turbidity and scouring, constraining primary production.

3. NUTRIENT UPTAKE AND RETENTION

Nitrogen and phosphorus uptake parameters (uptake distance and uptake rate) were measured using short-term continuous injections together with NaCl as a conservative tracer (Webster & Ehrmann, 1996). Because nitrogen concentrations were high in stream water, N uptake was determined for only 3 different stream types. Nutrient injections of 1-1.5 hr duration were run on stream reaches of 100-190 m in length. The study reaches had no substantial lateral surface water or groundwater inputs, and added nutrients increased background concentrations by a factor of 2 to 15. For each site, we corrected plateau concentrations against background levels and then fit the data to an exponential uptake model to calculate standard uptake parameters.

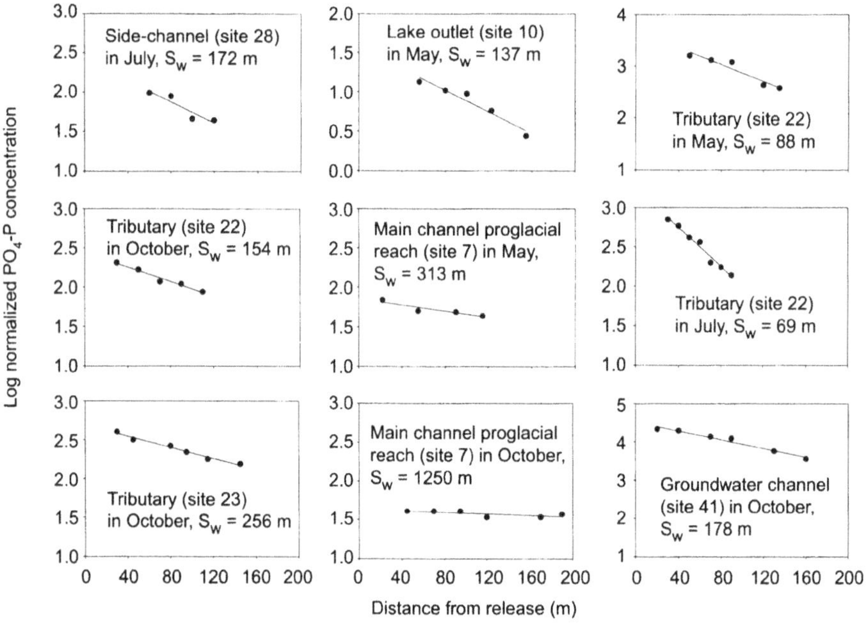

Figure 3. Phosphorus uptake lengths in various stream types in the Roseg flood plain. Each point represents an average of 5 samples collected at plateau concentrations after a set injection period and then fit to an exponential uptake model. The uptake length (S_w) is the inverse of the slope of the regression line.

As expected, nitrogen uptake lengths were much greater (2.7—30×) than phosphorus. Uptake lengths for nitrogen ranged from 0.4 km at the lake outlet to

15.4 km in the main glacial channel. Seasonally, N uptake length increased in the main channel from 15.4 km in spring to 45.4 km in autumn. Regardless, these data suggest that most nitrogen deposited in the flood plain is passed quickly through the system and certainly is non-limiting to ecosystem functioning.

Phosphorus, on the other hand, had relatively short uptake lengths, ranging from a low of 69 m in the springbrook to 1250 m in the main channel (Fig. 3). Two side-slope tributaries showed uptake lengths of 154 and 265 m. Phosphorus uptake at the glacial-influenced lake outlet was 137 m. Seasonally, P uptake length increased from ca. 313 m in spring to 1250 m in autumn in the glacial main channel.

A comparison of P uptake kinetics between groundwater-fed and glacial streams revealed that, although uptake lengths were longer in glacial channels, uptake rates (as mg m^{-2} h^{-1}) were actually higher in glacial channels (Fig. 4). We suspect the higher uptake rates of phosphorus in glacial channels were due to adsorption of P to particles of glacial flour. An exception to this pattern was a tributary fed by hillslope ground water (site 22) that had a high abundance of moss in the channel[1]: mosses are known to take up nutrients from the water through their leaves.

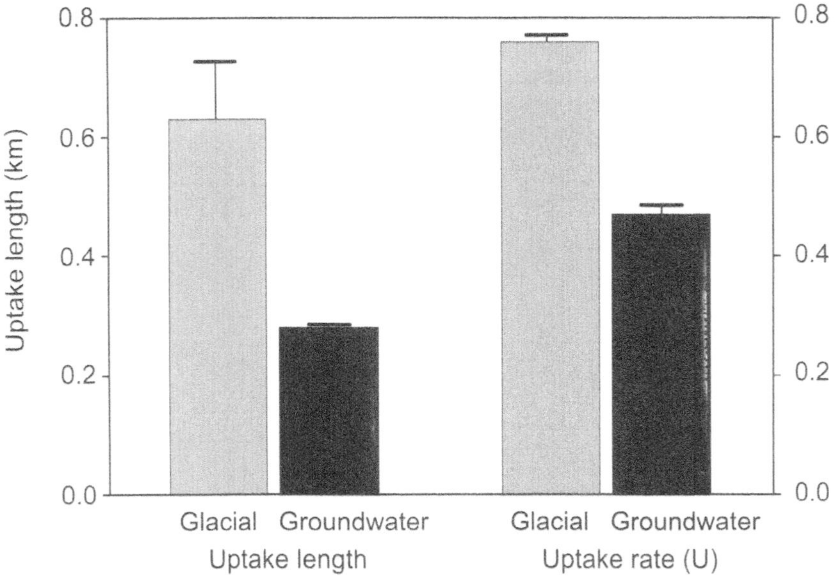

Figure 4. A comparison of phosphorus uptake kinetics between groundwater-fed streams glacial streams in the Roseg flood plain. Average (bar) + 1 standard error.

[1] Mosses were not normally encountered in the glacial floodplain of the Val Roseg, even in groundwater fed streams. The moss present was restricted to the experimental reach, which was outside of the actual floodplain.

4. WHOLE STREAM NUTRIENT ENRICHMENT

We added nutrients as slow-release fertilizer briquettes in summer to a 50-m reach of a side-tributary of the Roseg flood plain (site 14). A 50-m reach just upstream of the fertilized reach was used for comparison. A weir was temporarily installed directly upstream of the study section, enabling us to maintain relatively constant flow (10-20 L s^{-1}) during the study period. Unglazed quarry tiles were used as substrata for periphyton growth. Replicate samples were collected after 1 month, 2 months, and in the following spring after fertilization. Periphyton biomass on the tiles was assessed as chlorophyll *a* and ash-free dry mass (AFDM), but only the AFDM data are shown here.

Nutrient addition initially increased both phosphorus and nitrogen concentrations in the water. After about a month, N concentrations were again similar between the two reaches but dissolved phosphorus levels remained around 5 times higher in the treatment reach. Periphyton biomass increased within two months following nutrient addition, with chlorophyll levels being 13× and AFDM 5× greater in the fertilized reach than in the unfertilized reach (Fig. 5). However, the increase in periphyton biomass was not sustained over winter with levels being similarly low in both reaches. Ice cover had a significant effect on periphyton in the tributary stream, where freezing to the stream bottom was noted in some areas. Both nitrogen and phosphorus contents of the periphyton were higher in the fertilized reach, and nutrient contents were directly related to periphyton biomass.

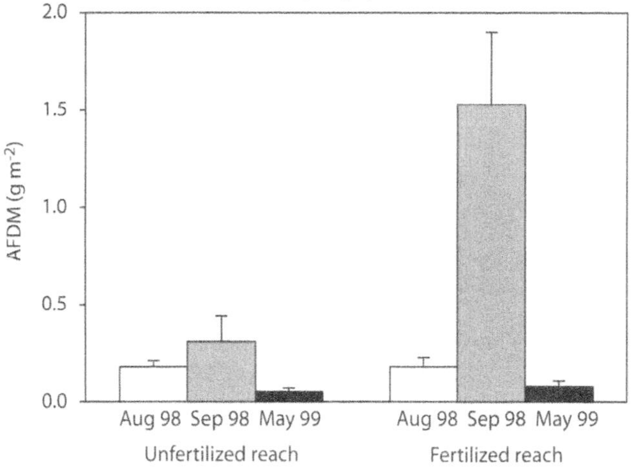

Figure 5. Mean (+1 SD, n = 7 to 10 per date) periphyton biomass as ash-free dry mass (AFDM) on quarry tiles placed in fertilized and unfertilized reaches of a tributary stream of the Roseg flood plain. The study began in late July with replicate samples taken 1, 2, and 10 months later. Modified from Robinson et al. (2002), Fig. 4, p. 269.

5. NUTRIENT ENRICHMENT AND LEAF BREAKDOWN

In this experiment we added nutrients (N and P) in the form of slow-release fertilizer briquettes to leaf packs and compared breakdown rates with unfertilized leaf packs. Fresh alder leaves (*Alnus viridis*) were used in the experiment and the study was terminated after 42 days. The fertilizer briquettes initially increased the phosphorus and, to a lesser extent, nitrogen (not shown) concentrations associated with the alder leaves, although this increase disappeared within 3 weeks (Fig. 6). Leaf nutrient concentrations were actually less in fertilized packs at the end of the experiment, reflecting the fact that little leaf material remained in these packs relative to unfertilized leaf packs.

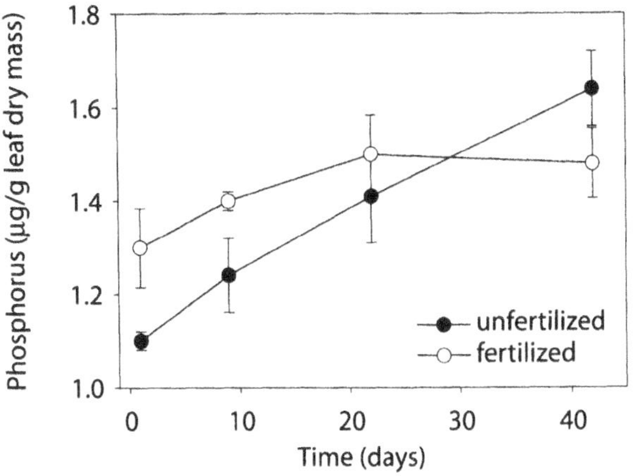

Figure 6. Mean (±1 standard error, n=4) phosphorus concentrations of leaves decomposing in an alpine springbrook in packs with or without nutrients added. Modified from Robinson & Gessner (2000), Fig. 2, p. 260.

The the breakdown of fertilized leaf packs ($k = 0.0284$ day^{-1}) was significantly faster than that of unfertilized packs ($k = 0.0137$ day^{-1}), resulting in a 30% higher leaf mass loss after 42 days in the fertilized leaf packs (Fig. 7). We expected that microbes would colonize the fertilized leaves to a greater extent than the unfertilized leaves. However, although fungal biomass was substantial on leaves (up to 55 mg dry mass per g leaf dry mass on day 42), and continued to increase over time, we found no differences between the two treatments. The number of fungal species recorded from the conidia produced from the different leaf packs was markedly lower on fertilized packs, especially after 42 days, although sporulation on both types of packs was dominated by a single species, *Flagellospora curvula* Ingold.

The primary reason for the greater breakdown of fertilized leaves was apparently the preferred colonization of these packs by shredding insects (Fig. 8). Indeed, there were 4 times the number of invertebrates associated with fertilized packs than with unfertilized packs by day 22, and about 50% of these animals were shredders.

Macroinvertebrate biomass showed a similar pattern as numbers (data not shown), with fertilized packs having twice as much biomass as unfertilized packs and shredders representing more than 66% of this biomass. The predominant shredders were nemourid stoneflies (*Nemoura* spp. and *Protonemura* spp.), and a few caddisflies.

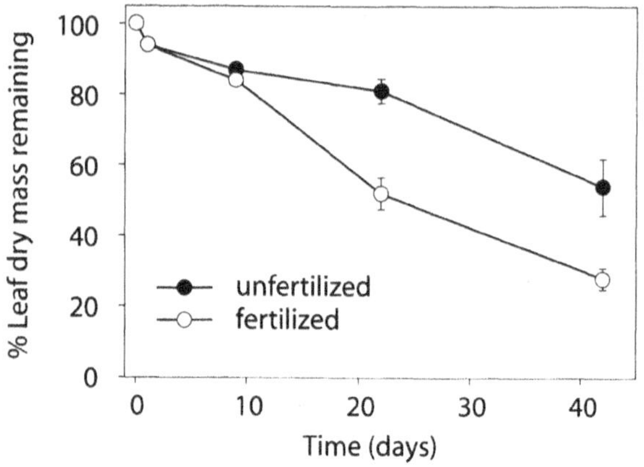

Figure 7. Mean (±1 standard error, n=4) percent dry mass remaining of alder leaves decomposing in an alpine springbrook in packs with and without nutrients added. Modified from Robinson & Gessner (2000), Fig. 1, p. 260.

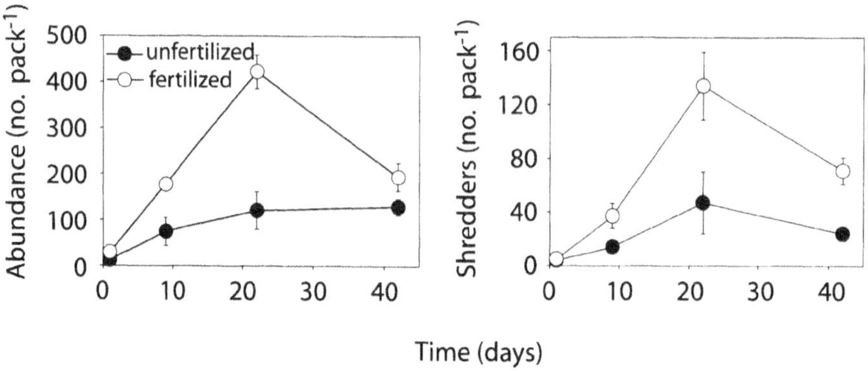

Figure 8. Mean (±1 standard error, n=4) total abundance (left) and number of shredding macroinvertebrates (right) associated with leaves decomposing in an alpine springbrook in packs with and without nutrients added. Modified from Robinson & Gessner (2000), Fig. 4, p. 261.

Overall, the results suggest that heterotrophic processes, such as leaf breakdown, are nutrient limited in the study system. Other researchers have observed that the degradative activity of aquatic fungi is stimulated by increased nutrient availability (Grattan & Suberkropp, 2001; Gulis & Suberkropp, 2003), suggesting that nutrient limitation of leaf breakdown may be a common phenomenon in streams. The greater abundance of shredders in the fertilized packs relative to unfertilized packs suggests that nutrient enrichment resulted in higher-quality resource patches, which these animals were able to exploit. The greater shredder abundances likely translate into enhanced feeding activity, thus accelerating leaf breakdown. Conceivably, this greater feeding activity masked the response of the aquatic fungi to nutrient enrichment, as any increase in fungal biomass would have been removed by the shredders, which select patches of leaves colonized by fungi (Arsuffi & Suberkropp, 1985). A treatment of fertilized packs with no access by macroinvertebrates is needed to test this idea.

6. SUMMARY AND CONCLUSIONS

The combination of approaches provided different insights into nutrient limitation of streams in the Val Roseg flood plain. One clear finding is that the nitrogen concentration in these streams is not limiting to primary producers. The long uptake lengths (kilometers) and the lack of periphyton response to nitrogen addition indicate nitrogen is in excess in these alpine streams. There was a seasonal difference in nitrogen uptake lengths, but this can be attributed to the differences in discharge.

Most sites exhibited some degree of phosphorus limitation in spring prior to snowmelt runoff. Phosphorus limitation varied by stream type and season, being most pronounced in glacial streams of the upper flood plain. The glacial sites lower in the flood plain showed less of a response to nutrient enrichment than upper flood plain sites, although some limitation was clearly evident, suggesting lateral inputs of groundwater and side-slope tributaries were altering nutrient dynamics in the longitudinal dimension. Mulholland & Rosemond (1992) found longitudinal depletion of phosphorus in a stream lacking substantial groundwater and side-slope inputs. These results suggest that the functional capacity of the glacial streams increases longitudinally in response to nutrient subsidies (particularly phosphorus) from tributaries and groundwater-fed springbrooks.

Seasonally, spring was the period when periphyton showed a major response to phosphorus enrichment, whereas overall primary production was high in autumn, regardless of nutrient enrichment, and may have swamped any nutrient effect. In the glacial streams, primary production in summer appears constrained more by physical factors (turbidity, scouring) than by nutrient limitation. For example, a nutrient diffuser study conducted in summer in some other glacial streams in the Swiss Alps showed no response to enrichment of phosphorus (Rinke et al., 2001). In autumn, glacial ablation is low, water clarity is high and phosphorus levels are typically above limiting concentrations in the glacial streams. The higher phosphorus levels may be related to the deposition of glacial flour enriched in apatite as discharge declines in late summer (Tockner et al., 2002), as particulate phosphorus levels were 4—6× greater in autumn than in spring. Others have shown that glacial flour provides biologically available phosphorus for primary producers (e.g., Bretschko, 1968), although most of the particulate form is not directly bioavailable. The uptake rate data also support the contention that phosphorus adsorbs onto glacial flour, perhaps making P less biologically accessible in summer but becoming available later once flows are reduced in the glacial channels.

The addition of nutrients to the side-slope tributary suggests seasonal dynamics in nutrient limitation is dependent on the type of stream being assessed in the flood plain. That experiment showed a clear response of periphyton to enrichment in autumn, a response not observed in the glacial streams for this season. This suggests that phosphorus may be limiting in clear water channels during most times of the year. However, most of the alpine springbrooks are small and become ice and snow covered in winter, thus over-riding any effect of added nutrients, as was demonstrated in the reduction of periphyton biomass by spring in the above experiment. Groundwater channels also had greater periphyton biomass than glacial channels in summer, and uptake lengths were shorter in groundwater and side-slope tributaries than in glacial channels, indicating that the former are more retentive. The diverse results among channel types emphasizes the spatial heterogeneity in nutrient dynamics among streams in the flood plain, indicating, as with other floodplain properties, that the channel networks constitute not only a structural mosaic but also a functional mosaic that enhances the overall biocomplexity of the floodplain system.

The leaf litter experiment demonstrated that heterotrophic properties also can be nutrient limited in these alpine streams. Leaf breakdown was markedly enhanced when nutrients were added, primarily reflecting the greater number of shredding insects in those leaf packs. Whether the leaves with nutrients provided a better food resource for the aquatic insects is not definitive but highly suggestive. The predominance of a single species of aquatic fungi on the nutrient enriched leaves, compared to the much more diverse community on unfertilized leaves, suggests that enrichment may change the composition of fungal communities. The functional capacity (i.e. litter breakdown rate) of this stream, however, was clearly constrained by the low supply of nutrients.

The combination of approaches has shown that these alpine flood plain streams are sensitive to changes in nutrients, although additional research is needed to better clarify the mechanisms causing the observed responses. Functional response trajectories to nutrient enrichment are complex, affecting not only microbes and algae but also macrozoobenthos and the interactions of these organisms. Alterations in resource supply has been shown to dramatically change trophic dynamics of streams in forested streams (Wallace et al., 1999) and those in arctic tundra (Peterson et al., 1993). Our results demonstrate that in spite of notably different physical features, alpine streams respond to changes in nutrient supply quickly and in a manner similar to streams in other biomes.

7. ACKNOWLEDGEMENTS

Most of the results presented in this chapter represent the thesis work of various individuals examining nutrient dynamics in alpine streams of Switzerland: F. Guidon, P. Schenkel, R. Skvarc, and K. Rinke. We gratiously thank them all.

Analytical assistance was provided by R. Illi, S. Meyns, B. Ribi, and G. Meier. Financial support from the Swiss National Science Foundation is greatly appreciated (no. 21-49243.96).

8. REFERENCES

Arsuffi, T. L, & Suberkropp, K. (1985). Selective feeding by stream caddisfly (Trichoptera) detritivores on leaves with fungal-colonized patches. Oikos 45: 50-58.

Bretschko, G. (1968). Zur Hydrobiologie zentralalpiner Gletscherabflüsse. *Verhandlungen der Deutschen Zoologischen Gesellschaft in Innsbruck*, 168, 741-750.

Francoeur, S. N., Biggs, B. J. F., Smith, R. A., & Lowe, R. L. (1999). The limitation of algal biomass accrual in streams: seasonal patterns and a comparison of methods. *Journal of the North American Benthological Society, 18*, 242-260.

Grattan R.M. & Suberkropp K. (2001). Effects of nutrient enrichment on yellow poplar leaf decomposition and fungal activity in streams. *Journal of the North American Benthological Society, 20*, 33–43.

Gulis, V., & Suberkropp, K. (2003). Leaf litter decomposition and microbial activity in nutrient-enriched and unaltered reaches of a headwater stream. *Freshwater Biology, 48*, 123–134.

Malard, F., Tockner, K., & Ward, J. V. (1999). Shifting dominance of subcatchment water sources and flow paths in a glacial floodplain, Val Roseg, Switzerland. *Arctic, Antarctic, and Alpine Research, 31*, 135-150.

Mulholland, P. J., & Rosemond, A. D. (1992). Periphyton response to longitudinal nutrient depletion in a woodland stream: evidence of upstream-downstream linkage. *Journal of the North American Benthological Society, 11*, 405-419.

Peterson, B. J., Deegan, L., Helfrich, J., Hobbie, J. E., Hullar, M., Moller, B., Ford, T. E., Hershey, A., Kipphut, G., Lock, M. A., Fiebig, D. M., McKinley, V., Miller, M. C., Vestal, J. R., Ventullo, R., & Volk, G. (1993). Biological responses of a tundra river to fertilization. *Ecology, 74*, 653-672.

Rinke, K., Robinson, C. T., & Uehlinger, U. (2001). A note on abiotic factors that constrain periphyton growth in alpine glacier streams. *International Review of Hydrobiologia, 86*, 361-366.

Robinson, C. T., & Gessner, M. O. (2000). Nutrient addition accelerates leaf breakdown in an alpine springbrook. *Oecologia, 122*, 258-263.

Robinson, C. T., Uehlinger, U., Guidon, F., Schenkel, P., & Skvarc, R. (2002). Limitation and retention of nutrients in alpine streams of Switzerland. *Verhandlungen der Internationalen Vereinigung für Theoretische und Angewandte Limnologie, 28*, 263-272.

Tockner, K., Malard, F., Burgherr, P., Robinson, C. T., Uehlinger, U., Zah, R., & Ward, J. V. (1997). Physico-chemical characterization of channel types in glacial floodplain ecosystem (Val Roseg, Switzerland). *Archiv für Hydrobiologie, 140*, 433-463.

Tockner, K., Malard, F., Uehlinger, U., & Ward, J. V. (2002). Nutrients and organic matter in a glacial river-floodplain system (Val Roseg, Switzerland). *Limnology and Oceanography, 47*, 266-277.

Wallace, J. B., Eggert, S. L., Meyer, J. L., & Webster, J. R. (1999). Effects of resource limitation on a detrital-based ecosystem. *Ecological Monographs, 69*, 409-442.

Webster, J. R., & Ehrmann, T. P. (1996). Solute dynamics. In F. R. Hauer & G. A. Lamberti (Eds.). *Methods in Stream Ecology* (pp.145-160). San Diego: Academic Press.

Zah, R., & Uehlinger, U. (2001). Particulate organic matter inputs to a glacial stream ecosystem in the Swiss Alps. *Freshwater Biology, 46*, 1597-1608.

CHAPTER 15

K. TOCKNER, C. T. ROBINSON & P. BURGHERR

DRIFT AND COLONIZATION DYNAMICS

1. INTRODUCTION

The downstream transport of benthic organisms in the water column is a global feature of lotic ecosystems (Waters, 1972; Brittain & Eikland, 1988) that serves as an important colonization mechanism (Williams & Hynes, 1976; Müller, 1982; Mackay, 1992). Concomitant with the drift of invertebrates, stream currents also transport large amounts of particulate matter (e.g., Waringer, 1992), the temporal patterns of which reflect functional attributes of lotic ecosystems. For example, differences in drift patterns of benthic invertebrates among river types and exchange patterns of organic matter across floodplain channels are associated with differences in system productivity (Hershey et al., 1993; Bilton et al., 2001).

Until recently, few studies have been undertaken on drift ecology in high alpine rivers (Ilg et al., 2001; Robinson et al., 2002; Hieber et al., 2003). Most published work is based on studies of single thread channels conducted during the summer period (but see Hieber et al., 2003). The research synthesized herein represents the first year-round study of drifting invertebrates in a high elevation flood plain that encompassed diel, seasonal, and spatial patterns of drifting invertebrates and organic matter (seston). The chapter begins with an analysis of the seasonal and diel patterns of drifting seston, dissolved organic carbon (DOC) and nitrate among different channel types in the glacial flood plain. This is followed by detailed analyses of the seasonal and diel periodicity of drifting invertebrates. Finally, we assess the role of the glacial flood plain as a sink or source for benthic invertebrates and particulate organic matter. We formulated three hypotheses before conducting the research:

H1: Benthic invertebrates exhibit high drift densities as a behavioral adaptation to inhabiting a harsh environment.
H2: Drift patterns are coupled to the expansion and contraction cycle of the channel network. High seasonal and diel periodicity of drifting invertebrates and seston organic matter will occur, particularly in those channels directly influenced by glacial runoff.
H3: Close functional interactions occur among different floodplain channels, in particular between groundwater channels and the main channel. Groundwater channels serve as primary sources for the recolonization of channels influenced by glacial melt water.

J.V. Ward & U. Uehlinger (eds.), Ecology of a Glacial Flood Plain, 243-257.
© 2003 *Kluwer Academic Publishers. Printed in the Netherlands.*

Seasonally, we focused on four channel types, including the main channel at the upstream end of the flood plain, the main channel at the downstream end of the flood plain, a groundwater channel, and a channel carrying mixed waters (see Channel typology: Chapter 4). These channels were sampled about monthly over a one-year period (10 dates). In addition, an intermittent tributary was sampled in spring and summer (6 dates), and the main channel downstream of the glacier snout was sampled five times over the annual cycle. The most information on diel drift patterns was available for the main channel at the downstream end of the flood plain (11 dates, 7 times per date over a diel cycle). Drift sampling methods are described in detail by Robinson et al. (2002) and Hieber et al. (2003).

In addition to dispersal by drift, short-term instream colonization was investigated in five channel types in the flood plain (upstream and downstream ends of the main channel, a side channel, an intermittently-connected channel, a groundwater channel, and a tributary). Colonization experiments were carried out four times from November 1998 to November 1999. On each sampling date, four benthic samples were collected at each site using a Hess sampler (standard sampling procedure as described in Chapter 10). Colonization by macroinvertebrates was evaluated using substrate-filled cages (12 x 17 x 30 cm, mesh size 1cm^2) placed in each stream on each date. The colonization cages were left in place for 6 days during each experiment. Similarities between benthic, drift, and cage assemblages among sites in the different seasons were assessed using Kendall's coefficient of concordance (tau), a nonparametric test based on relative taxon abundances (Zar, 1984).

2. SPATIO-TEMPORAL TRANSPORT OF SUSPENDED SOLIDS, PARTICULATE ORGANIC MATTER, DISSOLVED ORGANIC CARBON, AND NITRATE

In the main channel, the transport of suspended solids (SS) and particulate organic matter (expressed as AFDM) exhibited a distinct seasonality with highest concentrations in summer (Fig. 1). Diel variability (expressed as the standard deviation of seven samples taken at 3 hour intervals) was also highest in summer; in the main channel, variation in SS and AFDM was driven by the diel cycle of glacial melting and freezing. In all other channel types, seasonal and diel patterns in suspended solids and particulate organic matter (POM) were less distinctive (Tockner et al., 1997).

Seston coarse particulate organic matter (CPOM, >1.0 mm) was sampled simultaneously with drifting invertebrates by exposing drift nets (Robinson et al., 2002). CPOM displayed a clear seasonality in all channel types (Fig. 2). Results of a two-way ANOVA (season and channel type as factors) demonstrated significant differences among channel types ($F = 59.1$, $p < 0.001$), among dates ($F = 143.8$, $p < 0.001$), and significant interactions between channel type and season ($F = 40.8$, $p < 0.001$). Seasonality was more pronounced in the main channel, with a peak during the main ablation period resulting from CPOM released from the glacier and

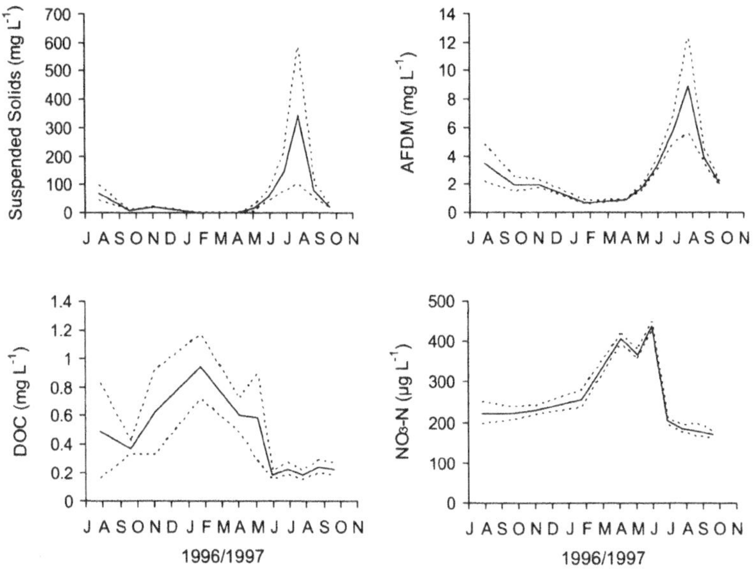

Figure 1. Seasonal patterns in suspended solids, AFDM, DOC, and nitrate in the main channel at the downstream end of the flood plain. Solid line: average; dashed line: ±1 SD. On each date, samples were taken from 6:00 to 24:00 at 3 hour intervals (n = 7 per date.)

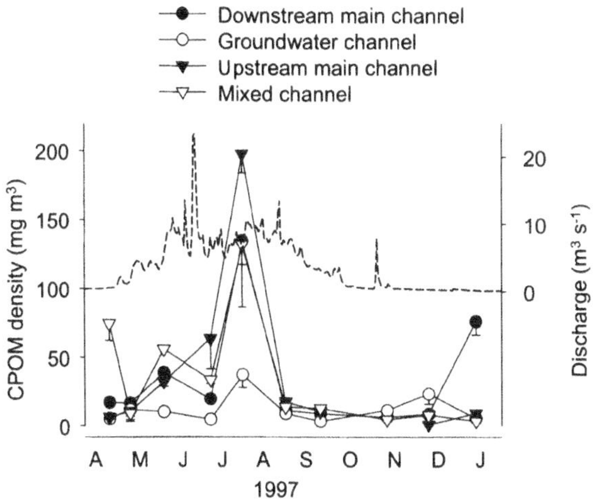

Figure 2. Seasonal drift concentration (average ± 1 SD) of CPOM (> 1.0 mm) in four different floodplain channels. Samples were taken around noon. Dashed line: daily discharge.

subsequently transported downstream (Teeri & Barrett, 1977; Tockner et al., 2002). The average maximum concentration was 200 mg m^{-3} in the main channel and 37 mg m^{-3} in the groundwater channel. Secondary peaks of CPOM occurred in April The average maximum concentration was 200 mg m^{-3} in the main channel and 37 mg m^{-3} in the groundwater channel. Secondary peaks of CPOM occurred in April and autumn when CPOM was primarily composed of needles and algae, respectively. At the downstream end of the main channel and in the groundwater channel, needles formed an important component of CPOM during the expansion period (e.g., 10 June 1997, main channel: 54 ± 26 % of CPOM). Pine needles were absent at the upstream end of the flood plain. In total, CPOM contributed less than 5 % to POM transport (see Chapter 6; Tockner et al., 2002).

3. SPATIO-TEMPORAL DRIFT OF INVERTEBRATES

3.1 Seasonal drift patterns

Drifting invertebrates displayed a high degree of seasonality in all channel types (Fig. 3). Drift densities were low at all sites during the main ablation period, even in channels that were not directly influenced by glacial runoff (e.g., the groundwater-fed channel). Based on a two-way ANOVA (season and channel type as factors), drift densities varied significantly among channel types ($F = 220.4$, $p < 0.001$) and among dates ($F = 191.0$, $p < 0.001$). In addition, significant interactions between dates and seasons occurred ($F = 35.2$, $p < 0.001$).

Channel location in the floodplain network played an important role in drift dynamics. The upper main channel site always exhibited the lowest drift densities compared with other channel types (annual average: 63±51.2, n=36). Groundwater channels, however, exhibited the highest drift densities during most sampling dates (annual average: 217±147, n=40). In the main channel, drift densities at the downstream end of the flood plain ranged from ~6 Ind. m^{-3} during the main ablation period in July to ~300 Ind. m^{-3} during the contraction period (beginning of November). In the groundwater-fed channel, a maximum drift density of 670 Ind. m^{-3} was observed in September. Seasonal heterogeneity, expressed as the coefficient of variation (CV), was high in all channels and ranged from 68% (groundwater channel) to 103% (mixed channel).

Compared to most headwater streams, drift densities were an order-of-magnitude higher in the Val Roseg flood plain. Hieber et al. (2003) compared four alpine stream types and found much lower drift densities than in the Val Roseg flood plain, with average densities ranging between 10 and 100 Ind. m^{-3}. In other high altitude streams and lake outlets, drift densities were even lower, mostly below 10 Ind. m^{-3} (Allan, 1987; Tilley, 1989; Kownacki et al., 1997; Matthaei et al., 1998; Ramirez & Pringle, 2001). In the Val Roseg flood plain, similar low densities occurred only during the main ablation period. However, drift densities were not directly comparable among studies as the mesh size of drift nets varied, and sampling intensity was generally much lower compared to the present study (e.g. Ilg et al., 2001; Saltveit et al., 2001).

Most drift occurred during the expansion and contraction periods. This high seasonality may reflect an "avoidance" strategy against harsh conditions during the main ablation period. In contrast to the Val Roseg, drift density peaks in most headwater streams in summer, and seasonal variation is lower than in the glacial flood plain (Matthaei et al., 1998; and references therein). In particular, high drift densities during the contraction period favored the recolonization of the main channel and other floodplain channels influenced by glacial meltwater. For instance, benthic invertebrate density in the main channel increased from about 3,000 Ind. m^{-2} to more than 12,000 Ind. m^{-2} between August and early autumn (Chapter 10).

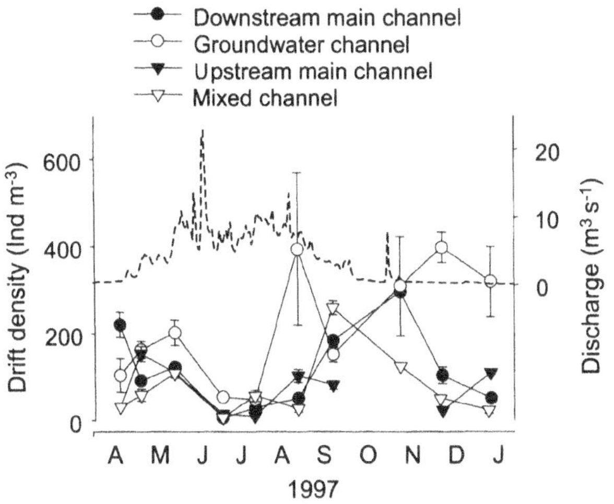

Figure 3. Seasonal density (average ± 1 SD) of drifting invertebrates in four different floodplain channels. Samples were taken around noon. Dashed line: average diel discharge.

3.2. Drift-Benthos relationships

Another index of mobility is the relative proportion of the benthic community actually found in the drift (P, %; Elliott, 1977). Overall, P-ratios were high in all channels with values >0.10 % during most of the time (Fig. 4). In the main channel, the estimated proportion of benthic individuals drifting in the water column over a 1 m^2 area at any instant in time ranged from <0.01 % (winter) to 1.99% (main ablation period) (Fig. 4). A P-ratio of 1.99 is around two orders-of-magnitude higher than P-ratios normally observed in streams, where ratios always were well below 0.01% (e.g., Danube: P = 0.0026-0.0064 %; Anderwald et al., 1991). The P-ratio differed markedly among channel types. In the main channel, the maximum ratio was found during the main ablation period. In the groundwater channel, however, the P-ratio was lowest during this period, and seasonal variation was less pronounced compared to other channel types (Fig. 4). In the groundwater channel, benthic invertebrate densities remained constant over seasons, although drift density decreased significantly in summer. Therefore, different processes are expected to control benthic and drifting invertebrate dynamics. Benthic invertebrates are mainly regulated by extrinsic factors (sediment movement, high turbidity) and the availability of resources (e.g., benthic algae), while the dynamics of drifting invertebrates can be explained by the interplay of extrinsic (environmental conditions) and intrinsic (life history strategies) variables.

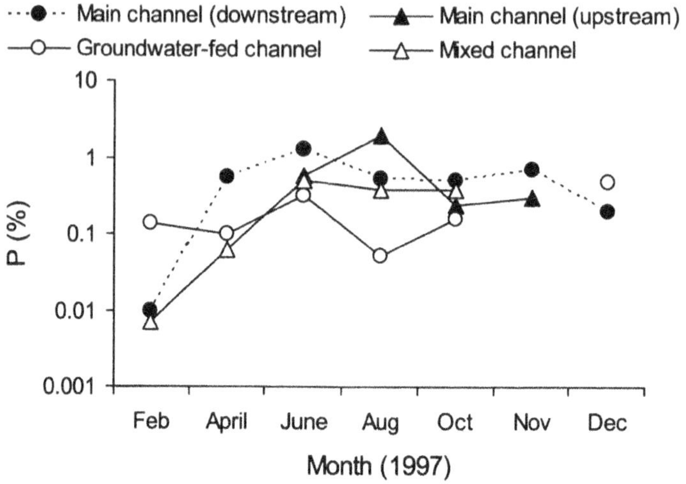

Figure 4. The proportion of benthic invertebrates in the drift (P, %; log-scale) in four channel types over an annual cycle. P was calculated after the formula of Elliott (1977).

3.3 Drift composition

In all channel types, Chironomidae was the most dominant group, averaging a relative contribution to drift from 68% (mixed channel) to 90% (main channel downstream of the glacial snout) (Table 1). Ephemeroptera were abundant only in the main channel at the downstream end of the flood plain (15%) and in the channel carrying water of different sources (mixed channel, 13%). Simuliidae (tributary, 7%, main channel, 6%) and Copepoda (groundwater-fed channel, 6%) were locally important. The two most common stoneflies (*Leuctra* spp. and *Rhabdiopteryx alpina*; Robinson et al., 2002) were observed at all sampling sites in low numbers, although their drift density was lowest during the main ablation period. Trichoptera were found only in low numbers in the drift. All other invertebrate groups were present only in small numbers, although seasonal peaks were observed for some groups (e.g., terrestrial invertebrates, Fig. 5).

Seasonal patterns of drift were influenced highly by the presence of particular taxa, although chironomids dominated the drift at all sites and most times (Fig. 5). In the groundwater channel, Chironomidae were most abundant after peak discharge, microcrustaceans (Copepoda, Cladocera, and Ostracoda) peaked in spring, terrestrial organisms (mainly Collembola and the terrestrial stages of aquatic insects) were most abundant in summer, and mayflies (mainly *Baetis* spp.) were predominant in

Table 1. Relative contribution (%) of invertebrate groups to total drift density (pooled data from all seasonal samples) and total number of invertebrates collected over an annual cycle. n = 4 per date and channel type.

	Main channel (downstream of glacier)	*Main channel (upstream of flood plain)*	*Main channel (downstream of flood plain)*	*Mixed channel*	*Groundwater-fed channel*	*Tributary*
Number of samples	20	36	40	40	40	24
Chironomidae	89.5	89.9	69.8	68.1	76.3	81.6
Simuliidae	0.2	0.4	6.2	5.6	0.6	7.1
Plecoptera	0.8	1.9	2.2	1.6	0.2	0.8
Ephemeroptera	0.5	3.1	14.9	13.3	6.5	1.2
Trichoptera	-	0.1	0.1	0.3	0.1	0.1
Oligochaeta	0.2	0.3	0.5	0.9	0.3	1.2
Copepoda	0.6	0.4	0.8	0.3	5.9	0.2
Cladocera	-	-	0.5	0.5	-	-
Ostracoda	-	-	-	0.3	-	-
Hydracarina	0.7	0.8	2.2	3.1	2.0	1.5
Nematoda	2.3	0.9	0.4	0.3	0.3	0.6
Collembola	3	0.8	0.9	2.9	1.5	4.6
Terrestrial div.	0.4	0.3	0.8	1.4	2.4	1.1
Others	1.9	0.9	0.7	1.4	3.7	-
Total individuals	4,086	5,197	16,511	6,354	11,621	7,364

late autumn. Two recent studies, Ilg et al. (2001) and Robinson et al. 2002), also found high drift densities in early autumn, and a pronounced peak in the density of *Baetis alpinus* (see: Fig. 5). In the Val Roseg flood plain, the contraction phase was the main dispersal period for benthic invertebrates such as mayflies and chironomids. During the main expansion period, however, the increase in the groundwater table led to a mass release of microcrustaceans into the drift. Even true groundwater species were encountered in the drift during the expansion period. Terrestrial organisms were transported primarily during the expansion and main ablation period. They were absent during winter.

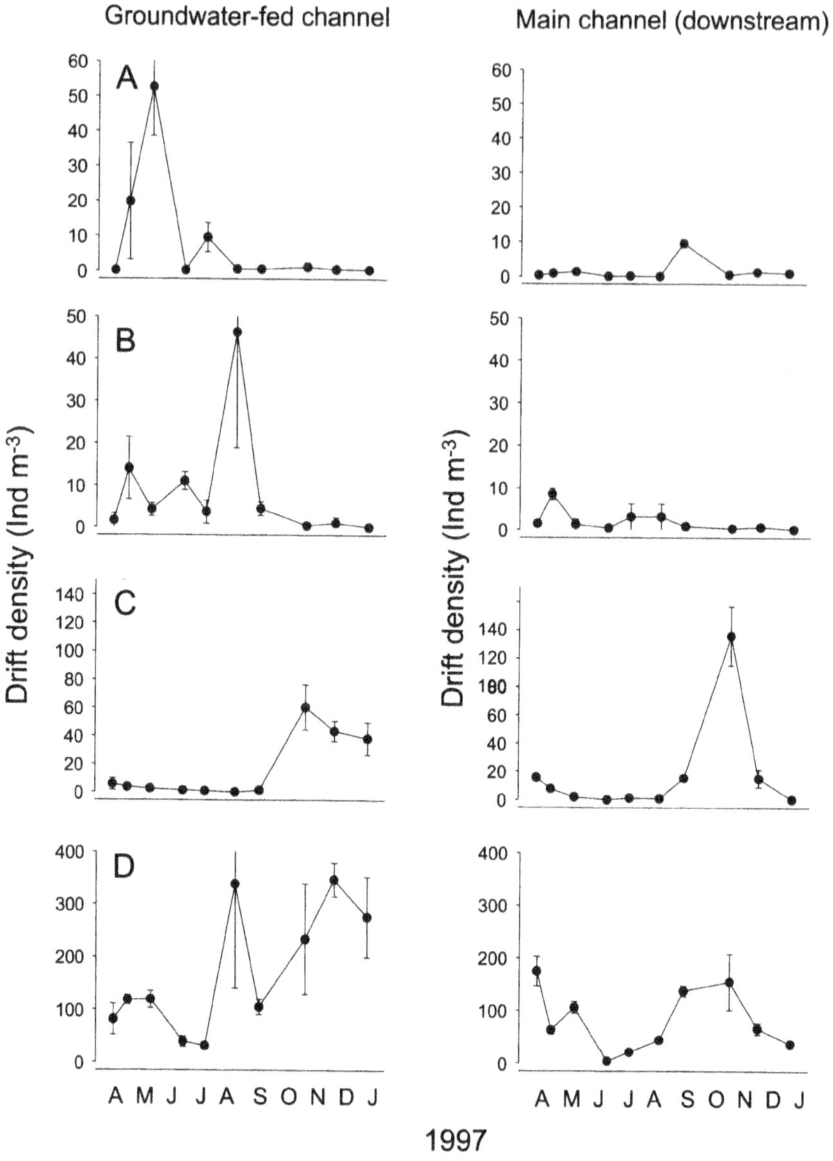

Figure 5. Seasonal drift patterns (average ± 1 SD) of A) microcrustaceans, B) terrestrial invertebrates, C) Ephemeroptera, and D) Chironomidae in a groundwater-fed channel and in the main channel at the downstream end of the flood plain. Samples were taken around noon (n = 4 per date and sampling site). Please note different scales of y-axis.

3.4. Diel drift patterns

A significant time of day effect on drift density occurred on all sampling dates, with densities being higher during the day than at dawn or night (Figs. 6 & 7). This diel pattern was observed in all channel types and during all seasons, although diel differences were most prominent in the main channel (Table 2). Highest diel periodicity was found during the expansion and contraction periods and in channels influenced by glacial melt water. In most rivers, drift displays peaks at dawn or during night, a periodicity thought to be a mechanism to reduce predation by drift-feeding fishes. Studies in fishless rivers found no evidence of distinct nocturnal drift patterns (Flecker, 1992; Pringle & Ramirez, 1998; Hieber et al., 2003). In glacial streams in Western Norway, Saltveit et al. (2001) also found peaks during daytime in summer, probably caused by passive drift in response to glacial melt. However, in the Val Roseg flood plain daytime peaks of drifting invertebrates occurred in all channel types and during all seasons. Therefore, factors other than predator avoidance or passive drift may cause diel periodicity in the glacial flood plain. Temperature may be an important factor that can lead to elevated moving and foraging activities of invertebrates during daytime. Armitage (1977) and Sagar & Glova (1992) have shown that low temperatures depressed drift rates, although the thermal influence was often superceded by flow and flood pulses. The diel maximum temperature in the Val Roseg flood plain ranged between 7.6-19.4°C (May-September) and 0.0-4.5°C (December-February). Diel temperature pulses were low in the main channel during summer, but high in floodplain channels with maximum diel pulses of up to 15°C (Uehlinger et al., 2003). Although temperature was expected to drive diel periodicity of drifting invertebrates in floodplain channels, flow pulses were expected to control drift patterns in the main channel. Hence, diel drift periodicity was a result of both active (floodplain channels and main channel during the low water period) and passive processes.

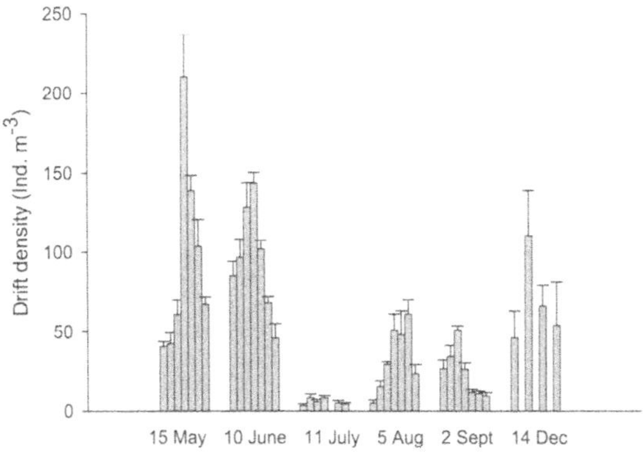

Figure 6. Diel drift patterns (average ± 1 SD) in the main channel at the downstream end of the flood plain. Samples were taken at 3 hour intervals between 06:00 and 24:00 (6 hour intervals in December).

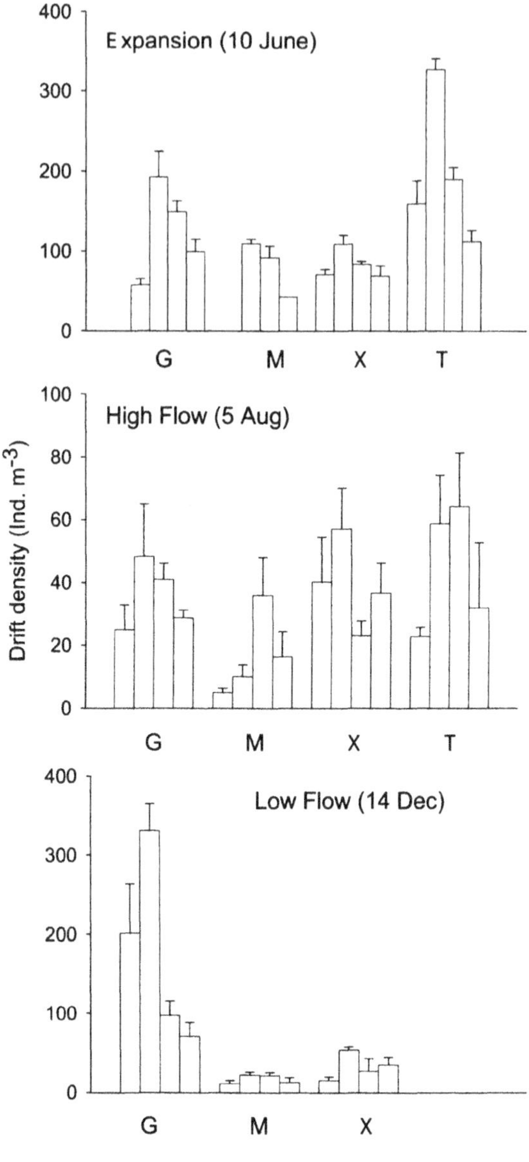

Figure 7. Diel drift patterns (average ± 1 SD) during the expansion, high flow and low flow periods in a groundwater-fed channel (G), the main channel at the upstream end of the flood plain (M), a channel carrying mixed water sources (X) and a tributary (T). Samples were taken at 6 hour intervals between 06:00 and 24:00. Please note different scales of y-axis.

Table 2. Results of one-way ANOVAs testing for differences in diel drift densities in the main channel at the downstream end of the flood plain (28 replicates per date).

Month	*SS*	*df*	*MS*	*F*	*P*
May	866×10^2	6	144×10^2	60.62	< 0.001
June	269×10^2	6	4487	37.1	< 0.001
July	176	6	29.4	11.4	< 0.001
August	8356	6	1393	15.9	< 0.001
September	5506	6	917.7	41.4	< 0.001
December	8626	3	2875	5.54	0.013

3.5. Drift rates and sink-source dynamics

A two-end member mixing model based on measured sodium (see Chapter 3) was used to determine whether the flood plain was acting as sink or source of organic matter and invertebrates. At the floodplain scale, the Val Roseg acted as both a sink and source for particulate organic matter (POM) and drifting invertebrates. The flood plain was a major sink for POM during the main ablation period and a source during the expansion period. During the contraction and low water periods, POM rates were too low to assess the role of the flood plain as a sink or source. For benthic invertebrates, the flood plain served as a source during the contraction and low water periods, however, it was a sink during the expansion and the main ablation period, except in August (Fig. 8). Based on this calculation, in early autumn more than 16 million invertebrates were exported every day from the flood plain to the river downstream.

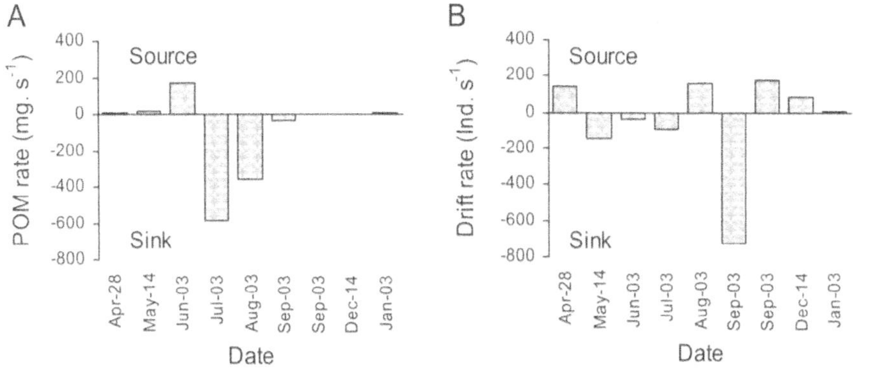

Figure 8. Seasonal sink and source patterns (expressed as mg AFDM and individuals sec^{-1}) of A) particulate organic matter, and B) drifting invertebrates. Calculations were derived from a mixing model based on the relative contribution of ground water and melt water to total floodplain discharge (see Tockner et al., 2002).

4. COLONIZATION PATHWAYS

The results of the cage experiments indicated that colonization dynamics by larvae differed substantially among sites and among seasons within a site. Assemblage composition strongly influenced larval colonization patterns within each stream, as inferred from the results of the Kendall's analyses (Table 3). For example, benthic assemblages were dominated by chironomids and baetids in the upper main channel in all seasons, both taxa being disturbance resistant and early colonizers of denuded substrates. Here there was a close association between benthic and drift assemblages, and cage assemblages were similar to both (Table 3). In contrast, benthic and drift assemblages were significantly different in the groundwater channel in all seasons, reflecting a higher proportion of non-drifting benthic invertebrates such as caddisflies, turbellarians, and microcrustaceans. Here cage assemblages were most similar to drift assemblages. These data strongly suggest that drift is the primary mechanism used by larvae of some aquatic insects in these streams to quickly disperse into vacant habitats. Lastly, drift and cage assemblages were similar in the lower main channel in spring and summer when chironomids and baetid mayflies also were predominant in the benthic assemblages (Burgherr et al., 2002).

Table 3. Similarity comparisons in the relative abundances based on the densities of macroinvertebrates between benthic, drift, and cage samples in each of the channel types for each date using Kendall's coefficient of concordance (tau). Significant P-values indicate difference in macroinvertebrate assemblage structure for a particular comparison (Bold= $p<0.05$). (Dates: November 1998, 1999; May 1999; August 1999).

Date	*Main channel (upstream)*		*Tributary*		*Mixed channel*		*Main channel (downstream)*		*Groundwater-fed channel*	
	tau	*p*	*tau*	*p*	*tau*	*p*	*tau*	*p*	*tau*	*p*
				Benthic versus Drift Macroinvertebrates						
N- 1998	0.06	0.24	NA	NA	0.12	0.11	0.14	0.08	0.19	**0.04**
M-1999	0.05	0.32	0.13	0.095	0.53	**0.001**	0.09	0.16	0.42	**0.003**
A-1999	0.13	0.10	0.14	0.08	0.03	0.41	0.17	0.06	0.18	**0.05**
N-1999	0.17	0.06	0.25	**0.02**	0.25	**0.02**	0.40	**0.004**	0.16	0.07
				Benthic versus Cage Macroinvertebrates						
N-1998	0.004	0.78	NA	NA	0.05	0.32	0.003	0.79	0.05	0.32
M-1999	0.007	0.70	0.04	0.36	0.00	1.00	0.14	0.08	0.22	**0.03**
A-1999	0.05	0.32	0.09	0.17	0.13	0.10	0.26	**0.02**	0.09	0.16
N-1999	0.02	0.56	0.05	0.32	0.08	0.19	0.22	**0.03**	0.19	**0.04**
				Drift versus Cage Macroinvertebrates						
N-1998	0.02	0.47	NA	NA	0.07	0.21	0.14	0.08	0.05	0.32
M-1999	0.06	0.26	0.13	0.096	0.36	**0.004**	0.01	0.70	0.07	0.21
A-1999	0.01	0.71	0.06	0.26	0.02	0.53	0.08	0.21	0.004	0.76
N-1999	0.09	0.16	0.02	0.56	0.13	0.096	0.13	0.096	0.06	0.25

5. SUMMARY

Our understanding of abiotic and biotic dynamics in high alpine streams has increased in recent years; however, most studies were limited to short periods of time (Ward, 1994, Milner et al., 2001). Drift related dispersal is seen as a pivotal functional process that controls gene flow, maintains ecological stability, and serves as a trophic pathway, both longitudinally and laterally. The present study contributed to the understanding of drift dynamics by assessing temporal and spatial drift patterns in different channel types of a glacial flood plain. Density of drifting invertebrates displayed a clear seasonality in all channel types, with highest densities during the expansion and contraction phases, and lowest densities during the period of maximum expansion. The results support the hypothesis (H1 in the introduction) that a high drift density in all channels during most of the year maintain ecosystem resilience in an otherwise harsh environment (e.g., Bilton et al., 2001; Chapter 10). Results also demonstrated that assemblage dynamics of macroinvertebrates were controlled by the expansion and contraction of the stream network over an annual cycle, in strong support of the second hypothesis (H2). Two factors played a predominant role in dispersal and colonization patterns within the flood plain: 1) location in respect to upstream sources of larvae and (2) the environmental harshness associated with a channel type (Table 4). The complex interplay between channel type and drift dynamics have important implications regarding the function of glacial networks of alpine flood plains over an annual cycle. Because of changes in the relative proportion of channel types during ecosystem expansion and contraction, the contribution of drift and seston by various channel types to floodplain function changes seasonally (Robinson et al., 2002). For example, recolonization of the main channel after harsh summer conditions primarily occurs via groundwater channels. In conclusion, the channel mosaic of the Val Roseg flood plain can be viewed as a distinct freshwater system that functions in a similar manner to a headwater drainage network. The assemblages in each channel type are strongly determined by dispersal movements among different channel types as stated in hypothesis 3 (H3).

Table 4: Dispersal patterns and processes over an annual cycle in the Val Roseg flood plain.

Pattern/Process	*Seasonal Period*			
	Low Flow	*Expansion*	*High Flow*	*Contraction*
Drift density	Variable	High	Low	High
Dominant taxa in drift	Chironomidae/ Ephemeroptera	Microcrustaceans/ Chironomidae	Chironomidae	Ephemeroptera
Recruitment/Dispersal	Production	Dispersal	Survival	Recolonization
Active/passive drift pattern	Passive/active	Active	Passive	Active
Sink/Source (Groundwater channel)	Source/Sink	Source	Sink	Source
Sink/Source (Main channel)	Sink/Source	Source/Sink	Sink	Sink

6. REFERENCES

Allan, J.D. (1987). Macroinvertebrate drift in a Rocky Mountain stream (Colorado, USA). *Hydrobiologia, 144,* 261-268.

Anderwald, P.H., Konar, M., & Humpesch, U.H. (1991). Continuous drift samples of macroinvertebrates in a large river, the Danube in Austria. *Freshwater Biology, 25,* 461-476.

Armitage, P.D. (1977). Invertebrate drift in the regulated River Tees, and an unregulated tributary Maize Beck, below Cow Green dam. *Freshwater Biology, 7,* 167-183.

Bilton, D.T., Freeland, J.R., & Okamura, B. (2001). Dispersal in freshwater invertebrates. *Annual Review of Ecology and Systematics, 32,* 159-181.

Brittain, J.E., & Eikeland, T.J. (1988). Invertebrate drift – A review. *Hydrobiologia, 166,* 77-93.

Burgherr, P., Robinson, C.T., & Ward, J.V. (2002) Seasonal variation in zoobenthos across habitat gradients in an alpine glacial flood plain (Val Roseg, Swiss Alps). *Journal of the North American Benthological Society, 21,* 561-575.

Elliott, J.M. (1977). *Some methods for the statistical analysis of samples of benthic invertebrates.* Scientific Publications of the Freshwater Biological Association No 25.

Flecker, A.S. (1992). Fish predation and the evolution of invertebrate drift periodicity: evidence from neotropical streams. *Ecology, 73,* 438-448.

Hershey, A.E., Pastor, J., Peterson, B.J., & Kling, G.W. (1993). Stable isotopes resolve the drift paradox for *Baetis* mayflies in an arctic river. *Ecology, 74,* 2315-2325.

Hieber, M., Robinson, C.T., & Uehlinger, U. (2003). Seasonal and diel patterns of invertebrate drift in different alpine stream types. *Freshwater Biology, 48,* 1078-1092.

Ilg, C., Castella, E., Lodz-Crozet, B., & Marmonier, P. (2001). Invertebrate drift and physico-chemical variables in the tributaries of the Mutt, a Swiss glacial stream. *Archiv für Hydrobiologie, 139,* 113-128.

Kownacki, A., Dumnicka, E., Galas, J., Kawecka, B., & Wojtan, K. (1997). Ecological studies of a high mountain lake-outlet stream (Tatra Mts, Poland). *Archiv für Hydrobiologie, 139,* 113-128.

Mackay, R.J. (1992). Colonization by lotic macroinvertebrates: a review of processes and patterns. *Canadian Journal of Fisheries and Aquatic Sciences, 49,* 617-629.

Matthaei, C.D., Werthmüller, D., & Frutiger, A. (1998). An update on the quantification of stream drift. *Archiv für Hydrobioloie, 143,* 1-19.

Milner, A. M., Brittain, J.E., Castella, E., & Petts, G. E. (1994). Trends of macroinvertebrate community structure in glacier-fed rivers in relation to environmental conditions: a synthesis. *Freshwater Biology 46,* 1833-1847.

Müller, K. (1982). The colonisation cycle of freshwater insects. *Oecologia, 52,* 202-207.

Pringle, C.M., & Ramirez, A. (1998). Use of benthic and drift sampling techniques to assess tropical stream invertebrate communities along an altitudinal gradient, Costa Rica. Freshwater Biology, 39, 359-373.

Ramirez, A., & Pringle, C.M. (2001). Spatial and temporal patterns of invertebrate drift in streams draining a Neotropical landscape. *Freshwater Biology, 46,* 47-62.

Robinson, C. T., Tockner, K. & Burgherr, P. (2001). Seasonal patterns in macroinvertebrate drift and seston transport in streams of an alpine glacial flood plain. *Freshwater Biology, 47,* 985-994.

Sagar, P.M., & Glova, G.J. (1992). Invertebrate drift in a large, braided New Zealand river. *Freshwater Biology, 27,* 405-416.

Saltveit, S.J., Haug, I., & Brittain, J.E. (2001). Invertebrate drift in a glacial river and its non-glacial tributary. *Freshwater Biology, 46,* 1777-1789.

Teeri, J. A., & Barrett, P. E. (1977). Detrital transport by wind in a high arctic terrestrial ecosystem. *Arctic and Alpine Research 7,* 387-391.

Tilley, L.J. (1989). Diel drift of Chironomidae larvae in a pristine Idaho (USA) mountain stream. *Hydrobiologia, 174,* 133-150.

Tockner, K., Malard, F., Burgherr, P., Robinson, C. T., Uehlinger, U., Zah, R., & Ward, J. V. (1997). Physico-chemical characterization of channel types in a glacial floodplain ecosystem (Val Roseg, Switzerland). *Archiv für Hydrobiologie, 140,* 433-463.

Tockner, K., Malard, F., Uehlinger, U., & Ward, J. V. (2002). Nutrients and organic matter in a glacial river-floodplain. *Limnology and Oceanography, 47,* 266-277.

Uehlinger, U., Malard, F., & Ward, J.V. (2003). Thermal patterns in the surface waters of a glacial river corridor (Val Roseg, Switzerland). *Freshwater Biology, 48,* 278-300.

Ward, J. V. (1994). Ecology of alpine streams. *Freshwater Biology, 32,* 277-294.
Waringer, J.A. (1992). The drifting of invertebrates and particulate organic matter in an Austrian mountain brook. *Freshwater Biology*, 27, 253-378.
Waters, T.F. (1972). The drift of stream insects. *Annual Review of Entomology, 17*, 253-272.
Williams, D.D., & Hynes, H.B.N. (1976). The recolonization mechanisms of stream benthos. *Oikos, 27*, 265-272.
Zar, J.H. (1984). *Biostatistical Analysis*. Englewood Cliffs (NJ, USA): Pentrice Hall.

CHAPTER 16

C. T. ROBINSON, P. BURGHERR, F. MALARD, K. TOCKNER &
U. UEHLINGER

SYNTHESIS AND PERSPECTIVES

1. INTRODUCTION

The findings presented in this book represent a holistic perspective on the structure and function of a high elevation glacial flood plain in the Swiss Alps, encompassing landscape features and their history, physical and chemical characteristics of surface waters and ground waters, aquatic and terrestrial flora, benthic and hyporheic biota, and mechanistic studies of ecosystem function (Fig. 1). The results demonstrate the inherent dynamic complexity of alpine glacial flood plains, a level of complexity typically attributed to their low elevation counterparts. An important finding was how sensitive these systems are to changes in environmental conditions, thus making them key indicators for assessing landscape transformation resulting from changes in climate or human demographics. Glacier retreat has accelerated globally, increasing the probability that fundamental ecological changes will occur in alpine environments, in particular those with alluvial flood plains and complex channel networks. As the glacial influence diminishes over time, expected changes in water sources and hydrologic regime will certainly affect the distribution and abundance of organisms and the resulting functional characteristics of alpine flood plains. This chapter synthesizes our current understanding of the Val Roseg and presents some perspectives that we believe are generally applicable to these spatio-temporally complex and dynamic ecosystems.

2. FEATURES OF AN ALPINE GLACIAL FLOOD PLAIN

Alpine glacial flood plains and their resident streams are highly complex and sensitive ecosystems. The ecology of glacial (kryal) streams, in particular, has attracted major attention in the past few years. Indeed, the pioneering works of Milner & Petts (1994) and Ward (1994) stimulated a number of studies of glacial streams across Europe (Brittain & Milner, 2001; Milner et al., 2001, Robinson et al., 2001; Schütz et al., 2001). Some characteristic features of glacial streams include: a) an extended seasonal flow pulse controlled by temperature and radiation, b) large diel flow pulses during summer, c) low temperatures, d) high turbidity during flow

J.V. Ward & U. Uehlinger (eds.), Ecology of a Glacial Flood Plain, 259-271.
© 2003 Kluwer Academic Publishers. Printed in the Netherlands.

pulses, and e) low channel stability mainly in recently deglacierized areas (Chapter 1). Most early studies of kryal streams were conducted on single channel streams and limited in temporal extent. Recent work has emphasized the habitat mosaic of channel types and seasonal expansion and contraction of the floodplain stream network (Tockner et al., 1997; Malard et al., 2000). The Roseg valley is a keen example as it is fed by two separate glacial systems, thus illustrating the composite features of a variety of alpine flood plains such as kryal and krenal streams, side-slope tributaries, an extensive alluvial aquifer, and a relatively large proglacial lake (Chapter 1). The stream network also is made up of a diversity of channel types, including a lake outlet and groundwater-fed streams. The terrestrial vegetation provides a mosaic of floral patches, each changing along a diverse array of successional trajectories.

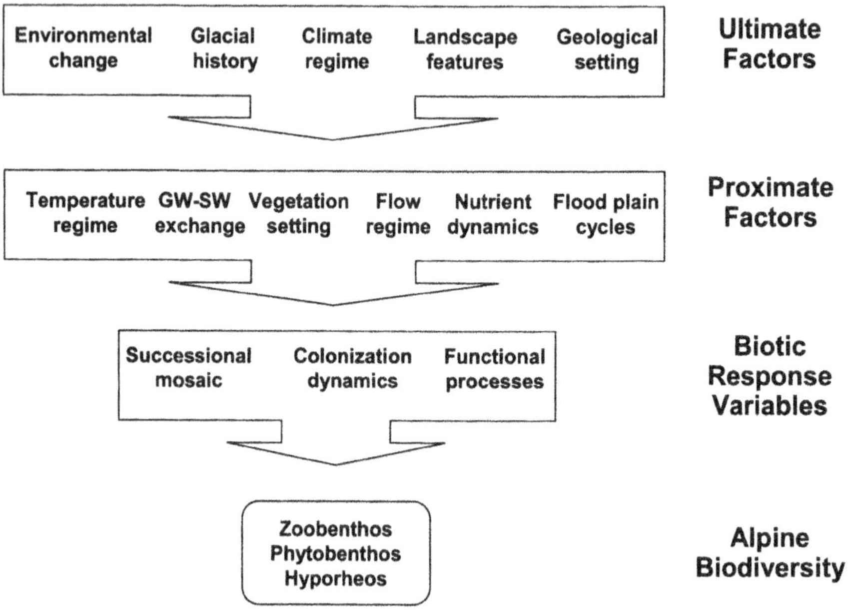

Figure 1. Conceptual diagram of forcing factors and biotic response variables structuring biodiversity patterns in alpine flood plains.

In the geologically young mountain range of the Alps, many headwater valleys are restricted in the lateral dimension, thus having relatively confined stream channels with little or no development of flood plains. Lateral expansion is generally limited to upland valleys where topography (knickpoints) facilitates the accumulation of alluvial sediments, so-called inter mountain flood plains (Stanford and Ward, 1993). In contrast to low elevation arctic regions (e.g., Iceland, Alaska, Arctic Canada) where wide glacio-fluvial flood plains develop in front of ice sheets and ice caps (Price, 1971; Church, 1972), glacial flood plains in the Alps are comparatively small or have been substantially modified by man. For example, human impacts from power generation, tourism, and habitat degradation or loss have influenced ca. 90% of streams and rivers in the Alps (CIPRA, 2001). An exception is the Val Roseg where an ancient landslide in the upper valley (Chapter 2) initiated the formation of a broad flood plain strongly influenced by glaciers and minimally affected by man. The Val Roseg study emphasizes the need for a better understanding of Alpine systems, as providers of ecosystem services and in sustaining biodiversity. Only recently (2001) has the Val Roseg been included in the 'Swiss Inventory of Wetlands of National Importance'. Indeed, this study has illustrated the importance of glacial history towards understanding present floodplain attributes and for predicting future possible directions in floodplain evolution (Chapter 2).

An evaluation of aerial photographs taken over a 52-year period showed that the annual channel turnover rate in the Val Roseg flood plain averages around 25% (Chapter 2). However, channels adjacent to the main channel changed the most, whereas the position of the main channel usually persisted for several years. The minimum age of the oldest floodplain vegetation, which may be considered a surrogate for turnover time (Hughes, 1997), is about 70 years. This turnover time is low compared to braided stream networks at lower elevations, and somewhat higher than low elevation meandering streams. This result is not surprising because the Roseg flood plain differs from low elevation systems in some important aspects. For example, during the seasonal flow pulse, water is confined to stream channels because of the steep slopes (range 1.9 - 4.8%). Only at discharges exceeding 25 m^3 s^{-1} (rare rainfall-induced spates) does water overflow the banks of the main channel forming a 30 – 60 m wide inundated corridor, which is a relatively small portion of the entire floodplain area. In contrast, bank overflows in low gradient systems typically inundate most of the flood plain, homogenizing physical and chemical differences among floodplain waterbodies. Moreover, low gradient flood plains often contain diverse waterbodies such as isolated ponds, scour pools, and oxbow lakes (e.g., Arscott et al., 2001; Ward et al. 2002), whereas lentic waters are few in glacial flood plains. Regardless, the interplay of glacial hydrology and sediment character result in a continuous turnover of floodplain area, thereby maintaining a heterogeneous mosaic of floodplain habitats of differing age and environmental conditions.

3. ENVIRONMENTAL DYNAMICS

The sources and flow paths of water in the flood plain displayed strong spatial and seasonal differences (Chapter 3). Although glacial water is clearly the primary source of water to the hydrological network of the flood plain, groundwater sources also played a substantial role in structuring the ecological conditions of surface waters. Glacial water showed high seasonality with summer dominated by glacial melt and winter by subglacial reservoirs. Consequently, the turbidity and chemistry of glacial water changes seasonally; although turbidity is substantially lower, subglacial water is chemically enriched relative to glacial meltwater. The seasonal switch in water sources occurs in late spring around mid-May as glacial melt commences and then again in autumn in October with the freezing of the glaciers. Indeed, natural tracer studies showed that ground water contributes almost 80% of the surface flow in winter but only 10% in summer. Glacial melt results in lower concentrations of base cations, bicarbonates and sulfates, with associated decreases in specific conductance. Groundwater sources showed higher ion concentrations than glacial meltwaters. Various types of ground water were distinguished: shallow alluvial ground water, hillslope ground water, and deep alluvial ground water (Chapter 3). Deep alluvial ground waters also upwell at locations of geological constrictions. These deep waters tended to be anoxic and much older (ca. 26 years) than the other ground waters (<1-4 years).

Additional study revealed that surface-subsurface connections were spatially distinct along the flood plain (Chapter 3). In the uppermost part of flood plain, the hyporheic zone was perched over ground water and with little or no hydrologic exchange between these layers, whereas hydrologic exchange was evident in the lower flood plain. Consequently, surface connections among channels in the lower and upper flood plains were lost when flows were reduced in late autumn/winter. In fact, in winter all water in the lower flood plain is derived from ground water as all surface water in the upper flood plain infiltrates to subsurface flow paths. This downwelling-upwelling phenomenon is likely a common attribute of most glacial flood plains with abrupt transitions in floodplain slope. Further, most of the groundwater channels tend to go dry by autumn, with only a small portion having permanent flow throughout the annual cycle. Essentially, floodplain waters are fed by snowmelt in spring, glacial ice-melt in summer, a mix of subglacial and glacial water in autumn, and primarily ground water in winter.

The Val Roseg flood plain exhibited a high heterogeneity of aquatic habitats with six distinct channel types being recognized: main glacial channel, side glacial channels, intermittently-connected channels, mixed source channels, groundwater channels, and tributaries (Chapter 4). Channels differed physically and chemically over an annual cycle, reflecting changes in water source and discharge. Viewed holistically, the different sources and pathways of water interact over time and space to maintain a mosaic of aquatic habitat patches within the catchment, the spatial heterogeneity of which, however, changes seasonally. Snow and ice meltwaters not only result in a distinct expansion of surface water habitats over the spring and summer ablation seasons but also replenish ground water and control the size of subglacial water stores. Thus, the importance of the summer recharge determines the extent of subglacial water-fed patches and groundwater-fed patches that persist during winter. In summer, glacial water expands throughout the catchment via surface and subsurface pathways. The simultaneous occurrence of these two hydrological pathways results in a dual flooding process which enables the development of a high diversity of aquatic habitat patches within the flood plain. Whereas the surface expansion of glacial water (overspill of main channel water) leads to the disappearance of most groundwater-fed patches that remained during winter in the lower flood plain, the subsurface expansion of glacial water (i.e., recharge of ground water by infiltrating glacial water) generates new groundwater-fed patches in the upper flood plain. This dual flooding process results in an almost complete spatial turnover of aquatic habitats within a very short period of time (less than 3 weeks). Only a small proportion of the total channel network (ca. 2.4 km of the 25 km of channels) maintains benign environmental conditions (permanent flow, high substrate stability and relatively warm temperature) during the entire hydrological cycle.

Geomorphic knickpoints were important factors controlling the spatial dynamics of floodplain channels; knickpoints were active areas of upwelling ground water. Consequently, channels in the upper flood plain experienced more extreme fluctuations in surface discharge patterns than those in the lower flood plain. Glacial melt in summer caused most channels to be turbid except for groundwater channels on the floodplain margins. Aquatic habitat heterogeneity tended to be highest in summer with high flows and more homogeneous in winter under low flow. Channels heavily influenced by glacial water tended to show the greatest temporal variability in physical and chemical properties, primarily due to the loss of glacial input in winter. Because the summer expansion of aquatic habitats is associated with high sediments loads, reduced water temperatures, and low nutrient concentrations in most channels, the Val Roseg flood plain lacks the "flood pulse advantage" typical for lowland flood plains (Bayley, 1991). However, the restriction of productivity imposed by the setting of harsh environmental conditions is compensated to some extent by the high spatial habitat heterogeneity resulting from the duality of the flooding process.

Flow and temperature are highly coupled in glacial streams, and air temperature and radiation control both. However, the complex dynamics of surface and subsurface flow paths also influenced thermal patterns and heterogeneity in streams of the flood plain (Chapter 5). Glacial melt kept summer stream temperatures low in some channels, although a small increase was visible in early spring. Temperatures declined rapidly in autumn from the decrease in air temperature and reduced solar radiation. The glacial influence on temperature decreases longitudinally because of heat exchange with the atmosphere and heat input by side-slope tributaries and springbrooks. The proglacial lake had a significant affect on water temperature of the outlet stream, which was warmer in summer and less variable than other glacial streams. The splitting of the main channel into smaller side-channels resulted in somewhat higher summer temperatures in the smaller channels, although overall floodplain temperatures remained low because most of the water is transferred via the main channel. Hyporheic temperatures were similar to surface temperatures in downwelling areas but less variable over the annual cycle. A contrasting pattern was observed in upwelling areas where little correlation was found between surface and hyporheic water temperatures. Interestingly, the influx of deep alluvial ground water had a warming influence on shallow subsurface waters in the proglacial reach, which reached temperatures >7°C in spring. A major finding was that thermal heterogeneity occurred along lateral, longitudinal, and vertical dimensions in the flood plain, with consequent effects on biodiversity (see below).

The seasonal shift in the relative contribution of different sources and flow paths of water produced a distinct sequence of concentration peaks in nutrients (Chapter 6). Melting of the snow pack on catchment slopes and the glacier surface resulted in a prolonged pulse of nitrate in surface waters (April – early June), although much of the nitrate was apparently released from soils. Melting of glaciers released significant amounts of particulate organic matter and mobilized fine subglacial sediments enriched in apatite, thereby resulting in elevated concentrations of particulate phosphorus. However, the reactivity of particulate phosphorous was presumably very low because the concentration of soluble reactive phosphorous remained below 3 μg/L during the ice ablation period. Moreover, ratios of N:P suggested phosphorus to be the limiting nutrient in channels of the flood plain, a conclusion supported by nutrient limitation experiments (see below). Groundwater dominance in winter was associated with high dissolved silica and dissolved organic carbon concentrations, although DOC may be generated during shallow subsurface flow paths through soils. Because of this sequence of concentration peaks in nutrients, resource availability (i.e., nutrient supply and availability of organic matter) coincided with benign environmental conditions (low flow velocity, turbidity, bedload transport and elevated temperatures) only during early spring and autumn. Both periods are of short duration (less than a month) but they offer favorable ecological windows that may well be essential for the accrual of benthic algae and the maintenance of invertebrate populations. Differences in nutrient concentrations between floodplain channels reflected the mixing of different water sources rather than the effect of autogenic processes operating within individual water bodies. As for turbidity and specific conductance, the relative contribution over space and time of different water sources was the main factor determining the shifting mosaic in nutrient levels. The flood plain acted primarily as a transfer area for nutrients because it was never a source or a sink of dissolved nutrients, particulate nitrogen, and carbon. Retention of particulate phosphorus reflected the sedimentation process of inorganic suspended solids which occurred as the main channel water was diverted into the network of distribution channels. The insignificant role of biogeochemical processes was mainly attributable to the low water retention capacity of the flood plain (most surface water resided in the flood plain for only about 40 min.).

4. BIODIVERSITY PATTERNS

The Roseg study illustrated the heterogeneous spatial and temporal patterns in floodplain flora (aquatic and terrestrial) and fauna (benthos and hyporheos). Terrestrial vegetation responded to two major processes, habitat turnover in the primary flood plain and successional processes following glacial retreat (Chapter 7). Habitat turnover in the flood plain resulted in a mixture of successional vegetative patches ranging from bare sediments virtually devoid of organic matter to islands of vegetation composed of alpine grasses and willow shrubs (*Salix* spp.). Trees currently are restricted to valley side slopes, although dense shrubs (*Alnus viridis, Salix*) were found in riparian areas of springbrooks and stable groundwater-fed channels. The spatial pattern in vegetation strongly influenced organic matter inputs into the flood plain (see Chapter 12). Successional patterns of trees follow glacial recession, although altitudinal constraints impose limitations on rates of development. These rates may change with expected increases in temperature. Recent models of glacial retreat in the western USA suggest dramatic elevation shifts in terrestrial vegetation with glacial retreat by 2100 (Hall & Fagre, 2003).

The distribution of aquatic fungi reflected the habitat template of the different channel types in the flood plain (Chapter 8). For instance, the proglacial reach generally had the least number of species and floodplain springbrooks the most. However, numerous taxa were present in other glacial channels, being similar to levels found in forested streams at lower elevations. This finding clearly indicated that aquatic hyphomycetes can function under low temperatures as found in glacial streams. Indeed, the functional attributes of fungal assemblages in the Roseg flood plain mirrored those of low elevation streams, although rates tended to be somewhat lower. We suspect that aquatic hyphomycetes may be limited by low organic matter inputs, but are able to quickly capitalize on these resources once available. The results certainly show that aquatic hyphomycetes are circum-global in distribution, probably dispersing in a similar fashion as bacteria and algae. Most fungal taxa identified in the Val Roseg have been found in a large number of other studies in different biomes (Ingold, 1975; Nilsson, 1964).

Patterns in surface water zoobenthos were much more variable than interstitial fauna and were quite distinctive among stream and channel types (Chapter 10). The spatial heterogeneity of surface waters enhanced the overall biodiversity of the flood plain, with some stream types potentially providing refugia during the harsh flow period in kryal streams in summer. In general, 146 taxa were identified with aquatic insects comprising 65% of these taxa. Chironomid midges were predominant with over 35 taxa, whereas most other insect groups had around 10 taxa. The abundance and diversity of surface zoobenthos tended to increase downstream and during more favourable periods in spring and late autumn/winter. Robinson et al. (2001) also noted increases in the abundance and richness of macroinvertebrates in autumn/winter in a number of other glacial streams in the Alps (also see Schütz et al., 2001). Indeed, more knowledge on the winter ecology of glacial flood plains is needed. The *Diamesa*-groups dominated upstream glacial streams, with Ephemeroptera (*Baetis alpinus* and *Rhithrogena* spp.) becoming more common in lower reaches.

Groundwater-fed channels had zoobenthic assemblages distinct from those in the glacial-fed streams (Chapter 10). During periods of higher channel stability additional taxa become abundant in upper glacial reaches, supporting the importance of substrate stability in assemblage structure in proglacial streams. The fact that certain groups of surface zoobenthos such as Ephemeroptera and Plecoptera occur closer to the glacier than predicted may reflect the thermal refugium provided by the hyporheic zone (Chapter 11). Indeed, a major finding was the overall role of habitat stability in dictating the presence or absence of taxa, both seasonally and spatially among channel types. For example, non-insecta taxa such as oligochaetes, nematodes and turbellarians were more prevalent in groundwater-fed channels than glacial channels, whereas some stoneflies and mayflies were more prevalent in the latter. The other channel type, the lake outlet, also contained an assemblage distinct from that of other proglacial sites due to the increased stability caused by the lake (Burgherr & Ward, 2000; Hieber et al., 2002). Although alpine streams are considered relatively harsh environments, the biota of the streams in the Val Roseg further reflected a two-dimensional habitat template of spatial and temporal heterogeneity (Chapter 10).

Interstitial fauna have been little studied in alpine environments, although hyporheic zones may be quite extensive both vertically and laterally (Chapter 11). Interstitial fauna were dominated by permanent aquatic taxa and the richness was quite high, totalling 72 taxa. Strong longitudinal gradients were evident in interstitial assemblages, reflecting hyporheic (groundwater) flow paths and, principally, water temperature. However, only two taxa preferentially colonized proglacial hyporheic zones; most others became common with the increase in groundwater inputs lower in the flood plain. Most true hyporheic and stygobitic groups were found in the lower flood plain. Temporal variability was low, reflecting the stable habitat conditions of the subsurface environment. Patterns in the distribution of some taxa suggested that hyporheic flow paths may act as major corridors for organism dispersal and migration (*sensu* Ward & Palmer, 1994). The results of the Val Roseg sampling lead to the development of a habitat template model with one axis represented by the upstream surface hydrological connection with glacial water and the other by the saturated hydrological connection with ground water (Chapter 11). This model generally fit the longitudinal and lateral distribution of interstitial fauna in the Val Roseg. Results strongly suggest that the distribution patterns of permanent aquatic interstitial fauna at the floodplain scale not only reflects the spatial heterogeneity of environmental conditions but also the constraints imposed by the channel network configuration and distribution of surface-subsurface hydrological linkages on the dispersal of organisms.

5. FUNCTIONAL PROCESSES

The primary source of energy in the flood plain was autochthonous (instream) via benthic algae, and was an important food resource for benthic invertebrates (Chapter 12). Higher aquatic plants were essentially lacking and mosses could be found only in some groundwater-fed tributaries. Terrestrial sources (allochthonous) of organic material were important only in some side-slope tributaries, springbrooks, and glacial channels along the floodplain margins. Because of the high seasonality in discharge and related changes in turbidity and substrate stability in glacial channels, periphyton abundance was high in spring and late autumn, but not in summer. In winter, periphyton could also attain high levels in the segments of glacial channels that remained open (i.e., lacked ice cover). Groundwater channels tended to have relatively high periphyton biomass throughout the year. The low accrual of periphyton in summer in glacial channels suggests the system may be energy limited at this time, although the diversity of channel types clearly enhances the overall productive capacity of the flood plain. Lateral inputs of terrestrial organic matter can be high on a seasonal basis, particularly in autumn, and terrestrial inputs increased longitudinally as the glacial stream became enveloped by forest in the lower parts of the Val Roseg (Chapter 12). Even so, terrestrial sources of organic matter were much lower than for streams at lower elevations. Organic matter in hyporheic sediments displayed a similar pattern with concentrations increasing longitudinally in the flood plain, although levels are much lower (2-4X) than found in forested streams. Present findings suggest that retention and storage of organic matter in the flood plain is quite low, further emphasizing the importance of autochthonous production as the driving energy source in most streams of alpine flood plains.

The functional potential of different streams in the flood plain was evaluated through studies of leaf breakdown (Chapter 13). It was initially thought that the harsh environment of glacial streams would constrain stream function, although the diversity of stream types inferred differences in biological activity among streams. It was found that leaves introduced to floodplain channels could be processed quite quickly and at rates similar to those observed in forested streams. Leaves were processed by physical fragmentation, microbes and invertebrates, the relative importance of each dependent on the particular channel. Fragmentation was a primary factor in kryal streams, as the low channel stability limited macroinvertebrates and fungi. Groundwater-fed springbrooks showed the highest biological activity, probably because of their inherent stability and greater inputs of terrestrial organic matter. An interesting finding was the interplay in resource limitation and channel stability on leaf processing as shown at the glacial lake outlet where introduced leaves were consumed rapidly by a resident trichopteran (Chapter 13). The stability of the outlet allowed for the presence of the trichopteran even though optimal food resources were in short supply. As such, the leaves acted as resource islands that attracted the larvae. As mentioned above, fungal activity on introduced leaves was high in most of the streams, being similar to those found in forested low elevation systems.

Leaf type played a major role in breakdown dynamics with alder leaves being processed much faster than the more recalcitrant larch needles (Chapter 13). Nutrient (N + P) concentrations also showed similar patterns, being greater on alder leaves. The aquatic fungi apparently took up phosphorus to a much greater extent than nitrogen, as inferred by the differences in nutrient availability in the flood plain. Insects quickly colonized leaf packs and consumed leaves, although the numbers and kinds of animals differed considerably among channels. Again, the more stable channels such as the springbrooks contained more invertebrates that readily consumed the leaf material. The findings suggest that functional processes in alpine floodplain streams were constrained physically by the harsh environment and by limited resource availability. However, the diversity of stream types enhanced the overall functional potential of the glacial flood plain, as orders of magnitude differences in processing rates were found among stream types in the flood plain. Lastly, the functional potential of streams in the flood plain probably is related to the successional stage of each stream. Kryal sites were the youngest and least stable and had the slowest breakdown, whereas springbrooks are older and quite stable and had the highest leaf breakdown rates. A similar relationship was observed for needle breakdown among terrestrial habitats, with the slowest breakdown found on young bare sediments and the fastest rates in older side-slope forests.

Nutrient limitation (phosphorus) was evident among streams in the flood plain on a seasonal basis (Chapter 14). Nitrogen was never limiting because of high atmospheric deposition that is characteristic over Europe, and especially the Alps.

Nutrient concentrations changed seasonally in response to changes in water source (Chapter 6). Water source also played a role in nutrient dynamics among channel types and longitudinally in main floodplain channels. Limitation was most pronounced in spring and typically in channels higher in the flood plain; groundwater channels showed no limitation. Although nutrient limitation may occur in summer, it is overridden during that period by other physical factors (turbidity, scour) in the glacial streams. The high periphyton biomass in autumn in all sites suggested that nutrients were not limiting at this time of year. Nutrient uptake rates were much greater for phosphorus than nitrogen and differed among channel types. Shortest uptake lengths were found in groundwater systems and longer lengths in the main glacial channel. Findings also revealed that much phosphorus may be adsorbed onto particles of glacial flour, thus causing uptake rates to be higher in glacial channels than in channels without glacial flour. Experimental nutrient addition clearly stimulated periphyton production in a side-slope tributary and enhanced leaf decomposition in a groundwater-fed springbrook, thus indicating limitation (Chapter 14). The fast processing rates in the fertilized leaf packs were attributed to greater colonization by invertebrates, probably because of a higher quality food resource resulting from the uptake of nutrients by microbes. Inputs of nutrients from side-slope tributaries and groundwater systems influenced nutrient dynamics longitudinally in the flood plain. In general, the high heterogeneity of stream types clearly enhanced the functional potential of the floodplain network.

Drifting in the water column is a major avenue of dispersal and colonization by many invertebrates that can potentially influence functional processes among the different channel types in the flood plain (Chapter 15). As expected, the diversity and density of drifting invertebrates displayed a clear seasonality among channel types, with highest densities found during the expansion (spring) and contraction (autumn) phases, and lowest densities during the period of maximum expansion in summer. Groundwater channels showed the lowest among-season variability in drift parameters, acting as continual sources of colonists to downstream channels. Groundwater channels also were apparent refugia for some macroinvertebrate groups, sustaining populations during periods of extreme harshness in other channel types, glacial-influenced channels in particular. Application of a water mixing model (see Chapter 4) to drift patterns indicated that the entire flood plain served primarily as a sink for invertebrates during the snowmelt and glacial melt periods and as a source of organisms from autumn to early spring. Two factors appeared to play major roles in the dispersal and colonization dynamics among channels in the active flood plain: 1) location in respect to upstream sources of larvae; and 2) the environmental harshness associated with a particular channel type.

In general, drift densities were quite high (up to 500 ind/m^3) in the different channel types relative to low elevation temperate streams. However, drift peaks occurred during the day, perhaps in response to the diurnal discharge maxima in glacial channels or the lack of fish predators. For instance, the greatest diel periodicity in drift occurred in the main channel during glacial ablation, coinciding with transport of suspended solids. Drift patterns of individual taxa also were associated with seasonal changes in floodplain expansion and contraction. For example, upwelling ground water in spring resulted in a dramatic peak in microcrustacean drift in groundwater channels. Further, terrestrial invertebrates contributed significantly to drift during the main ablation period in glacial channels and mayflies (primarily *Baetis* spp.) were common drifters during the contraction phase in autumn. In summary, drift and associated colonization patterns reflected the strong interaction of biotic properties in response to seasonal changes in environmental conditions among the different channel types.

6. POTENTIAL EFFECTS OF CLIMATE CHANGE

During the last 10,000 years, glaciers in the Alps have advanced and receded within a relatively narrow range (Fig. 3 in Chapter 2). Since the end of the Little Ice Age, the glacierized area of the Bernina massif (Val Roseg) has decreased by more than 20%,with an accelerated pace during the last 20 years. The expected loss of the main valley glaciers in the Val Roseg under different scenarios of climate change is depicted in Figure 6 of Chapter 2. The rise of the snow line and loss of ice are anticipated to markedly influence the flow regime and the sediment balance of the upper Roseg catchment. The increasing proglacial areas comprise large amounts of unconsolidated sediment susceptible to fluvial transport, below the Tschierva Glacier in particular, whereas the proglacial Roseg lake may continue to act as a sink for coarse sediments. Snow melt is expected to have a stronger influence on the shape of the hydrograph than today; annual discharge may decline after a transient increase, and rain induced floods will presumably occur more frequently in late June and July (Kuhn & Batlogg, 1999). The reduction of permafrost areas may increase the inputs of lateral sediments into the flood plain. One can only speculate how these changes will affect the turnover rates of floodplain habitats. An expected consequence of the glacier recession is that kryal reaches will move upstream and temperature will increase in surface-connected channels. Within the next 20 years, large amounts of sediment exposed by the retreat of the Tschierva glacier may be transported into the flood plain either continuously (transient increase in summer discharge) or during high discharge floods (heavy rain storms). At least temporarily, the increased sediment delivery rate will cause the floodplain channel network to become more braided. This will promote the lateral extension of glacial water, thereby restricting the proportion of groundwater-fed habitat patches within the flood plain. However, on a longer time scale (e.g., 100 years), the Roseg flood plain will necessarily shift from a glacier- to a snow-dominated ecosystem if the proportion of glaciated area in the catchment continues to decrease at the present rate. A possible increase in the presence of intermittent and groundwater (snowmelt fed) streams is highly plausible.

Alpine stream ecosystems act as integrators and centers of organization within the landscape, affecting nearly all aspects of the natural environment and human culture (Füreder, 1999). However, they are highly sensitive to the effects of climate change and human impacts (McGregor et al., 1995). Although climate change models predict that ca. 95% of Alpine glacier mass could disappear by 2100 (Watson et al., 1997), the ecological effects on alpine streams are difficult to predict because many significant consequences will result at the scale of small catchments, a level of resolution as yet unresolved by Global Circulation Models. For example, lower channel stability could result in reduced abundances of benthic macroinvertebrates; increased water temperatures could lead to extinction of cold-water species and colonization by new species from lower elevations. Glacial retreat may fragment populations of aquatic invertebrates with variable affects on population persistence and genetic structure (Monaghan et al., 2002). Further, the zoobenthos may respond to indirect effects. Glacier retreat may promote the enlargement or formation of proglacial lakes, the outlets of which are characteristically stable habitats for zoobenthos (Burgherr & Ward, 2001; Hieber et al., 2001).

Mountains are insular environments and glacial retreat an imminent phenomenon in many alpine regions globally. Consequently, the insularity of high mountain areas will increase and freshwater habitats probably become more fragmented or temporary. Previous kryal systems will follow expected successional changes in habitat properties with concomitant changes in biodiversity and ecosystem function. The present mosaic of stream habitats in the flood plain represents various stream types differing in their successional stage of development. As glacial retreat continues upslope, the mosaic successional stages will change longitudinally and laterally. Longitudinally, new or younger streams will be found at higher elevations and will exhibit a long-term colonization pattern. Colonization of these new streams will be constrained by biogeographic and local factors such as distance, elevation, channel stability, and stream temperature, and biotic factors such as species specific life histories and dispersal behaviors. As streams move headward, the lateral dimension may become more confined unless an alpine plateau is present. These alpine plateaus may develop new high alpine flood plains with, as of yet, undetermined stream structure and function. One thing is certain, the ecosystems of high mountain glacial flood plains are in a state of transition with many promising and exciting avenues for research and ecological understanding.

7. REFERENCES

Arscott, D. B., Tockner, K. & Ward, J.V. (2001). Thermal heterogeneity along a braided floodplain river in the Alps. *Canadian Journal of Fisheries and Aquatic Sciences, 58*, 2358-2373.

Bayley, P. B. (1991). The floodpulse advantage and the restoration of river-floodplain systems. *Regulated Rivers, 6*, 75-86.

Brittain, J. E., & Milner A. M. (2001). Ecology of glacier-fed rivers: current status and concepts. *Freshwater Biology, 46*, 1571-1578.

Burgherr, P. & Ward, J. V. (2000). Zoobenthos of kryal and lake outlet biotopes in a glacial floodplain. *Verhandlungen der Internationalen Vereinigung für Theoretische und Angewandte Limnologie, 27*, 1587-1590.

Burgherr, P. & Ward, J. V. (2001). Longitudinal and seasonal distribution patterns of the benthic fauna of an alpine glacial stream (Val Roseg, Swiss Alps). *Freshwater Biology, 46*, 1705-1721.

Church, M. A. (1972). Baffin Island sandurs: a study of Arctic fluvial processes. *Geological Survey of Canada Bulletin 216.*

CIPRA (Internationale Alpenschutzkommission) (2001). *2. Alpenreport: Daten, Fakten, Probleme, Lösungsansätze.* Berne, Switzerland: Haupt-Verlag.

Füreder, L. (1999). High alpine streams: cold habitats for insect larvae. In R. Margesin & F. Schinner (Eds.), *Cold adapted organisms: Ecology, physiology, enzymology and molecular biology* (pp 181-196). Berlin: Springer.

Hall, M. H. P. & Fagre, D. B. (2003). Modeled climate-induced glacier change in Glacier national Park, 1850-2100. *BioScience, 53*, 131-140.
Hieber, M., Robinson, C. T., Rushforth, S. R. & Uehlinger, U. (2001). Algal communities associated with different alpine stream types. *Arctic, Antarctic, and Alpine Research, 33*, 447-456.
Hieber, M., Robinson, C. T., Uehlinger, U. & Ward, J. V. (2002). Are alpine lake outlets less harsh than other alpine streams. *Archiv für Hydrobiologie, 154*, 199-223.
Hughes, F. M. R. (1997). Floodplain biogeomorphology. *Progress in Physical Geography 214*.
Ingold, C. T. (1975). An illustrated guide to aquatic and waterborne hyphomycetes (Fungi Imperfecti) with notes on their biology. *Freshwater Biological Association, Scientific Publication, 30*, 1-96.
Kuhn, M., & Batlogg, N. (1999). Modellierung der Auswirkung von Klimaänderung auf verschiedene Einzugsgebiete in Österreich. Page 94 In: *Schriftenreihe der Forschung im Verbund. Österreichische Elektrizitätswirschaft-AG.*
Malard, F., Tockner, K. & Ward, J. V. (2000). Physico-chemical heterogeneity in a glacial riverscape. *Landscape Ecology, 15*, 679-695.
McGregor, G., Petts, G. E., Gurnell, A. M. & Milner, A. M. (1995). Sensitivity of alpine stream ecosystems to climate change and human impacts. *Aquatic conservation: Marine and Freshwater Ecosystems, 5*, 233-247.
Milner, A. M., Brittain J. E., Castella E., & Petts, G. E. (2001). Trends of macroinvertebrate community structure in glacier-fed rivers in relation to environmental conditions: a synthesis. *Freshwater Biology, 46*, 1833-1847.
Milner, A. M., & Petts, G. E. (1994). Glacial rivers: physical habitat and ecology. *Freshwater Biology, 32*, 295-307.
Monaghan, M. T., Spaak, P., Robinson, C. T. & Ward, J. V. (2002). Population genetic structure of 3 alpine stream insects: influences of gene flow, demographics, and habitat fragmentation. *Journal of the North American Benthological Society, 21*, 114-131.
Nilsson, S. (1964). Freshwater hyphomycetes: taxonomy, morphology and ecology. *Symbolae Botanicae Upsaliensis, 18*, 5-130.
Price, R. J. (1971). The development and destruction of a sandur, Breidmerkurjökull. *Arctic and Alpine Research, 3*, 225-237.
Robinson, C. T., Uehlinger, U., & Hieber, M. (2001). Spatio-temporal variation in macroinvertebrate assemblages of glacial streams in the Swiss Alps. *Freshwater Biology, 46*, 1663-1672.
Schütz, C., Wallinger, M. Burger, R., & Fureder, L. (2001). Effects of snow cover on the benthic fauna in a glacier-fed stream. *Freshwater Biology, 46*, 1691-1704.
Stanford, J. A., & Ward J. V. (1993). An ecosystem perspective of alluvial rivers: connectivity and the hyporheic corridor. *Journal of the North American Benthological Society, 12*, 48-60.
Tockner, K., Malard, F., Burgherr, P, Robinson, C. T., Uehlinger, U., Zah, R. & Ward, J. V. (1997). Physico-chemical characterization of channel types in a glacial floodplain ecosystem (Val Roseg, Switzerland). *Archiv für Hydrobiologie, 140*, 433-463.
Ward, J. V. (1994). Ecology of alpine streams. *Freshwater Biology, 32*, 277-294.
Ward, J. V. & Palmer, M. A. (1994). Distribution patterns of interstitial meiofauna over a range of spatial scales, with emphasis on alluvial river-aquifer systems. *Hydrobiologia, 287*, 147-156.
Ward, J.V., Tockner, K., Arscott, D.B., & Claret, C. (2002) Riverine landscape diversity. *Freshwater Biology*, 47, 517-539.
Watson, R. T., Zinyowera, M. C. & Moss, R. H. (1997). *The regional impacts of climate change: an assessment of vulnerability. A special report of IPCC Working Group II.* Cambridge: Cambridge University Press.

APPENDIX I

1. SITE MAPS

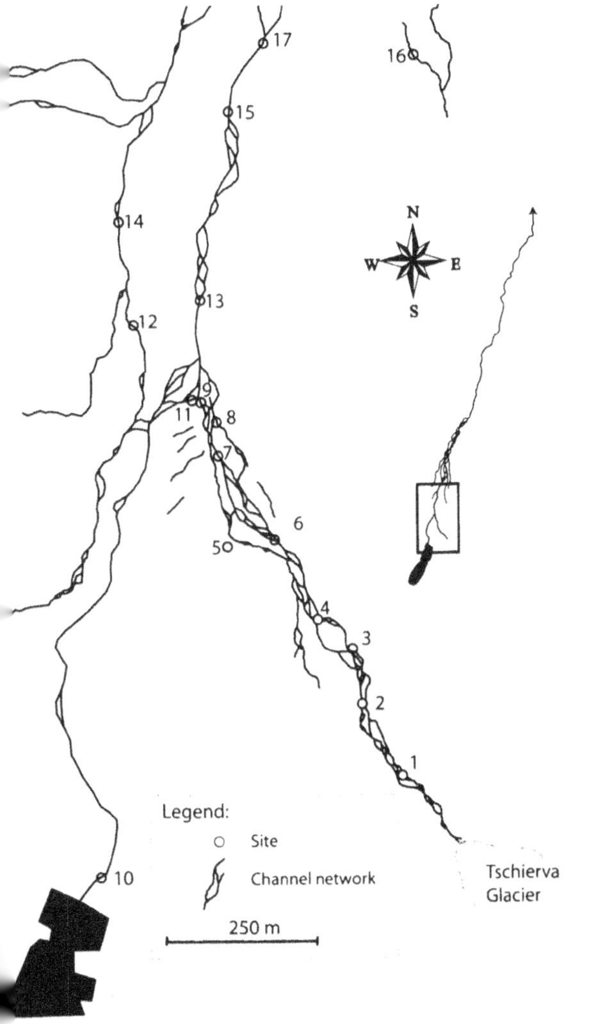

n the proglacial reach (1-9), lake outlet (10 and 11), and incised). The inserted overview map shows the location of these reaches Roseg River corridor. Sites 12, 14, and 16 are tributaries.

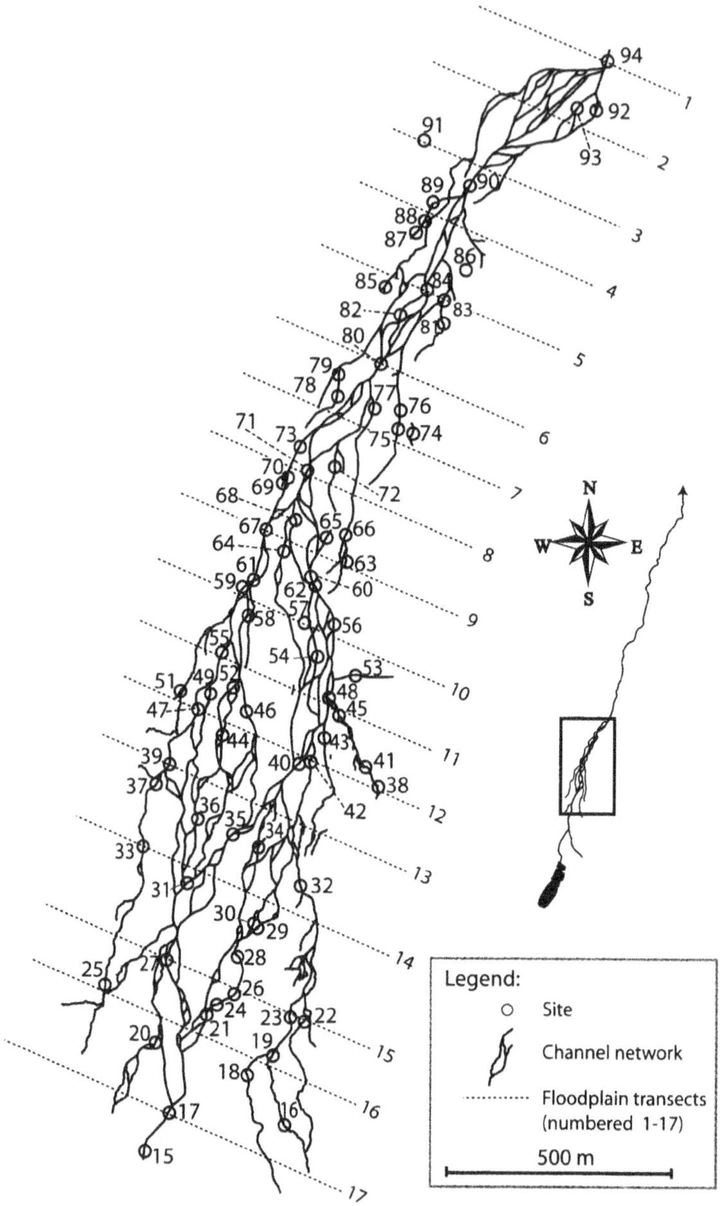

Sampling sites within the flood plain and in the adjacent area. Sites16, 18, 19, 25, 33, 37, and 53 are outside of the flood plain. Site 91 is a groundwater well. The inserted overview map shows the location of the floodplain within the Roseg River corridor. Temperature, turbidity and specific conductance were measured along the transects in monthly intervals in 1997 and 1999.

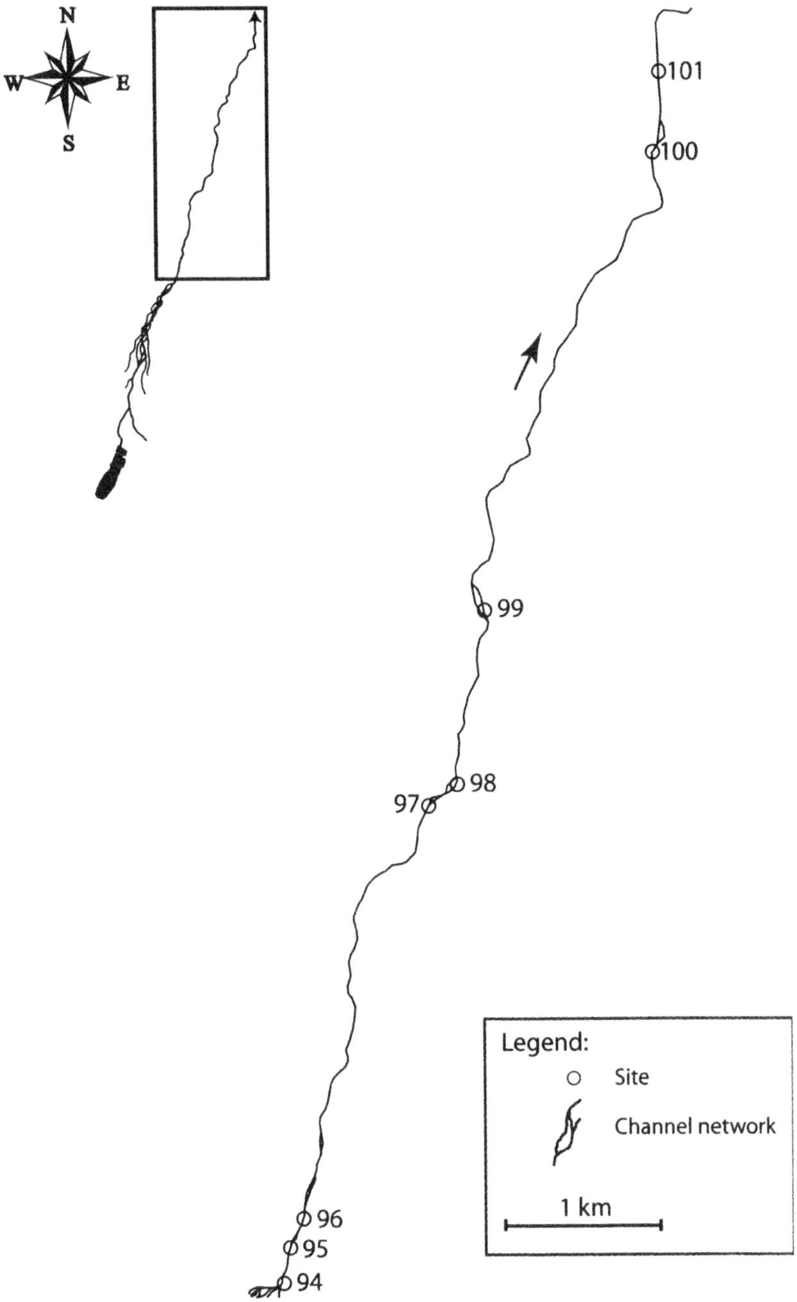

Sampling sites along the constrained reach. The inserted overview map shows the location of the constrained reach within the Roseg River corridor.

2. SAMPLING SITES AND PARAMETERS

This list describes where water, algae, invertebrates and sediment organic matter were collected, where temperature was continuously recorded and sediment respiration measured, and where nutrient uptake and leaf decomposition experiments were performed.

No. = site number, TS = surface temperature continuously recorded, TH = hyporheic temperature continuously recorded. CS = surface water chemistry, CH = hyporheic water chemistry, AL = benthic algae, BI = benthic invertebrates, HI = hyporheic invertebrates, NU = nutrient uptake experiments, LD = leaf decomposition experiments, RE = sediment respiration and sediment particulate organic matter. + indicates sampling.

Reach	No.	TS	TH	CS	CH	AL	BI	HI	NU	LD	RE
Proglacial reach	1					+	+				
	2	+		+	+	+	+	+			
	3			+							
	4			+		+					
	5						+				
	6		+	+							+
	7			+	+	+	+	+	+	+	
	8	+	+	+			+				
	9	+				+					
Lake outlet	10	+		+			+		+	+	+
	11	+		+		+					
Incised reach and adjacent area	12								+		
	13			+					+		
	14								+		
	15	+			+	+	+	+			
Flood plain and adjacent area	16						+				
	17			+	+			+			
	18			+							
	19			+							
	20	+		+	+	+		+			

Reach	No.	TS	TH	CS	CH	AL	BI	HI	NU	LD	RE
Flood plain and adjacent aarea	21										+
	22								+	+	
	23								+		
	24	+		+	+			+			
	25			+							
	26					+					+
	27			+							
	28	+									
	29			+							
	30			+							
	31			+							
	32	+		+	+	+		+			+
	33			+	+			+			
	34			+							
	35			+	+	+		+			
	36			+							
	37										+
	38			+	+		+	+			
	39			+	+			+			
	40	+									+
	41					+			+		+
	42			+	+	+		+			+
	43			+	+			+			+
	44			+	+			+			+
	45										+
	46	+	+	+	+	+	+	+			+
	47			+							
	48										+
	49			+	+			+			
	50										
	51			+							
	52					+					
	53	+		+							
	54			+	+			+			
	55										+
	56				+		+	+			
	57			+							+
	58			+							
	59								+	+	
	60								+	+	

Reach	No.	TS	TH	CS	CH	AL	BI	HI	NU	LD	RE
Flood plain and adjacent aarea	61			+	+			+			+
	62	+	+	+		+					
	63						+				
	64	+	+	+	+	+	+	+			+
	65										+
	66			+	+	+		+			+
	67	+				+					
	68				+			+			
	69	+		+	+		+	+			
	70					+					
	71										+
	72			+	+			+			
	73	+	+	+			+				
	74	+		+			+				
	75			+	+			+	+		
	76										+
	77	+									
	78			+	+			+			
	79										+
	80		+	+	+	+	+	+			
	81			+			+				+
	82	+		+	+	+		+			
	83	+	+	+	+			+			
	84										+
	85			+							
	86		+	+	+						
	87	+									
	88										+
	89			+	+			+			
	90			+	+			+			+
	91				+						
	92			+							+
	93										+
	94	+		+		+					

Reach	No.	TS	TH	CS	CH	AL	BI	HI	NU	LD	RE
Constrained reach	95			+							
	96				+		+	+			
	97			+	+	+	+	+			
	98			+							
	99			+	+		+	+			
	100			+							
	101	+		+	+	+	+	+			

APPENDIX II

COMMON TERRESTRIAL PLANT TAXA OF THE TSCHIERVA GLACIER FORELAND AND THE ROSEG FLOOD PLAIN

Life forms after Lauber & Wagner (1998)[1]: T = therophyte, G = geophyte, H = hemicryptophyte, kCH = herbaceous chamaephyte, hCH = woody chamaephyte, snP= summer-green nanophanerophyte, inP = evergreen nanophanerophyte, sP=summergreen phanerophyte, iP = evergreen phanerophyte.

Family/*Species*	Life form
Aspidiaceae	
Dryopteris filix-mas (L.) Schott	H
Gymnocarpium dryopteris (L.) Newman	G
Polystichum lonchitis (L.) Roth	H
Asteraceae	
Achillea erba-rotta All. ssp. *moschata* (Wulf.) Vacc.	H
Achillea millefolium agg. L.	H
Achillea nana L.	H
Adenostyles leucophylla (Willd.) Rchb.	H
Antennaria dioica (L.) Gaertn.	kCh
Artemisia umbelliformis Lam.	kCh
Carduus defloratus ssp. *rhaeticus* (DC.) Rothm.	H
Doronicum clusii (All.) Tausch	H
Erigeron acer s.l. L.	H
Erigeron alpinus L.	H
Gnaphalium sylvaticum L.	H
Hieracium bifidum Hornem.	H
Hieracium intybaceum All.	H
Hieracium pilosella L.	H
Hieracium staticifolium All.	H
Hieracium villosum agg. Jacq.	H
Homogyne alpina (L.) Cass.	H
Leontodon autumnalis L.	H
Leontodon hispidus s.l. L.	H

[1] Lauber, K., & Wagner, G. (1998). *Flora Helvetica* (2 ed). Bern: Paul Haupt.

J.V. Ward & U. Uehlinger (eds.), Ecology of a Glacial Flood Plain, 281-286.
© 2003 *Kluwer Academic Publishers. Printed in the Netherlands.*

Leucanthemopsis alpina (L.) Heyw.	H
S. virgaurea ssp. *minuta* (L.) Arcang.	H
Senecio doronicum L.	H
Senecio incanus ssp. *carniolicus* (Willd.) Br.-Bl.	H
Taraxacum officinale Web.	H
Taraxacum alpinum (Hoppe) Hegetschw.	H
Tussilago farfara L.	G
Athyriaceae	
Athyrium filix-femina (L.) Roth	H
Cystopteris fragilis (L.) Bernh.	H
Betulaceae	
Alnus viridis (Chaix) DC.	snP
Betula pendula Roth	P
Boraginaceae	
Myosotis alpestris F.W. Schmidt	H
Brassicaceae	
Arabis alpina L.	kCh
Biscutella laevigata (L.)	H
Cardamine resedifolia (L.)	H
Campanulaceae	
Campanula cochleariifolia Lam.	H
Campanula scheuchzeri Vill.	H
Caprifoliaceae	
Lonicera caerulea (L.)	snP
Caryophyllaceae	
Cerastium arvense ssp. *strictum* (Koch) Sch. et K.	kCh
Cerastium fontanum ssp. *vulgare* (Hartm.)	kCh
Cerastium pedunculatum Gaudin	kCh
Cerastium uniflorum Clairv.	kCh
Minuartia verna (L.) Hiern	kCh
Sagina saginoides (L.) Karsten	kCh
Silene exscapa All.	Ch
Silene nutans s.str. L.	H
Silene vulgaris ssp. *galerosa* (Jord.) Marsd.-Jon. et Turill	Ch
Cistaceae	
Helianthemum nummularium ssp. grandiflorum (Scop.) Sch. et Th.	HCh
Crassulaceae	
Sedum alpestre Vill.	Ch
Sempervivum arachnoideum L.	kCh
Sempervivum montanum L.	kCh

Cupressaceae	
Juniperus communis ssp. *nana* Syme	inP
Cyperaceae	
Carex bicolor All.	G
Carex flava s.str. L.	H
Carex frigida All.	G
Carex hostiana DC.	H
Carex leporina L.	H
Carex nigra (L.) Reichard	G
Equisetaceae	
Equisetum variegatum Schl.	H
Ericaceae	
Arctostaphyllos uva-ursi (L.) Sprengel	hCh
Rhododendron ferrugineum L.	inP
Vaccinium myrtillus L.	hCh
Vaccinium uliginosum L.	hCh
Fabaceae	
Anthyllis vulneraria ssp. *alpestris* (Schult.) A. et G.	H
Lotus alpinus (DC.) Ramond	H
Trifolium badium Schreber	H
Trifolium pallescens Schreber	H
Trifolium pratense ssp. *nivale* (Koch) Arcang.	H
Trifolium repens L.	H
Gentianaceae	
Gentiana nivalis L.	T
Juncaceae	
Juncus alpino-articulatus Chaix	H
Juncus jacquinii L.	H
Juncus trifidus L.	H
Luzula multiflora (Retz.) Lej.	H
Luzula spicata (L.) DC.	H
Lamiaceae	
Thymus polytrichus Kerner	kCh
Lycopodiaceae	
Huperzia selago (L.) Schr. et Mart.	kCh
Onagraceae	
Epilobium alsinifolium Vill.	H
Epilobium angustifolium L.	H
Epilobium fleischeri Hochst.	kCh
Ophioglossaceae	
Botrychium lunaria (L.) Sw.	G

Oxalidaceae	
Oxalis acetosella L.	G
Pinaceae	
Larix decidua Miller	snP
Pinus cembra L.	iP
Plantaginaceae	
Plantago serpentina All.	H
Poaceae	
Agrostis rupestris All.	H
Agrostis schraderiana Becherer	H
Agrostis stolonifera L.	H
Anthoxanthum alpinum Löve et Löve	H
Avenella flexuosa (L.) Parl.	H
Calamagrostis villosa (Chaix) Gmelin	H
Deschampsia caespitosa (L.) P.B.	H
Festuca halleri All.	H
Festuca quadriflora Honckeny	H
Festuca rubra s.l. L.	H
Nardus stricta L.	H
Phleum alpinum L.	H
Phleum rhaeticum (Humph.) Rausch.	H
Poa alpina L.	H
Poa chaixii Vill.	H
Poa nemoralis L.	H
Poa pratensis L.	H
Poa supina Schrader	H
Poa trivialis L.	H
Poa violacea Bell.	H
Polygalaceae	
Polygala alpina (DC.) Steudel	H
Oxyria digyna (L.) Hill.	H
Polygonum viviparum L.	H
Rumex scutatus L.	H
Pyrolaceae	
Pyrola minor L.	H
Ranunculaceae	
Aconitum napellus (Rchb.) Gayer	H
Ranunculus aconitifolius L.	H
Ranunculus acris L.	H
Rosaceae	
Alchemilla decumbens Buser	H

Alchemilla fissa Gnth. et Schumm.	H
Alchemilla flabellata Buser	H
Alchemilla xanthochlora agg. Rothm.	H
Dryas octopetala L.	hCh
Geum reptans L.	H
Potentilla aurea L.	H
Potentilla crantzii (Crantz) Beck	H
Rubus idaeus L.	snP
Sibbaldia procumbens L.	H
<u>Rubiaceae</u>	
Galium anisophyllon Vill.	H
<u>Salicaceae</u>	
Populus tremula L.	sP
Salix appendiculata Vill.	snP
Salix caprea L.	snP
Salix daphnoides Vill.	sP
Salix foetida DC.	snP
Salix hastata L.	snP
Salix helvetica Vill.	snP
Salix herbacea L.	H
Salix nigricans var. *alpicola* Buser	snP
Salix pentandra L.	sP
Salix purpurea ssp. *gracilis* (Wimmer) Buser	snP
Salix reticulata L.	hCh
Salix retusa L.	hCh
Salix serpyllifolia Scop.	hCh
<u>Santalaceae</u>	
Thesium alpinum L.	H
<u>Saxifragaceae</u>	
Parnassia palustris L.	H
Saxifraga aizoides L.	kCh
Saxifraga bryoides L.	kCh
Saxifraga exarata Vill.	kCh
Saxifraga oppositifolia L.	kCh
Saxifraga paniculata Miller	kCh
Saxifraga stellaris L.	kCh
<u>Scrophulariaceae</u>	
Euphrasia minima Schleicher	T
Euphrasia rostkoviana Hayne	T
Linaria alpina (L.) Miller	kCh
Rhinanthus glacialis Personn.	T
Veronica alpina L.	H

Veronica fruticans Jacq.	hCh
Tamaricaceae	
Myricaria germanica (L.) Desv.	inP

APPENDIX III

SPECIES LIST OF IDENTIFIED AQUATIC HYPHOMYCETES FROM VARIOUS STREAM HABITATS IN THE ROSEG FLOOD PLAIN

Identifications of the aquatic hyphomycetes are based on conidia released during laboratory incubations from alder (*Alnus viridis* (Chaix.) Dc.) or larch (*Larix decidua* Mill.) leaves placed in five different stream types and retrieved periodically for analysis.

Species
Alatospora acuminata Ingold
Anguillospora longissima (Sacc. et Syd.) Ingold
Articulospora tetracladia Ingold
Clavariopsis aquatica de Wild.
Clavatospora longibrachiata (Ingold) Marvanová et Nilsson
Flagellospora curvula Ingold
Heliscella stellata (Ingold et Cox) Marvanová et Nilsson
Heliscus lugdunensis Sacc. et Thérry
Lemonniera aquatica de Wild.
Lemonniera terrestris Tubaki
Taeniospora gracilis Marvanová
Tetrachaetum elegans Ingold
Tetracladium marchalianum de Wild.
Tricladium angulatum Ingold
Tricladium curvisporum Descals
Tumularia aquatica Descals et Marvanová

J.V. Ward & U. Uehlinger (eds.), Ecology of a Glacial Flood Plain, 287.
© 2003 *Kluwer Academic Publishers. Printed in the Netherlands.*

APPENDIX IV

LIST OF ALGAL TAXA FOUND IN THE ROSEG RIVER

Bacillariophyta
Achnanthes spp.
A. bioretii Germain
A. linearis (W. Smith) Grunow
A. minutissima Kützing
A. oblongella Oestrup
Amphora ovalis Kützing
Caloneis silicula (Ehrenb.) Cleve
Cocconeis placentula Ehrenb.
Cyclotella spp.
Cymbella affinis Kützing
C. cistula (Ehrenb.) Kirchner
C. gracilis (Ehrenb.) Kützing
C. minuta Hilse
C. sinuata Gregory
C. ventricosa Kützing
Denticula tenuis Kützing
D. elegans Kützing
Diatoma anceps (Ehrenb.) Kirchner
D. hiemalis (Roth) Heiberg
D. mesodon (Ehrenb.) Kützing
D. vulgaris Bory
Diploneis spp.
Eunotia spp.
Eunotia exigua (Brébisson) Rabenh.
Fragilaria spp.
F. arcus (Ehrenb.) Cleve
F. arcus var. *Amphioxys* Rabenh.
F. arcus var *linearis* Holmboe
F. capucina Desmazières
F. construens (Ehrenb.) Grunow
F. construens f. binodis (Ehrenb.) Hustedt
F. pinnata Ehrenb.
F. ulna (Nitzsch) Lange-Bertalot
Frustulia rhomboides (Ehrenb.) De Toni
Gomphonema spp.

J.V. Ward & U. Uehlinger (eds.), Ecology of a Glacial Flood Plain, 289-290.
© 2003 *Kluwer Academic Publishers. Printed in the Netherlands.*

Bacillariophyta (cont.)
G. parvulum Kützing
*G. angustatum (*Kützing) Rabenh.
G. olivaceum (Hornemann) Brébisson
Navicula spp.
N. cryptocephala Kützing
N. radiosa Kützing
Melosira varians Agardh
Meridion circulare (Greville) Agardh
Nitzschia spp
N. alpina Hustedt
N. paleacea Grunow
Rhoicosphenia spp.
Stauroneis anceps Ehrenb.
Tabellaria flocculosa (Roth) Kützing
Chrysophyta
Hydrurus foetidus (Villars) Trevisan
Chlorophyta
Closterium spp.
Cosmarium spp.
Microspora spp.
Mougeotia spp.
Pleurotaenium spp.
Scenedesmus opoliensis Richter
Spirogyra spp.
Staurastrum spp.
Stigeoclonium spp.
Ulothrix spp.
Zygnema spp.
Cyanobacteria
Calothrix spp.
Chamaesiphon spp.
C. confervicola (Syn. curvatus) Gutwinski
C. incrustans Grunow
C. polonicus (Rostaf.) Hansgirg
Homoeothrix janthina (Bron. et Flah.) Starm.
Lyngbya spp.
Nodularia spp.
Phormidium autumnale (Agardh) Gomont
P. incrustatum (Näg.) Gomont.
Rivularia spp.
Schizothrix spp.
Scytonema spp.
Tolypothrix spp.

APPENDIX V

LIST OF BENTHIC AND HYPORHEIC TAXA

Inventory of benthic and hyporheic macroinvertebrate taxa recorded in the Val Roseg river corridor.

Taxon	**Benthic layer**	**Hyporheic zone**
Phylum PLATHELMINTHES		
Class **Turbellaria**		
Order **Seriata**		
Family Planariidae		
Crenobia alpina (Dana, 1766)	+	+
Phylum NEMATHELMINTHES		
Class **Nematoda**	+	+
Phylum MOLLUSCA		
Class **Bivalvia**		
Order **Eulamellibranchiata**	+	
Phylum ANNELIDA		
Class **Polychaeta**		
Family Nerillidae		
Troglochaetus cf. *beranecki* Delachaux, 1921		+
Class **Aphanoneura**		
Family Aeolosomatidae		
Aeolosoma hyalinum Bunke, 1967	+	+
Class **Clitellata**		
Order **Oligochaeta**		
Family Naididae		
Nais alpina Sperber, 1948	+	
Nais bretscheri Michaelsen, 1899		+
Nais communis Piguet, 1906	+	+
Nais elinguis Müller, 1773	+	
Nais simplex Piguet, 1906	+	
Nais variabilis Piguet, 1906	+	+

J.V. Ward & U. Uehlinger (eds.), Ecology of a Glacial Flood Plain, 291-297.
© 2003 *Kluwer Academic Publishers. Printed in the Netherlands.*

Family Enchytraeidae		
Cernosvitoviella atrata (Bretscher, 1903)	+	+
Cernosvitoviella carpatica Nielsen & Christensen, 1959		+
Cognettia cognettii (Issel, 1905)	+	+
Cognettia glandulosa Michaelsen, 1888		+
Fredericia sp. Michaelsen, 1889		+
Hemienchytraeus sp.		+
Henlea ventriculosa (d'Udekem, 1854)	+	+
Marionina argentea (Michaelsen, 1889)	+	+
Mesenchytraeus sp. Eisen, 1878	+	
Mesenchytraeus armatus (Levinson, 1883)	+	+
Lumbricillus rivalis Levinson, 1883	+	
Family Lumbriculidae		
Stylodrilus heringianus Claparède, 1862	+	+
Family Tubificidae		
Tubifex tubifex (Müller 1874)	+	
Family Lumbricidae		
Eiseniella tetraedra (Savigny, 1826)	+	
Family Haplotaxidae		
Haplotaxis gordioides (Hartman, 1921)	+	+
Family Dorydrilidae		
Dorydrilus michaelseni Piguet, 1913		+
Family Propappidae		
Propappus volki Michaelsen, 1916	+	+
Phylum TARDIGRADA	+	
Phylum ARTHROPODA		
Class **Arachnidae**		
Order **Acari**		
Hydrachnellae		+
Family Feltriidae		
Feltria minuta Koenike, 1896	+	
Feltria cf. *rubra* Piersig, 1898	+	
Feltria setigera Koenike, 1896	+	
Family Hygrobatidae		
Atractides loricatus Piersig, 1898	+	
Atractides cf. *vaginalis* (Koenike, 1905)	+	
Atractides walteri (K. Viets, 1925)	+	
Hygrobates norvegicus (Thor, 1897)	+	
Family Lebertiidae		

Lebertia cf. *extendens* Walter,1922	+	
Lebertia cf. *glabra* Thor, 1914	+	
Lebertia sefvei Walter, 1911	+	
Lebertia tuberosa Thor, 1914	+	
Family Hydryphantidae		
Panisopsis cf. *curvifrons* (Walter, 1907)	+	
Panisus michaeli Koenike, 1896	+	
Family Sperchonidae		
Sperchon mutilus Koenike, 1895	+	
Sperchon squamosus Kramer, 1879	+	
Sperchon thienemanni Koenike, 1907	+	
Family Proziinae gen. sp.	+	
Class **Crustacea**		
Subclass **Branchiopoda**		
Order **Cladocera**		
Chydorus sphaericus (O.F. Müller, 1785)		+
Subclass **Ostracoda**		
Family Cyprididae		
Candona neglecta Sars, 1887	+	+
Candona candida (Müller, 1776)	+	+
Fabaeformiscandona tyrolensis (Löffler, 1963)	+	+
Fabaeformiscandona cf. *lapponica (*Ekman, 1908)	+	+
Cypria opthalmica Jurine, 1820	+	+
Cyclocypris ovum (Jurine, 1820)	+	+
Eucypris pigra Fischer, 1851	+	+
Potamocypris sp. 1	+	+
Potamocypris sp. 2	+	+
Cavernocypris subterranea Wolf 1921	+	+
Subclass **Copepoda**		
Order **Cyclopoida**		
Eucyclops serrulatus (Fischer, 1851)		+
Acanthocyclops robustus (G. O. Sars, 1863)		+
Diacyclops bisetosus (Rehberg, 1880)		+
Diacyclops languidus (G. O. Sars, 1863)		+
Diacyclops cf. *disjunctus* (Thallwitz, 1927)		+
Order **Harpactoida**	+	
Moraria (Moraria) alpina Stoch, 1998		+
Bryocamptus (Arcticocamptus) cuspidatus (Schmeil, 1893)		+
Bryocamptus (Arcticocamptus) raethicus (Schmeil, 1893)		+

Bryocamptus (Rheocamptus) zschokkei (Schmeil, 1893)		+
Bryocamptus (Limocamptus) echinatus (Mrazek, 1893)		+
Maraenobiotus insignipes insignipes (Lilljeborg, 1902)		+
Parastenocaris glacialis Noodt, 1954		+
Syncarida		
Family Bathynellidae		+
Amphipoda		
Family Niphargidae		
Niphargus cf. *tatrensis* Wrzesniowsky, 1933		+
Class **Insecta**		
Order **Ephemeroptera**		
Family Baetidae		+
Baetis alpinus Pictet, 1843	+	
Baetis rhodani Pictet, 1843	+	
Family Heptageniidae		+
Rhithrogena loyolea Navas 1922	+	
Rhithrogena nivata (Eaton, 1871)	+	
Rhithrogena degrangei Sowa, 1969	+	
Rhithrogena alpestris Eaton, 1885	+	
Ecdyonurus picteti (Meyer-Dür, 1864)	+	
Ecdyonurus helveticus (Eaton, 1885)	+	
Order **Plecoptera**		
Family Taenopterygidae		
Rhaphidiopterix alpina Kühtreiber, 1934	+	
Family Nemouridae		+
Amphinemura spp.	+	
Nemoura spp.	+	
Protonemura spp.	+	
Family Leuctridae		+
Leuctra spp.	+	
Family Capniidae		
Capnia vidua Klapalek, 1904	+	
Family Perlodidae		+
Isoperla spp.	+	
Perlodes intricata (Pictet, 1841)	+	
Dictyogenus alpinus (Pictet, 1841)	+	
Family Chloroperlidae		+
Siphonoperla spp.	+	
Order **Coleoptera**		

Family Dytiscidae		
Agabus bipustulatus (Limnaeus, 1767)	+	+
Agabus sturmi (Gyllenhal, 1808)	+	+
Hydroporus marginatus (Duftschmid, 1805)	+	
Oreodytes borealis (Gyllenhal, 1827)	+	+
Family Helophoridae		
Helophorus glacialis Villa, 1883	+	
Helophorus nivalis Giraud, 1851	+	
Family Hydrophilidae		
Laccobius striatulus (Fabricius, 1801)	+	
Family Curculionidae		
Trachyploeus bifoveolatus (Beck, 1817)	+	
Family Haliplidae	+	
Order **Trichoptera**		
Family Rhyacophilidae		
Rhyacophila intermedia McLachlan, 1868	+	
Rhyacophila hirticornis McLachlan, 1879	+	
Family Limnephilidae		+
Acrophylax zerberus Brauer, 1867	+	
Drusus biguttatus (Pictet, 1834)	+	
Drusus discolor (Rambur, 1842)	+	
Pseudopsilopteryx zimmeri (McLachlan, 1876)	+	
Allogamus uncatus (Brauer, 1857)	+	
Melampophylax melampus (McLachlan, 1867)	+	
Melampophylax mucoreus (Hagen, 1861)	+	
Anomalopterygella chauvinana (Stein, 1874)	+	
Order **Diptera**		
Family Empididae		
Wiedemannia spp.	+	+
Clinocera spp.	+	
Hemerodromia spp.	+	
Family Tipulidae		
Tipula spp.	+	
Family Tabanidae	+	+
Family Syrphidae	+	
Family Limoniidae		
Dicranota spp.	+	+
Pedicia sp.	+	
Delichopeza sp.	+	

Rhabdomastix sp.	+	+
Rhypholophus spp.	+	+
Elaeophila sp.	+	
Limonia sp.		+
Family Culicidae		
Culex sp.	+	
Family Stratiomyidae	+	
Family Simuliidae		
Simulium argenteostriatum (Strobl, 1898)	+	
Simulium argyreatum Meigen, 1838	+	+
Simulium carthusiense Grenier & Dorier, 1959	+	
Simulium cryophilum (Rubzov, 1959)	+	
Simulium trifasciatum Curtis, 1839	+	
Simulium tuberosum (Lundström, 1911)	+	
Prosimulium latimucro (Enderlein, 1925)	+	
Prosimulium rufipes (Meigen, 1830)	+	
Family Chironomidae		
Subfamily Diamesinae		+
Diamesa sp.	+	
Diamesa steinboecki-group	+	+
Diamesa latitarsis-group	+	+
Diamesa bertrami-group	+	
Diamesa cinerella-zernyi-group	+	
Diamesa cf. *incallida* (Walker, 1856)	+	
Diamesa bertrami Edwards, 1935	+	
Pseudodiamesa branickii (Nowicki, 1873)	+	
Subfamily Prodiamesinae		
Prodiamesa olivacea (Meigen, 1818)	+	
Subfamily Orthocladiinae		
Brillia modesta (Meigen, 1830)	+	
Chaetocladius piger (Goetghebuer, 1913)	+	
[1]COP	+	
Coryoneura sp.	+	+
Eukiefferiella brevicalcar (Kieffer, 1911)	+	
Eukiefferiella minor/fittkaui	+	

[1] COP = larvae of either genus *Cricotopus*, *Orthocladius* or *Paratrichocladius* that cannot be separated by morphological characteristics

Eukiefferiella sp.	+	
Heleniella sp.	+	
Krenosmittia camptophleps (Edwards, 1929)		+
Limnophyes sp.	+	
Orthocladius frigidus (Zetterstedt, 1838)	+	+
Orthocladius cf. *obumbratus* Johannsen, 1905	+	
Orthocladius rivicola Kieffer, 1921	+	
Orthocladius rivulorum Kieffer, 1909	+	
Orthocladius gr. *thienemanni* Kieffer, 1906	+	
Parametriocnemus stylatus (Kieffer, 1924)		+
Paraphaenocladius pseudirritus Strenzke, 1950	+	
Paratrichocladius cf. *rufiventris* (Meigen, 1830)	+	
Paratrichocladius nivalis Goetghebuer, 1938	+	
Paraorthocladius nudipennis (Kieffer, 1908)	+	
Rheocricotopus effusus (Walker, 1856)	+	
Stilocladius montanus Rossaro, 1979	+	
Synorthocladius semivirens (Kieffer, 1909)	+	
Thienemaniella sp.	+	
Tvetenia sp.	+	
Tvetenia bavarica (Goetghebuer, 1934)	+	
Subfamily Chironominae		
Tribe Tanytarsini non det.	+	+
Microspectra sp	+	
Family Ceratopogonidae	+	+
Order **Collembola**	+	+
TOTAL	**146**	**72**

The following taxonomic specialists provided support for species identifications of several groups of aquatic organisms:

- Dr. M. Lafont, Lyon, France: Oligochaeta
- Dr. R. Gerecke, Tübingen, Germany: Hydrachnellae
- Dr. P. Marmonier, Lyon, France: Ostracoda
- Dr. D. Galassi, Aquila, Italy: Copepoda
- Dr. J. Waringer, Vienna, Austria: Trichoptera
- Dr. R. Glatthaar, Würenlos, Switzerland: Simuliidae
- Dr. B. Janecek, Vienna, Austria;
- Dr. B. Lods-Crozet, Geneva, Switzerland: Chironomidae
- Dr. P. Richoux: Coleoptera
- Dr. W. Riss, Münster, Germany: Limoniidae

INDEX

D

E

F

G

H

I

J

K

L

Z

GPSR Compliance

The European Union's (EU) General Product Safety Regulation (GPSR) is a set of rules that requires consumer products to be safe and our obligations to ensure this.

If you have any concerns about our products, you can contact us on ProductSafety@springernature.com

In case Publisher is established outside the EU, the EU authorized representative is:

Springer Nature Customer Service Center GmbH
Europaplatz 3
69115 Heidelberg, Germany

Batch number: 10755466

Printed by Printforce, the Netherlands